Methods in Molecular Biology™

Series Editor
John M. Walker
School of Life Sciences
University of Hertfordshire
Hatfield, Hertfordshire, AL10 9AB, UK

For further volumes:
http://www.springer.com/series/7651

Proteomics for Biomarker Discovery

Methods and Protocols

Edited by

Ming Zhou and Timothy Veenstra

Laboratory of Proteomics and Analytical Technologies, SAIC-Frederick, Inc., Frederick National Laboratory for Cancer Research, Frederick, MD, USA

Humana Press

Editors
Ming Zhou
Laboratory of Proteomics and Analytical Technologies, SAIC-Frederick, Inc.
Frederick National Laboratory for Cancer Research
Frederick, MD, USA

Timothy Veenstra
Laboratory of Proteomics and Analytical Technologies, SAIC-Frederick, Inc.
Frederick National Laboratory for Cancer Research
Frederick, MD, USA

ISSN 1064-3745 ISSN 1940-6029 (electronic)
ISBN 978-1-62703-359-6 ISBN 978-1-62703-360-2 (eBook)
DOI 10.1007/978-1-62703-360-2
Springer New York Heidelberg Dordrecht London

Library of Congress Control Number: 2013934219

Printed on acid-free paper

Humana Press is a brand of Springer
Springer is part of Springer Science+Business Media (www.springer.com)

Preface

With the advent of proteomics came the development of technologies, primarily mass spectrometry, which allowed high-throughput identification of proteins in complex mixtures. While the mass spectrometer resides at the heart of proteomics, its ability to characterize biological samples is only as good as the sample preparation and data analysis tools used in any study. Not only has proteomics increased our capacity to identify proteins, it has enabled other characteristics of proteomes to be measured. Of utmost interest has been the development of techniques for measuring posttranslational modifications such as phosphorylation and glycosylation. Since the amount of any specific protein within a cell is important to its function, methods to quantitate protein levels have also been developed. These quantitative methods include both label-free approaches and those that utilize stable isotopes incorporated both during cell growth or added via a chemical reaction once the proteome is extracted from the cell.

The purpose of this book is to provide the student and researcher in the fields of Biochemistry, Biomedicine, Molecular and Cellular Biology, and Bioinformatics a detailed description of many of the different sample preparation and data analysis tools used in proteomics today. The editors are indebted to each of the authors for providing their time and expertise in making this edition an invaluable resource to anyone involved or interested in proteomics.

Frederick, MD, USA

Ming Zhou
Timothy Veenstra

Contents

Contributors

ALEX APFFEL • *Agilent Laboratories, Santa Clara, CA, USA*
SABINE BAHN • *University of Cambridge, Cambridge, UK*
MATTHEW BELLGARD • *Centre for Comparative Genomics, Murdoch University, Perth, WA, Australia*
JOSIP BLONDER • *Frederick National Laboratory for Cancer Research, Frederick, MD, USA*
JON BURROWS • *Oncoplex Diagnostics, Rockville, MD, USA*
NATALIE CASTELLANA • *University of California San Diego, La Jolla, CA, USA*
ANGELA CHAMBERY • *Second University of Naples, Caserta, Italy*
KING C. CHAN • *Laboratory of Proteomics and Analytical Technologies, Advanced Technology Program, SAIC-Frederick, Inc., Frederick National Laboratory for Cancer Research, Frederick, MD, USA*
BRETT CHAPMAN • *Centre for Comparative Genomics, Murdoch University, Perth, WA, Australia*
SIMONA COLANTONIO • *Protein Chemistry Laboratory, Advance Technology Program, SAIC-Frederick, Frederick National Laboratory for Cancer Research, Frederick, MD, USA*
BRETT COOKE • *Macquarie University, North Ryde, Australia*
GRANT R. CRAMER • *University of Nevada, Reno, NV, USA*
SURENDRA DASARI • *Vanderbilt University Medical Center, Nashville, TN, USA*
LEROI V. DESOUZA • *Department of Chemistry, York University, Toronto, ON, Canada*
MICHAEL R. EMMERT-BUCK • *National Cancer Institute, Bethesda, MD, USA*
ANNARITA FARINA • *Geneva University, Geneva, Switzerland*
JESSICA M. FAUPEL-BADGER • *National Cancer Institute, Rockville, MD, USA*
LIQIAN GAO • *National University of Singapore, Singapore, Republic of Singapore*
RYAN GHAN • *University of Nevada, Reno, NV, USA*
DAVID M. GOOD • *Karolinska Institute, Stockholm, Sweden*
SCOTT P. GOULDING • *University of Pittsburgh School of Medicine, Pittsburgh, PA, USA*
PAUL C. GUEST • *University of Cambridge, Cambridge, UK*
DAVID K. HAN • *University of Connecticut Health Center, Farmington, CT, USA*
JEFFREY HANSON • *National Cancer Institute, Bethesda, MD, USA*
PAUL A. HAYNES • *Macquarie University, North Ryde, NSW, Australia*
JINTANG HE • *Department of Surgery, University of Michigan Medical Center, Ann Arbor, MI, USA*
JERRY D. HOLMAN • *Vanderbilt University Medical Center, Nashville, TN, USA*
SUN-II HWANG • *Carolinas Medical Center, Charlotte, NC, USA*
HALEEM J. ISSAQ • *Laboratory of Proteomics and Analytical Technologies, Advanced Technology Program, SAIC-Frederick, Inc., Frederick National Laboratory for Cancer Research, Frederick, MD, USA*
JULIAN A.J. JAROS • *University of Cambridge, Cambridge, UK*
DONALD J. JOHANN JR. • *National Cancer Institute, Bethesda, MD, USA*

TIM KEIGHLEY • *Macquarie University, North Ryde, NSW, Australia*
DAVID B. KRIZMAN • *Oncoplex Diagnostics, Rockville, MD, USA*
DAISUKE KUBOTA • *Department of Orthopedic Surgery, Juntendo University School of Medicine, Tokyo, Japan*
YASHU LIU • *Department of Surgery, University of Michigan Medical Center, Ann Arbor, MI, USA*
DAVID M. LUBMAN • *Department of Surgery, University of Michigan Medical Center, Ann Arbor, MI, USA*
BRIAN T. LUKE • *Frederick National Laboratory for Cancer Research, Frederick, MD, USA*
MICHAEL J. MACCOSS • *University of Washington, Seattle, WA, USA*
DANIEL MARTINS-DE-SOUZA • *Department of Chemical Engineering and Biotechnology, University of Cambridge, Cambridge, UK; Lab. de Neurociencias (LIM-27), Inst. de Psiquaitria, Faculdade de Medicina da Universidade de Sao Paulo, Sao Paulo, Brazil*
ZHAOJING MENG • *Laboratory of Proteomics and Analytical Technologies, Advanced Technology Program, Frederick National Laboratory for Cancer Research, SAIC-Frederick, Inc., Frederick, MD, USA*
SUMANA MUKHERJEE • *National Cancer Institute, Bethesda, MD, USA*
KARLIE A. NEILSON • *Macquarie University, North Ryde, NSW, Australia*
DIANNE L. NEWTON • *Frederick National Laboratory for Cancer Research, Frederick, MD, USA*
BENJAMIN C. ORSBURN • *LMIV Molecular Pathogenesis and Biomerkers, Frederick, MD, USA*
DANA PASCOVICI • *Macquarie University, North Ryde, NSW, Australia*
DARUE A. PRIETO • *Frederick National Laboratory for Cancer Research, Frederick, MD, USA*
THIERRY RABILLOUD • *Chemistry and Biology of Metals Grenoble, CEA Grenoble, Grenoble, France*
JAIME RODRIGUEZ-CANALES • *National Cancer Institute, Bethesda, MD, USA*
DOROTHEA RUTISHAUSER • *Science for Life Laboratory, Stockholm, Sweden*
TSUYOSHI SAITO • *Department of Human Pathology, Juntendo University School of Medicine, Tokyo, Japan*
VALERIA SEVERINO • *Second University of Naples, Caserta, Italy*
K.W. MICHAEL SIU • *Department of Chemistry, York University, Toronto, ON, Canada*
ROBERTA M. SMITH • *Pathology/Histology Laboratory, SAIC-Frederick Inc., Frederick National Laboratory for Cancer Research, Frederick, MD, USA*
APURVA K. SRIVASTAVA • *Laboratory of Proteomics and Analytical Technologies, Advanced Technology Program, Frederick National Laboratory for Cancer Research, SAIC-Frederick, Inc., Frederick, MD, USA*
LUKE H. STOCKWIN • *Frederick National Laboratory for Cancer Research, Frederick, MD, USA*
YOSHIYUKI SUEHARA • *Department of Orthopedic Surgery, Juntendo University School of Medicine, Tokyo, Japan*
HONGYAN SUN • *City University of Hong Kong, Hong Kong, China*
DAVID L. TABB • *Vanderbilt University Medical Center, Nashville, TN, USA*
MICHAEL A. TANGREA • *National Cancer Institute, Bethesda, MD, USA*
SARAH TRIBOULET • *Université Joseph Fourier, Grenoble, France*
MAHESH UTTAMCHANDANI • *DSO National Laboratories, Singapore, Republic of Singapore*
STEVEN C. VAN SLUYTER • *Macquarie University, North Ryde, NSW, Australia*
TIMOTHY VEENSTRA • *Laboratory of Proteomics and Analytical Technologies, SAIC-Frederick, Inc., Frederick National Laboratory for Cancer Research, Frederick, MD, USA*

SÉBASTIEN N. VOISIN • *Department of Chemistry, York University, Toronto, ON, Canada*
TIMOTHY J. WAYBRIGHT • *Frederick National Laboratory for Cancer Research, Frederick, MD, USA*
CHRISTINE C. WU • *University of Pittsburgh School of Medicine, Pittsburgh, PA, USA*
JING WU • *Department of Surgery, University of Michigan Medical Center, Ann Arbor, MI, USA*
ZHEN XIAO • *Frederick National Laboratory for Cancer Research, Frederick, MD, USA*
XIA XU • *Frederick National Laboratory for Cancer Research, Frederick, MD, USA*
SHAO Q. YAO • *National University of Singapore, Singapore, Republic of Singapore*
XIAOYING YE • *Frederick National Laboratory for Cancer Research, Frederick, MD, USA*
LI-RONG YU • *National Center for Toxicological Research, FDA, Jefferson, AR, USA*
MING ZHOU • *Laboratory of Proteomics and Analytical Technologies, SAIC-Frederick, Inc., Frederick National Laboratory for Cancer Research, Frederick, MD, USA*

Chapter 1

Affinity Depletion of Plasma and Serum for Mass Spectrometry-Based Proteome Analysis

Julian A.J. Jaros, Paul C. Guest, Sabine Bahn,
and Daniel Martins-de-Souza

Abstract

Protein biomarker discovery in blood plasma and serum is severely hampered by the vast dynamic range of the proteome. With protein concentrations spanning 12 orders of magnitude, conventional mass spectrometric analysis allows for detection of only a few low-abundance proteins. Prior depletion of high-abundant proteins from the sample can increase analytical depth considerably and has become a widely used practice. We describe in detail an affinity depletion method that selectively removes 14 of the most abundant proteins in plasma and serum.

Key words Plasma, Serum, Depletion, High abundant, Analytical depth, Removal, Biomarker, Mass spectrometry, CSF, Cerebrospinal fluid, Agilent, Sigma

1 Introduction

Blood is a typical source of protein biomarkers for a number of reasons. It is easily accessible, enabling repeated sampling with minimal inconvenience to the donor. This feature is important especially for clinical studies and the translation of findings into the clinic. Blood is in molecular exchange with most tissues and organs of the body and transports molecules that regulate a range of body functions. Therefore, it can reflect the physiological and pathological status of an individual (1). Furthermore, this body fluid is relatively uncomplicated to work with compared to tissue extracts or cellular homogenates. Blood proteins appear to suffer only moderate degradation when exposed to room temperature even for long periods, in contrast to other biomaterials (2). Further issues with cell and tissue lysates include protein extraction and solubility, mass spectrometry (MS)-incompatible buffer components (e.g., detergents, chaotropic agents, salts) that are needed to solubilize hydrophobic proteins, and, as a consequence, the necessity of numerous sample preparation and clean-up steps. These potential problems

Ming Zhou and Timothy Veenstra (eds.), *Proteomics for Biomarker Discovery: Methods and Protocols*, Methods in Molecular Biology, vol. 1002, DOI 10.1007/978-1-62703-360-2_1, © Springer Science+Business Media, LLC 2013

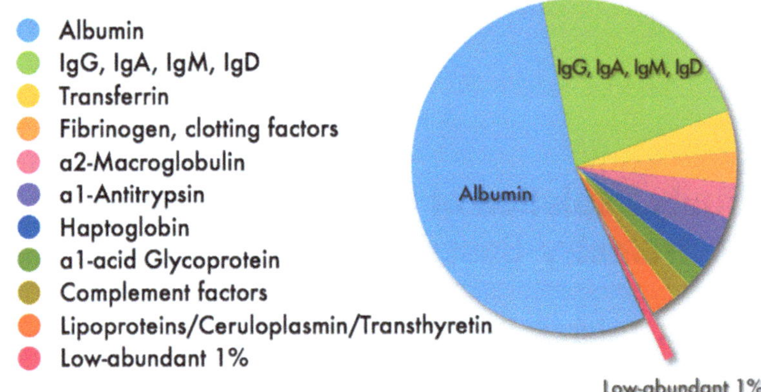

Fig. 1 The composition of the plasma/serum proteome. Albumin and the immunoglobulins account for approximately 75 % of the total protein weight, whereas 20 additional protein species make up most of the remaining weight. Low-abundant proteins correspond to only approximately 1 % of the protein weight in plasma/serum

are less pronounced or nonexistent when working with blood. Because of these characteristics, blood is a common choice for biomarker discovery.

Plasma is derived from crude blood through removal of blood cells by centrifugation in a tube containing an anticoagulant such as ethylenediaminetetraacetic acid (EDTA). In case of serum, neat blood samples are typically allowed to clot for 2 h at room temperature (21 °C) and then transferred into centrifugation tubes, leaving the clots behind. A centrifugation step serves to pellet any remaining blood cells, cellular material, and clots. The resulting serum supernatant does not contain blood cells and lacks most of the clotting factors. Plasma and serum contain up to 80 mg of protein per mL along with various small molecules such as salts, lipids, amino acids, and sugars (3). More than 10,000 different proteins in concentrations stretching over 12 orders of magnitude may be present in plasma and serum (4–6), resulting in one of the highest dynamic range for proteomes studied to date (5). Most of these proteins are present at relatively low concentrations. Contrarily, the major protein constituents of plasma and serum such as albumin, immunoglobulins, transferrin, haptoglobin, and lipoproteins are present at very high concentrations compared to the majority of proteins found in circulation. Approximately 99 % of the protein weight of plasma and serum consist of only 22 different protein species and their isoforms (6) (see Fig. 1).

This variance in protein concentrations poses a significant challenge to analysis using proteomic techniques such as two-dimensional gel electrophoresis or MS. As most MS-based quantitative platforms have a dynamic range of only 3–4 orders of magnitude (7), these methods are of limited use for direct plasma and serum analysis. Firstly, the limited loading capacity of analytical columns coupled to MS instruments is occupied primarily by high-abundant proteins. Secondly, high-abundant proteins are likely to mask the less concentrated proteins and prevent their detection. Masking of the small amount of low-abundant proteins occurs due to effects such as ion suppression (8). Efforts to overcome the dynamic range problem have led to the development of protein depletion solutions such as the ProteoExtract™ Albumin/IgG removal kit by Merck, the POROS® Affinity Albumin/IgG depletion cartridges by Applied Biosystems, the ProteoPrep® plasma immunodepletion kit by Sigma, and the Multiple Affinity Removal System (MARS®) by Agilent Technologies. The latter two products remove a dozen or more high-abundant proteins in addition to albumin and IgG and represent the state-of-the-art in this respect. They are available in the form of spin cartridges, compatible with standard bench top centrifuges, and/or liquid chromatography (LC) columns coupled to a high-performance liquid chromatography (HPLC) system (see Note 1). The two products contain resins with covalently linked polyclonal antibodies, which target some of the most abundant proteins in plasma and serum (see Table 1). Processing a plasma/serum sample using these columns leads to the removal of approximately 94 % (Agilent Hu14) and 97 % (Sigma Top20) of the total protein content, respectively.

Depletion of high-abundant proteins followed by multidimensional protein fractionation can increase the analytical depth of plasma and serum analysis (9, 10). It has become a common practice for biomarker discovery in both plasma and serum (11–15), increasingly so in cerebrospinal fluid (CSF) (16–19), and is also being explored as a method for the analysis of urine (20). However, the depletion of high-abundant proteins also removes some of the less abundant proteins. This effect is not necessarily due to nonspecific binding to the depletion columns, but more likely because some of the targeted proteins such as albumin and transferring have the biologically important capability of binding and transporting other proteins and small molecules. Due to these protein–protein interactions, considerable numbers of lower-abundant proteins are retained on the columns and found in the high-abundant fraction (21, 22). Traces of high-abundant proteins, on the other hand, are found in the low-abundant protein fraction, because the depletion is not quantitative (efficiency is in the range of 92–99 % depending on the protein). Evaluations of the different depletion solutions and further considerations are available (21–25) and should be taken into account in the study design.

In this chapter we present a detailed protocol using the Agilent's MARS Hu14 LC column (see Note 2) in combination

Table 1
List of all targeted proteins by the Agilent MARS Hu14 and the Sigma ProteoPrep top20 columns

Swissprot accession	Recommended name	Agilent	Sigma
P01009	Alpha-1-antitrypsin	x	x
P02763	Alpha-1-acid glycoprotein	x	x
P01023	Alpha-2-macroglobulin	x	x
P02768	Albumin	x	x
P02647	Apolipoprotein A I	x	x
P02652	Apolipoprotein A II	x	x
P01024	Complement C3	x	x
P02671, -75, -79	Fibrinogen alpha/beta/gamma chain	x	x
P00738	Haptoglobin	x	x
P01876, -77	IgA (IgAlpha1/2 chain C region)	x	x
P01857, -59, -60	IgG (IgGamma1/2/3 chain C region)	x	x
P01871	IgM (IgMu chain C region)	x	x
P02766	Transthyretin (Prealbumin)	x	x
P02787	Transferrin	x	x
P04114	Apolipoprotein B		x
P00450	Ceruloplasmin		x
P0C0L4, -5	Complement C4		x
P02745, -46, -47	Complement C1q		x
P01880	IgD (IgDelta chain C region)		x
P00747	Plasminogen		x

with a the LC system the ÄKTA purifier from GE Health-care. This procedure is routinely used for the depletion of high-abundant proteins in clinical plasma and serum samples, and is also suitable for depletion of the high-abundant proteins in CSF.

2 Materials

1. Human plasma or serum samples.
2. Agilent MARS Hu14 column (4.6 mm inner diameter, length 100 mm, 1.66 mL bed volume, capacity of 40 μL of plasma/serum).

3. Buffer solution A (1 L bottle, Agilent): A neutral salt solution containing phosphate, pH 7.4 (see Note 3).
4. Buffer solution B (1 L bottle, Agilent): A concentrated urea solution, pH 2.2 (see Note 3).
5. Filter centrifuge tubes with 0.22 μm pore size (available from various manufacturers including Agilent) for removing particulates in the samples before the depletion.
6. Concentration centrifuge tubes with a 5 kDa molecular weight cutoff (available from various manufacturers including Agilent), 4 mL capacity.
7. HPLC system (we use an ÄKTA™ purifier UPC 10, GE Healthcare, Unicorn control software (v5.11 build 407), pump P-900, Frac-920 fraction collector, and a manual injection system using a Hamilton syringe).
8. Microcentrifuge and larger centrifuge for up to 15 mL capacity centrifuge tubes.
9. Eppendorf LoBind tubes (0.5 mL).
10. Colorimetric protein assay.

3 Methods

All procedures can be carried out at room temperature (recommended) or at 4 °C in cases for which minimization of protein degradation is needed (HPLC carried out in a cold room, see Note 4).

3.1 Sample Preparation

1. Dilute the plasma/serum samples four times with buffer solution A in a 0.5 mL LoBind Eppendorf tubes (e.g., add 40 μL of plasma/serum to 120 μL of buffer solution A) (see Note 5).
2. Pre-wet the 0.22 μm filter centrifuge tubes with 500 μL of buffer solution A, and centrifuge and discard the filtrate using a pipette. The tubes are now ready to filter the diluted samples.
3. Transfer the entire sample (160 + 10 = 170 μL, see Note 5) into the filter tubes, and centrifuge at 16,000 × g for 1 min or until the samples have completely passed through the filter.
4. Prepare the samples immediately before the depletion procedure or keep the diluted and filtered samples on ice until use.

3.2 HPLC System and MARS Column Preparation

1. Set up buffer solution A and buffer solution B as the only mobile phases of the HPLC system (see Note 6).
2. Purge HPLC lines with buffer solution A and buffer solution B at a flow rate of 1 mL/min for 10 min.

Table 2
LC timetable for the MARS Hu14 column

Phase	Time (min)	% Buffer B	Flow rate (mL/min)	Accumulated volume (mL)	Max press. (MPa)
Sample loading	0.00	0	0.125	0	6
	18.00	0	0.125	2.25	6
Washing	18.01	0	1.0	2.25	6
	20.00	0	1.0	4.25	6
Elution	20.01	100	1.0	4.25	6
	27.00	100	1.0	11.25	6
Regeneration	27.01	0	1.0	11.25	6
	38.00	0	1.0	22.25	6

3. Set up the LC method (see Table 2 for details) and run a blank method by injecting 200 μL of buffer solution A without a column. Ensure proper sample loop size (see Note 7).
4. Connect the Agilent MARS Hu14 column as per usual with the HPLC system. Make sure to do a drop-to-drop connection in order to avoid air bubbles in the system.
5. Equilibrate the column with buffer solution A at a flow rate of 1 mL/min for 4 min. Prior to injection of the first sample, and after the last sample of the session, run a blank method by injecting 200 μL of buffer solution A (see Note 8).

3.3 Load Sample onto Column and Run Method

1. Inject the diluted, filtered serum into the HPLC sample loop (see Note 9).
2. Begin the depletion method. Collect the fractions as they come off the column (see Note 10). The high-abundance proteins targeted for depletion will bind to the column, whereas all other proteins should flow through within the first 5 mL/20 min of the method (see Fig. 2 for a representative chromatogram).
3. Collect and evaluate the eluted fractions. Use sodium dodecyl sulfate-polyacrylamide gel electrophoresis (SDS-PAGE) analysis for aliquots of this evaluation and the flow-through fractions to evaluate column efficiency and capacity (see Note 11).

3.4 Buffer Exchange and Concentration of Collected Fractions

1. Wash concentration centrifuge tubes (5 kDa molecular weight cutoff, Agilent) by adding 2 mL of 50 mM ammonium bicarbonate buffer solution (natural pH ~ 7.5) and centrifuging at 4,000 × g for 12 min or until remaining volume is 100 μL (see Note 12).

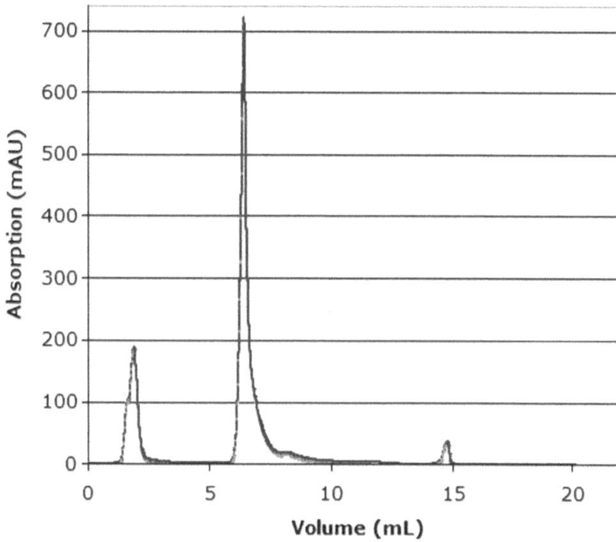

Fig. 2 Representative chromatograms of three serum sample depletion runs on the MARS Hu14 column (curves are superimposed). The *first peak* contains the low-abundant proteins (flow-through), the *second peak* contains the targeted high-abundant proteins (eluate), and the *third peak* is caused by the absorption of the buffer solutions as stated by Agilent's customer support

2. Add the flow-through and elution fractions to separate concentration centrifuge tubes and centrifuge at 4,000 × *g* for 12 min or concentrated to 200 μL.
3. Add 2 mL of 50 mM ammonium bicarbonate buffer solution and pipette the solution up and down to wash proteins off the filter membrane.
4. Centrifuge at 4,000 × *g* for 12 min or until concentrated to 200 μL.
5. Repeat steps 3 and 4 twice to reduce ionic strength and replace the MARS buffer solutions (see Note 13). It may be necessary to discard the filtrate several times during the process.
6. Concentrate the fraction to a volume 200 μL during the final centrifugation step. Add buffer to a final volume of 200 μL.
7. Pipette the concentrated fraction up and down several times to resuspend the proteins, and transfer 200 μL of the buffer-exchanged, concentrated fractions into Eppendorf tubes.
8. Measuring the protein concentrations using colorimetric protein assay should be carried out with bovine serum albumin as a standard ranging from 0.1 to 2 mg/mL (see Note 14) for further use (e.g., MS analysis).
9. Digest the fractions immediately using a protease (typically trypsin) and then store at −80 °C until MS analysis.

3.5 Column Care and Storage

1. After a blank run, disconnect and store the depletion column at 4 °C in buffer solution A.
2. Keep a column usage log. The manufacturer guarantees unchanged chromatographic performance for 200 runs (see Note 15).
3. Flush the HPLC system with water for 30 min at 1 mL/min, and then with 20 % ethanol/isopropanol in water for 30 min at 1 mL/min.

4 Notes

1. We have tested both spin cartridges and LC columns. While the former were significantly cheaper, the latter achieved a much more reproducible depletion performance (smaller coefficient of variation) and can be operated in fully automated mode, depending on the HPLC system used.
2. In our laboratory, we found the Agilent MARS Hu14 LC column to be the best combination of performance and price. The Sigma ProteoPrep 20 products remove an additional 3 % of the total protein weight from the sample (depletion of the 20 most abundant proteins versus the 14 most abundant proteins for the MARS Hu14 product), but at a significantly higher cost.
3. The exact composition of these buffer solutions is unknown to the authors and was not specified further by the manufacturer upon request.
4. The manufacturer recommends operating the MARS Hu14 column at room temperature (21 °C). However, the depletion column worked well for us using an HPLC system in a cold room (4 °C). No differences in separation performance were detected in an evaluation of the chromatograms and by SDS-PAGE testing of the fractions (data not shown). We also carried out tests for which we reduced the recommended flow rate by a factor of 2 to accommodate for the reduced temperature. This did not lead to different results in separation performance, but consequently doubled the required time to deplete a single sample. The Agilent customer support team was unable to comment on this matter. Agilent does not guarantee proper function at temperatures other than recommended.
5. Add an extra 10 μL of buffer A to enable the uptake of a full 160 μL later, or dilute more sample (e.g., 45 μL of sample with 135 μL of buffer solution A) and then inject 160 μL only.
6. The buffer solutions do not need any filtering or degassing. Purge the HPLC system with water first if it has been stored in 20 % ethanol/isopropanol in water, then with the buffer solutions. The direct change from an alcohol solution to the buffer solutions might cause salt precipitation within the system.

7. The sample loop volume should be at least twice the volume of the injected sample. We use a 500 μL loop for a sample volume of 160 μL.
8. This ensures complete elution of any residual proteins bound to the resin and maximum capacity of the column.
9. HPLC systems with autosamplers do not require manual steps. A manual injection with a Hamilton syringe, as fitted in our system, requires precautions specific for the system to avoid entrance of air into the sample loop. Before removal of the Hamilton syringe from the holder, switch the sample valve to the *inject* position. Then remove the syringe, wash it with buffer solution A to avoid cross-contamination between two samples, and take up the sample without any air bubbles in the syringe. Place the syringe back in the syringe support, switch the injection valve back to *load*, and inject the sample from the syringe into the sample loop now. Start your method, and leave the syringe in the holder during the run. Repeat these steps for each sample.
10. Use low-protein-binding tubes for fraction collection that do not contaminate the sample. Some tubes, e.g., made of polyethylene glycol (PEG), release material into the sample, which leads to a strong interfering signal during MS analysis. We use rimless polypropylene (PP) tubes with a capacity of 10 mL and, in combination with the Eppendorf protein LoBind tubes (0.5 mL, also made of PP), we do not experience such contamination.
11. Even if the high-abundant fraction is not analyzed immediately, we always perform the buffer exchange and desalting steps and store this sample at −80 °C for potential later studies. Even though current MS instruments are limited in their resolution power at present, future improvements might enable a deeper analysis of the proteins in this fraction. Furthermore, established approaches such as enzyme-linked immunosorbent assay (ELISA) or selective reaction monitoring (SRM) can be used for targeted analysis of this fraction.
12. Check the specifications of the concentration centrifuge tubes to adhere to the maximum allowed *g*-force. With higher *g*-forces, some proteins might leak trough the filter membrane. We have observed only minimal leakage at 4,000 × *g* with Agilent's concentration centrifuge tubes. The buffer exchange can be done in a refrigerated centrifuge at 4 °C to minimize protein degradation, as the procedure can take several hours to complete. Furthermore, buffer solutions other than the suggested ammonium bicarbonate buffer can be used if a particular downstream process (e.g., further protein fractionation) requires this. Check protein precipitation/solubility in this case.

13. For the flow-through fraction (low-abundant proteins), a single buffer-exchange step has worked for us, too. LC-MS systems fitted with a trap column wash the sample automatically as an additional desalting step prior to ionisation/analysis.
14. The expected protein concentration in the low-abundant fraction in a final volume of 200 μL is between 1–1.8 μg of protein per μL. Variation between samples from different donors is normal and depends on the hydration state of the individual at the time of blood sampling. This is true for the high-abundant fraction as well. In the latter, a typical protein concentration is 10–20 times higher than in the low-abundant protein fraction.
15. We keep a log for every depletion column. Even though Agilent guarantees no more than 200 runs without deterioration of the separation performance, we have seen little to no changes after 250 or even 300 runs per column in several cases. We use only the manufacturer's buffer solutions, filter every sample as recommended, and store the column correctly to prolong its life. We recommend keeping a large stock of dedicated quality control plasma/serum samples, and carrying out a test depletion of an aliquot of this at least every 10–20 runs to assess performance over the entire lifetime of a column.

Acknowledgments

This work was supported by the Stanley Medical Research Institute (SMRI, USA), Psynova Neurotech (UK), and European Union FP7 SchizDX research program (grant reference 223427).

JAJJ and DMS declare no conflict of interest. PCG and SB are consultants for Myriad-RBM although this does interfere with Springer Science policies with regards to sharing of data or materials.

References

1. Zhou M, Lucas DA, Chan KC et al (2004) An investigation into the human serum "interactome". Electrophoresis 25:1289–1298
2. Rai AJ, Gelfand CA, Haywood BC et al (2005) HUPO Plasma Proteome Project specimen collection and handling: towards the standardization of parameters for plasma proteome samples. Proteomics 5:3262–3277
3. Burtis CA, Ashwood ER (2001) Tietz fundamentals of clinical chemistry, 5th edn. Saunders Company, Philadelphia, PA
4. Wrotnowski C (1998) The future of plasma proteins. Gen Eng News 18:14
5. Anderson NL, Anderson NG (2002) The human plasma proteome: history, character, and diagnostic prospects. Mol Cell Proteomics 1:845–867
6. Tirumalai RS, Chan KC, Prieto DA, Issaq HJ, Conrads TP, Veenstra TD (2003) Characterization of the low molecular weight human serum proteome. Mol Cell Proteomics 2:1096–1103
7. Hoffman SA, Joo WA, Echan LA, Speicher DW (2007) Higher dimensional (Hi-D) separation strategies dramatically improve the potential for cancer biomarker detection in

serum and plasma. J Chromatogr B Analyt Technol Biomed Life Sci 849:43–52
8. Annesley TM (2003) Ion suppression in mass spectrometry. Clin Chem 49:1041–1044
9. Faca V, Pitteri SJ, Newcomb L et al (2007) Contribution of protein fractionation to depth of analysis of the serum and plasma proteomes. J Proteome Res 6:3558–3565
10. Smith MPW, Wood SL, Zougman A et al (2011) A systematic analysis of the effects of increasing degrees of serum immunodepletion in terms of depth of coverage and other key aspects in top-down and bottom-up proteomic analyses. Proteomics 11:2222–2235
11. Tang H-Y, Beer LA, Barnhart KT, Speicher DW (2011) Rapid verification of candidate serological biomarkers using gel-based, label-free multiple reaction monitoring. J Proteome Res 10:4005–4017
12. Zhao L, Liu Y, Sun X, Peng K, Ding Y (2011) Serum proteome analysis for profiling protein markers associated with lymph node metastasis in colorectal carcinoma. J Comp Pathol 144:187–194
13. Liao Q, Zhao L, Chen X, Deng Y, Ding Y (2008) Serum proteome analysis for profiling protein markers associated with carcinogenesis and lymph node metastasis in nasopharyngeal carcinoma. Clin Exp Metastasis 25:465–476
14. Hamrita B, Chahed K, Trimeche M et al (2009) Proteomics-based identification of alpha1-antitrypsin and haptoglobin precursors as novel serum markers in infiltrating ductal breast carcinomas. Clin Chim Acta 404: 111–118
15. Kaur P, Reis MD, Couchman GR, Forjuoh SN, Greene JF, Asea A (2010) SERPINE 1 links obesity and diabetes: a pilot study. J Proteomics Bioinform 3:191–199
16. Fratantoni SA, Piersma SR, Jimenez CR (2010) Comparison of the performance of two affinity depletion spin filters for quantitative proteomics of CSF: evaluation of sensitivity and reproducibility of CSF analysis using GeLC-MS/MS and spectral counting. Proteomics Clin Appl 4:613–617
17. Siegmund R, Kiehntopf M, Deufel T (2009) Evaluation of two different albumin depletion strategies for improved analysis of human CSF by SELDI-TOF-MS. Clin Biochem 42: 1136–1143
18. Zuberovic A, Hanrieder J, Hellman U, Bergquist J, Wetterhall M (2008) Proteome profiling of human cerebrospinal fluid: exploring the potential of capillary electrophoresis with surface modified capillaries for analysis of complex biological samples. Eur J Mass Spectrom 14:249–260
19. Thouvenot E, Urbach S, Dantec C et al (2008) Enhanced detection of CNS cell secretome in plasma protein-depleted cerebrospinal fluid. J Proteome Res 7:4409–4421
20. Sigdel TK, Lau K, Schilling J, Sarwal M (2008) Optimizing protein recovery for urinary proteomics, a tool to monitor renal transplantation. Clin Transplant 22:617–623
21. Zolotarjova N, Martosella J, Nicol G, Bailey J, Boyes BE, Barrett WC (2005) Differences among techniques for high-abundant protein depletion. Proteomics 5:3304–3313
22. Koutroukides TA, Guest PC, Leweke FM et al (2011) Characterization of the human serum depletome by label-free shotgun proteomics. J Sep Sci 34:1621–1626
23. Fania C, Vasso M, Torretta E et al (2011) Setup for human sera MALDI profiling: the case of rhEPO treatment. Electrophoresis 32:1715–1727
24. Martosella J, Zolotarjova N, Liu H, Nicol G, Boyes BE (2005) Reversed-phase high-performance liquid chromatographic prefractionation of immunodepleted human serum proteins to enhance mass spectrometry identification of lower-abundant proteins. J Proteome Res 4:1522–1537
25. Björhall K, Miliotis T, Davidsson P (2005) Comparison of different depletion strategies for improved resolution in proteomic analysis of human serum samples. Proteomics 5:307–317

Chapter 2

Tissue Sample Preparation for Biomarker Discovery

Yoshiyuki Suehara, Daisuke Kubota, and Tsuyoshi Saito

Abstract

Global protein expression studies, an approach known as "proteomics," can offer important clues for understanding tumor biology that cannot be obtained by other approaches. Proteomic studies have provided protein expression profiles of tumors that can be used to develop novel diagnostic and therapeutic biomarkers. In this chapter, we describe the strategy and design of proteomic studies, as well as the protocols for tissue sample collection and preparation for biomarker discovery, especially tumor biomarkers, followed by a few examples of our recent proteomic studies.

Key words Proteomics, 2D-DIGE, Tissue samples preparation, Soft tissue sarcomas

1 Introduction

The strategies and protocols used for human tissue sample preparation for developing biomarkers, especially tumor biomarkers, by proteomic approaches are described in this chapter. Proteomics is the study of global protein expression, and the phrase "expression profile" refers to the expression of thousands of individual proteins simultaneously in a given tissue sample (1–3). Unlike studies of a single protein or pathway, proteomic methods enable a systematic overview of the expressed protein profiles, which, in the case of tumors, can ultimately improve the diagnosis, prognosis, and management of patients by revealing the protein interactions affecting the overall tumor progression (1–3). Furthermore, a differential protein expression analysis can be used to compare tumors with normal tissues or high-grade with low-grade malignant tumors, which may implicate a range of protein biomarkers potentially indicative of disease and the prognosis of the disease (1–3).

Recent developments in high-throughput screening techniques, such as array-based comparative genomic hybridization analysis and cDNA microarray technology, now allow for the screening of several thousand DNA and mRNA sequences in tumors at once, thereby making it possible to identify the genes

Ming Zhou and Timothy Veenstra (eds.), *Proteomics for Biomarker Discovery: Methods and Protocols*, Methods in Molecular Biology, vol. 1002, DOI 10.1007/978-1-62703-360-2_2, © Springer Science+Business Media, LLC 2013

relevant to the histological diagnosis and also the clinical features of the tumors (1–3). However, DNA sequencing or measurement of the mRNA expression alone cannot detect the posttranslational modifications of proteins, such as phosphorylation, glycosylation, or differences in protein stability, and these factors play important roles in the malignant behavior of tumors (1–4). Global protein expression studies, known as proteomics, have a big advantage over global DNA or mRNA expression studies for detecting the post-translational modifications (1). Furthermore, the results obtained from proteomic studies will be more easily applicable to the clinical field, because the protein expression levels can be easily evaluated and confirmed using antibodies, such as Western blotting and immunohistochemistry, and these tests are both useful and convenient in the clinical setting (1). The proteomic approaches have already been used to develop molecular subclassifications and diagnostic biomarkers for several kinds of cancers (1–3). In addition, our proteomic technologies have identified candidate proteins associated with a differential diagnosis (3–6), predictive prognosis, (7–9) and the response to chemotherapy (10) in bone and soft tissue sarcomas.

With regard to human tissue sample preparation for biomarker discovery, the most important matters are that investigators must take special care to (1) use high-quality samples and/or proteins (avoiding degradation, dephosphorylation, and contamination) and (2) ensure that the sample received an exact and correct diagnosis by experts (histology, grade, and stage) and (3) that the clinical information can be followed completely (time of metastasis, time of death and disease status). To ensure that these three points are all met, we believe that collaboration between researchers and clinicians, including pathologists and surgeons, is critical for studies involving biomarker discovery and development (Fig. 1). Researchers may know the details of the techniques and skills for proteomics and how to extract proteins, while pathologists can collect surgical tissue samples and diagnose them, and surgeons can collect not only the samples but also the required clinical information. Furthermore, clinicians have a better understanding of what may improve the clinical outcome for the patients, and they can suggest the types of biomarkers that can be used to optimize existing therapeutic protocols based on their experience. Therefore, the collaboration of basic and clinical scientists can allow for the evaluation of the results from a more relevant clinical perspective or can therefore help to more effectively select biomarkers that can best benefit patients.

To discover useful biomarkers from surgical tissue samples by global protein expression studies (proteomics), it is necessary to conduct high-integrity and reliable studies which consist of three sets (Fig. 2): (1) a discovery set that tries to identify the candidate biomarkers from the global protein expression profiles of tissue

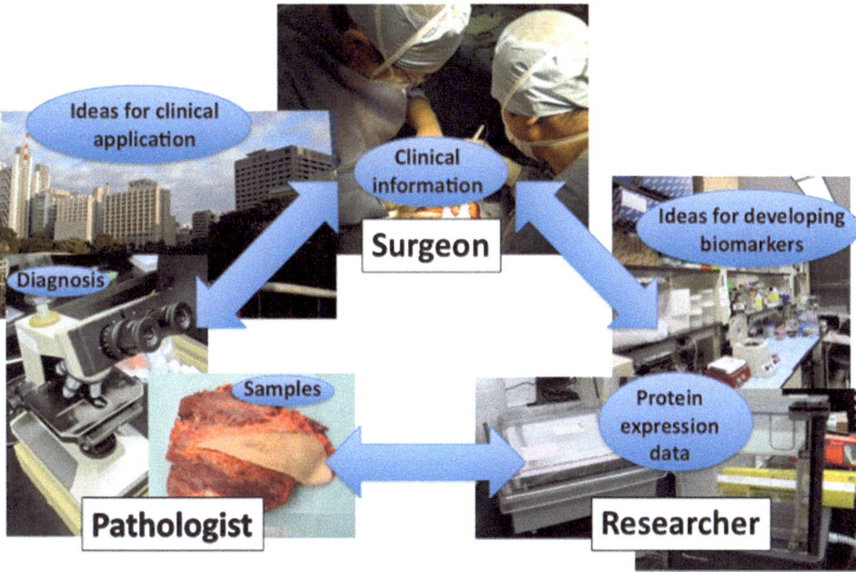

Fig. 1 Collaboration between basic researchers, pathologists, and surgeons for developing biomarker. The collaboration between researchers and clinicians, including pathologists and surgeons, is critical during biomarker discovery development. While researchers may know the details of the techniques and skills for proteomic studies, the pathologist can collect surgical tissue samples and diagnose the samples, and surgeons can collect samples and the required clinical information. Furthermore, clinicians often have a better understanding of what may improve the clinical outcome for the patients

samples (in our studies, we usually use two-dimensional difference gel electrophoresis (2D-DIGE) and mass spectrometry (MS) for these studies), (2) a confirmation set that is used to confirm the protein expression differences identified in the discovery set using other proteomic tools (in our studies, we usually use a Western blot analysis), and (3) a validation set that is used to verify the power and ability of a biomarker on a large scale using numerous samples, to develop a biomarker for clinical applications (in our studies, we generally use immunohistochemistry and Western blot analyses).

The first procedure involving the discovery set is to acquire protein expression profiles from tissue samples by 2D-DIGE, to list the candidate protein spots, and so on based on the expression data, and to identify proteins by MS. Second, the confirmation set is to substantiate the protein expression levels in the results of the discovery set. A Western blotting analysis employing specific antibodies against the candidate proteins is a useful and efficient confirmation tool for this step. Finally, the purpose of the validation set is to prove and establish the abilities and capabilities (as well as limitations) of biomarkers, in order to develop useful

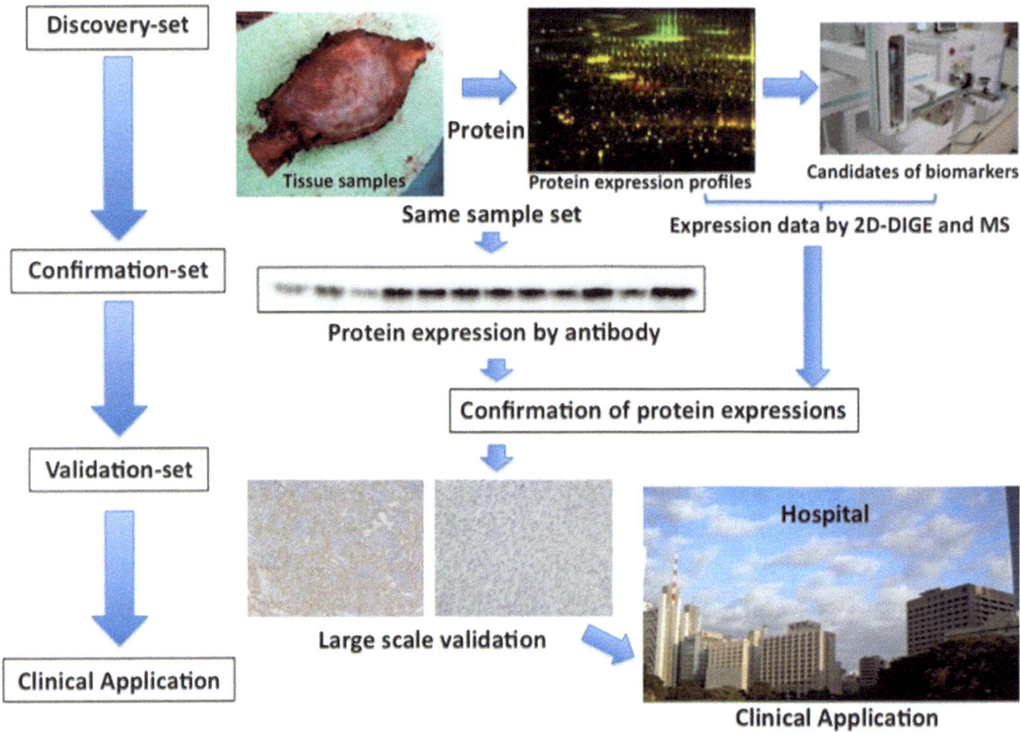

Fig. 2 The strategy for developing biomarkers using tissue samples. To discover and identify useful biomarkers using surgical tissue samples by proteomics studies, it is necessary to conduct high-integrity and reliable studies which consist of three sets: (1) a discovery set, (2) a confirmation set, and (3) a validation set. In our study, the discovery set is used to try to identify candidate biomarkers from global protein expression profiles of tissue samples using 2D-DIGE and MS. The confirmation set is used to try to confirm the protein expression of the discovery set by other techniques, such as a Western blot analysis. The validation set is used to verify the power and ability of a biomarker for predicting the disease using a large number of samples (e.g., FFPE samples), and often involves immunohistochemistry

biomarkers for clinical applications. As the validation set should include a large number of samples, we usually use immunohistochemistry for the confirmation of the expression changes. The reasons for this are (1) that the hospitals usually stock and store many more formalin-fixed paraffin-embedded (FFPE) than frozen tissue samples and (2) pathologists use FFPE samples to diagnose the patient's disease as part of their routine clinical assessment.

To provide an example of such studies, we used 2D-DIGE to investigate the tissue profiles and identify novel prognostic biomarkers in patients with gastrointestinal mesenchymal tumors (GISTs), which is one of the mesenchymal tumors of the gastrointestinal tract and has been shown to exhibit a broad spectrum of clinical behaviors (Fig. 3) (7). As our discovery set, we conducted

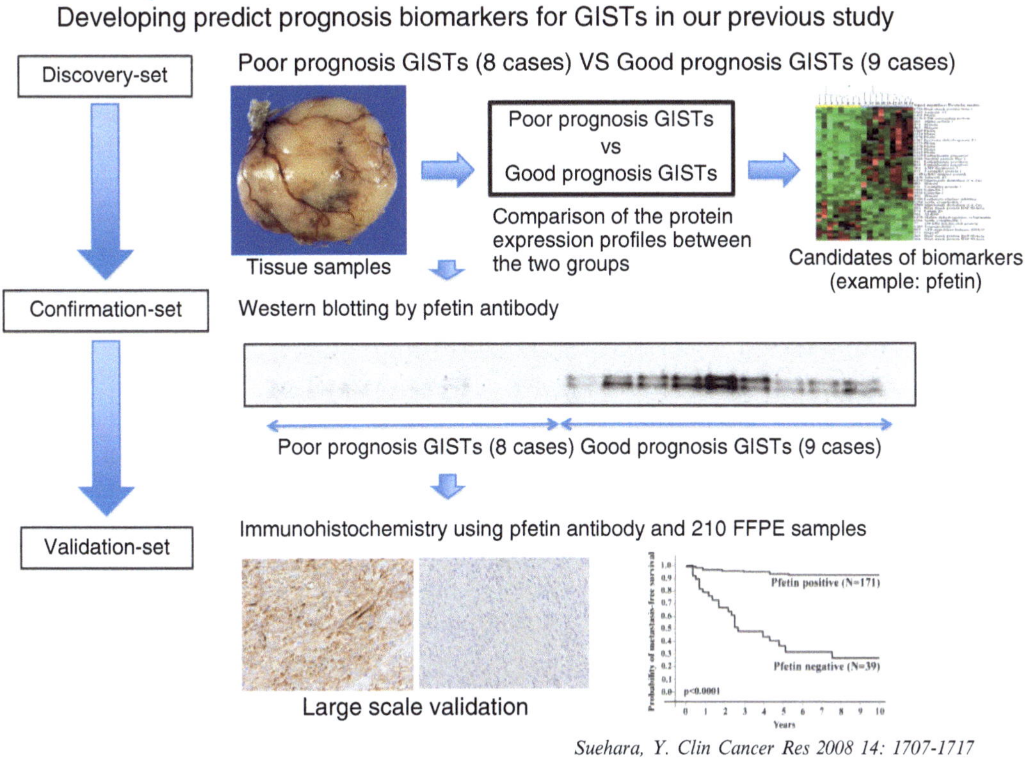

Fig. 3 The development of prognostic biomarker for GISTs (7). We previously performed a study to investigate the protein expression profiles and identify novel prognostic biomarkers in GISTs. As our discovery set, we conducted a proteomic study of GISTs using 2D-DIGE, and extracted proteins from surgical tissues samples. To identify the protein expression profiles that correlated with the prognosis of GISTs, we compared the protein expression profiles between two groups (eight poor prognosis patients vs. nine good prognosis patients) using 2D-DIGE. We identified that the pfetin protein was more highly expressed in samples from the good prognosis group. As our confirmation set, we confirmed the pfetin expression using the Western blot analyses. As our validation set, these results were validated based on immunohistochemical studies of the pfetin expression in 210 FFEP samples. There were statistically significant differences in the overall survival of GISTs patients between those who were pfetin positive and those who were negative ($p < 0.0001$)

a quantitative expression study of the proteins in GIST samples using 2D-DIGE. These protein samples were extracted from frozen tissues of GIST surgical samples. To identify the protein expression profiles that correlated with the prognosis of the GISTs, 17 GIST samples were divided into two groups (eight poor prognosis patients and nine with a good prognosis). We then compared the protein expression profiles between the two groups using 2D-DIGE. We identified that the pfetin protein had different intensities in the two groups of samples, with higher intensity expression in the good prognosis group. As our confirmation set, we confirmed that the pfetin expression was higher in the good prognosis group than in the poor prognosis group using a Western blot analysis.

Table 1
Optimal samples for proteomics research to develop biomarkers

Sample type	Tool	Frequency in use
Frozen tissue sample	2D-DIGE, MS, ELISA, WB	Discovery set > confirmation set > validation set
Formalin-fixed paraffin-embedded (FFPE)	IHC (includes TMA)	Validation set > confirmation set > discovery set

2D-DIGE two-dimensional difference gel electrophoresis, *MS* mass spectrometry, *ELISA* enzyme-linked immunosorbent assay, *WB* Western blot, *IHC* immunohistochemistry, *TMA* tissue microarray

Furthermore, as our validation set, the results were validated based on the immunohistochemical studies of the pfetin expression in 210 FFEP samples of GIST.

For proteomics studies intended to develop biomarkers for useful clinical applications, we employ frozen tissue samples as well as FFPE samples (Table 1) (1, 3–11). The frozen tissue samples are most suitable for acquiring the global protein expression profiles in the discovery set, as the tissue samples can provide high-quality proteins, which yield more accurate protein profiles. The extracted proteins are also used for the confirmation set. In the validation set, frozen tissue samples can sometimes be employed; however, it is generally difficult to prospectively collect a sufficiently large number of frozen tissue samples for large-scale analyses required for validation. On the other hand, FFPE samples are accumulated during routine clinical work for diagnosis, and a sufficient number of samples are likely to be stored to allow for a retrospective analysis. Therefore it is easier to collect FFPE than frozen tissue samples making FFPE samples the most commonly used type of sample for the validation set.

In the studies of biomarker discovery for humans, it is necessary to employ human tissue samples, especially patient samples, which can be obtained from resected surgical or biopsy materials (Fig. 4). In order to obtain high-quality proteins, it is necessary to recover the surgical material promptly from the operating room, and to collect the tissue samples from the fresh surgical materials before fixation with formalin. If tissues cannot be recovered promptly, the materials should be preserved at an optimal temperature (we recommend 4 °C). As, naturally, the clinical diagnosis for patients has first priority in the samples, it is necessary to collect representative parts of the materials after the pathologist has already secured sufficient information and samples for the diagnosis. It is also helpful to work with the pathologist to ensure that high-quality and representative samples of the tissue are collected. Therefore, investigators should be in close collaboration with pathologists.

With respect to sampling, all studies have to be permitted by the institutional review board. When collecting tissue samples from

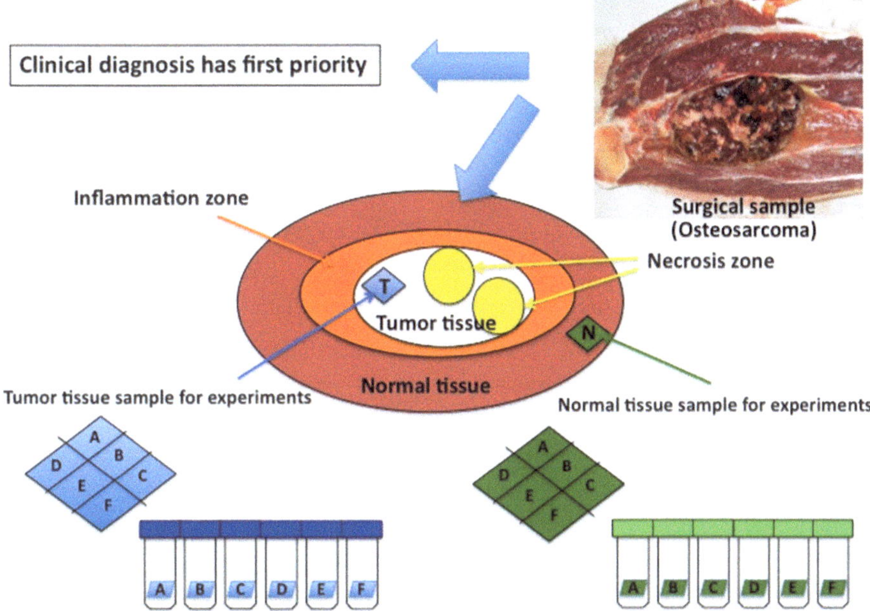

Fig. 4 Tissue sample acquisition from surgical samples and sample storage. Naturally, the clinical diagnosis for patients has first priority with regard to the tissue samples. During sampling, researches have to collect representative parts of the materials and different tumor locations, and avoid acquiring samples from normal tissues, necrotic areas, and reactive (inflammation) zones. The samples should be cut into 5 mm blocks, and divided samples should be placed in individual tubes with detailed labels

surgical materials, it is necessary to collect representative tumor sections, and to avoid acquiring samples from normal tissue, or necrotic and reactive (inflammatory) zones. Sampling errors are the most frequent cause of noise contamination of data, and interfere with the analyses and identification of proteins specifically expressed in the tumor. If stored samples that were collected previously will be used, it is necessary to check the sample quality and morphology before using the tissue for any analysis. Additionally, when developing biomarkers, it is sometimes necessary to compare tumor samples and normal samples. Therefore, if permission is given by the patient and the institution, then both normal and tumor samples should be collected and stored. For sample storage, the samples should be cut into 5 mm blocks, and divided samples should be placed in the each labelled tube separately. When the tissue samples have contamination due to either blood or other materials, then all samples should be rinsed in clean PBS prior to storage. The tubes should be stored in liquid nitrogen or −80 °C until they are used. Additionally, special care should be taken to avoid contamination during all of the steps in each of the procedures.

Table 2
Proteomics for developing biomarkers

Purpose of experiment	Tool(s)	Sample source(s)
To acquire global protein expression profiles	2D-DIGE, MS	Frozen tissue sample
To confirm expression of identified proteins	ELISA, WB	Frozen tissue sample
To confirm expression of identified proteins	IHC (includes TMA)	Formalin-fixed paraffin-embedded (FFPE)

2D-DIGE two-dimensional difference gel electrophoresis, *MS* mass spectrometry, *ELISA* enzyme-linked immunosorbent assay, *WB* Western blot, *IHC* immunohistochemistry, *TMA* tissue microarray

FFPE samples are usually managed and stored by a pathologist. Pathologists usually have detailed information about the FFPE samples. Therefore, supervision by a pathologist is necessary when an investigator has to choose a representative FFPE specimen from all the available tumor specimens. Unstained slides of the representative sample at an optimal thickness (we recommend 3–5 μm sections) should therefore be prepared to conduct more efficient validation studies.

To obtain global protein expression profiles (discovery studies), proteomic studies employ electrophoresis, mass spectrometry, and protein microarrays for the characterization of proteins (Table 2). These proteomic tools have their own individual advantages and limitations, affecting their ability to assess the protein profile. With regard to proteomic analyses to identify biomarkers as part of cancer research, electrophoresis (specifically 2D-DIGE) and MS (especially surface-enhanced laser desorption/ionization time-of-flight mass spectrometry (SELDI-TOF-MS), matrix-assisted laser desorption/ionization time-of-flight mass spectrometry (MALDI-TOF-MS), and liquid chromatography–mass spectrometry (LC/MS)) have mainly been used to obtain protein expression profiles. In our studies, we primarily employ 2D-DIGE, because this technology is the most frequently used method for examining protein expression profiles in proteomic studies of biomarkers (1, 3–11). When a specific target has already been identified (in the confirmation and discovery studies), proteomic studies employing a Western blot analysis, enzyme-linked immunosorbent assay (ELISA), and immunohistochemistry are common. These tools use specific antibodies against the target protein(s). In our studies, 2D-DIGE, MS, WB, and ELISA all detect proteins that are extracted from frozen tissue samples, while immunohistochemistry is used mainly for FFPE samples.

The procedure used for protein extraction from tumor or normal tissue is important to obtain precise and intact protein expression profiles for the proteomic analysis (1–3, 12). To prevent the degradation and dephosphorylation of the protein, it is essential to preserve the analytical quality during extraction. It is also necessary to maximize the reproducibility and to minimize waste. Containment of the sample is important to avoid the risk of contamination to and from each sample. Either the ionic detergent SDS or the chaotrope, urea, provides efficient ways to solubilize water-insoluble protein subunits.

2 Materials

2.1 Tissue Sample Preparation

1. Surgical materials: The use of surgical specimens must be approved by the institution's ethics committee, and written informed consent obtained from all of the patients.
2. Multi-beads shocker (Yasui Kikai, Osaka, Japan).
3. High speed refrigerated microcentrifuge MX-100 (TOMY Digital Biology, Tokyo, Japan).

3 Methods

3.1 Protocol for Sampling and Storage

1. Weigh the frozen clinical tissue samples.
2. Wash these samples twice with PBS at room temperature (see Note 1).
3. Cut the tissues into appropriately small pieces (see Note 2).
4. Place the cut pieces in tubes (see Note 3).
5. Store tissues in liquid nitrogen or −80 °C until they are used.

3.2 Protocol for Protein Extraction

1. Transfer the frozen sample to a new special collection tube that is provided with the Multi-beads shocker (Fig. 5a) (see Note 4).
2. Crush the frozen tissue to powder with the Multi-beads shocker after cooling with liquid nitrogen at 250× *g* for 20 s (Fig. 5b) (see Note 5).
3. Add urea-lysis buffer (6 mol/L urea, 2 mol/L thiourea, 3 % CHAPS, and 1 % Triton X-100) to the liquid nitrogen pools (Fig. 5c, d) (see Note 6).
4. Incubate on ice for 30 min (Fig. 5e) (see Note 7).
5. Centrifuge at 17,400× *g* for 30 min at 4 °C (Fig. 5f, g) (see Note 8).
6. Collect the supernatant, and place it in a new collection tube (Fig. 5h).
7. Determine the concentration of the total proteins (see Note 9).

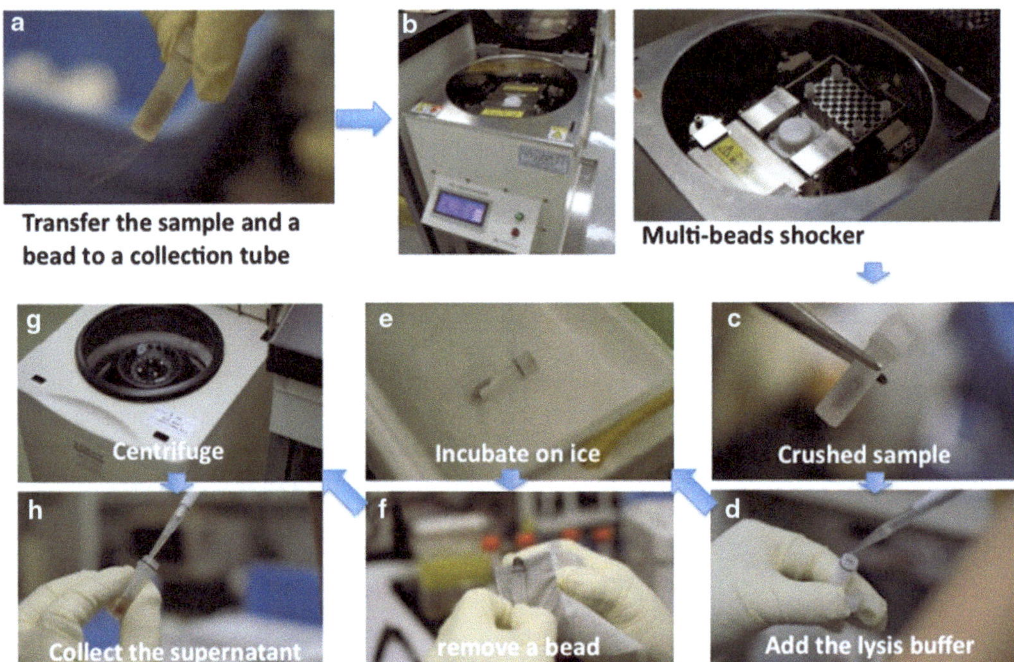

Fig. 5 Protocol for protein extraction. (**a**) The frozen sample is transferred with a bead to a new special collection tube that is provided with the Multi-beads shocker under cooling with liquid nitrogen. (**b**) The frozen tissue is crushed to powder with the Multi-beads shocker under cooling with liquid nitrogen at 250× *g* for 20 s. (**c**) Crushed sample. (**d**) The urea lysis buffer is added to the liquid nitrogen pools. (**e**) Incubate on ice for 30 min. (**f**) The bead is removed from the tube. (**g**) The tubes are centrifuged at 17,400× *g* for 30 min at 4 °C. (**h**) The supernatant is collected and placed in a new collection tube

4 Notes

1. The purpose of such washing is to remove any excess serum from the harvested samples by this procedure.
2. Usually, the sample should be about the size of a grain of rice (5 mm in diameter).
3. We recommend the use of tubes designed for the storage of biological materials, human or animal cells, at temperatures as low as −190 °C (e.g., Cryotubes, Cryogenics). Label the details of the samples on these tubes. Make a list and data sheet for the samples to organize them during both storage and analysis.
4. Be careful to prevent any potential contamination. Place an optimal amount of frozen tissue in each tube.
5. During crushing, be careful not to dissolve the frozen sample. Be careful to avoid potential contamination caused by the cracking of the tubes.

6. The amount is adjusted according to the size of the sample (50–500 μL). Be careful to avoid contamination.
7. Mix gently two or three times during this incubation. Wait until the sample is completely dissolved, even if it takes more than 30 min. Be careful to avoid potential contamination.
8. Remove the bead from the tube before the procedure of centrifuge.
9. If the concentration is too high (over 50 μg/μL), dilute using urea-lysis buffer.

Acknowledgments

This work was supported by a grant from the Japan Society for the Promotion of Science (JSPS), a science Grant-in-Aid for Young Scientists B, No-22791405. The authors appreciate critical comments and support from Dr. Makoto Endo and Dr. Eisuke Kobayashi.

References

1. Suehara Y (2011) Proteomic analysis of soft tissue sarcoma. Int J Clin Oncol 16:92–100
2. Kondo T, Hirohashi S (2006) Application of highly sensitive fluorescent dyes (CyDye DIGE Fluor saturation dyes) to laser microdissection and two-dimensional gel electrophoresis (2D-DIGE) for cancer proteomics. Nat Protoc 1:2940–2956
3. Suehara Y, Kondo T, Fujii K et al (2006) Proteomic signatures corresponding to histological classification and grading of soft-tissue sarcomas. Proteomics 6:4402–4409
4. Kawai A, Kondo T, Suehara Y, Kikuta K, Hirohashi S (2008) Global protein-expression analysis of bone and soft tissue sarcomas. Clin Orthop Relat Res 466:2099–2106
5. Suehara Y, Kikuta K, Nakayama R et al (2009) Anatomic site-specific proteomic signatures of gastrointestinal stromal tumors. Proteomics Clin Appl 3:584–596
6. Suehara Y, Kikuta K, Nakayama R et al (2009) GST-P1 as a histological biomarker of synovial sarcoma revealed by proteomics. Proteomics Clin Appl 3:623–634
7. Suehara Y, Kondo T, Seki K et al (2008) Pfetin as a prognostic biomarker of gastrointestinal stromal tumors revealed by proteomics. Clin Cancer Res 14:1707–1717
8. Kikuta K, Tochigi N, Shimoda T et al (2009) Nucleophosmin as a candidate prognostic biomarker of Ewing's sarcoma revealed by proteomics. Clin Cancer Res 15:2885–2894
9. Suehara Y, Tochigi N, Kubota D et al (2011) Secernin-1 as a novel prognostic biomarker candidate of synovial sarcoma revealed by proteomics. J Proteomics 74:829–842
10. Kikuta K, Tochigi N, Saito S et al (2010) Peroxiredoxin 2 as a chemotherapy responsiveness biomarker candidate in osteosarcoma revealed by proteomics. Proteomics Clin Appl 4:560–567
11. Kikuta K, Kubota D, Saito T et al (2012) Clinical proteomics identified ATP-dependent RNA helicase DDX39 as a novel biomarker to predict poor prognosis of patients with gastrointestinal stromal tumor. J Proteomics 75:1089–1098
12. Ericsson C, Nister M (2011) Protein extraction from solid tissue. Methods Mol Biol 675:307–312

Chapter 3

Subcellular Fractionation for Identification of Biomarkers: Serial Detergent Extraction by Subcellular Accessibility and Solubility

Sun-Il Hwang and David K. Han

Abstract

Cellular localization of proteins is one of the most valuable sources of information regarding spatiotemporal biological events involved in human disease. This information is sometimes enhanced by carrying out protein isolation using a process known as subcellular fractionation. This involves the sequential extraction of proteins from specific compartments and/or organelles within the cell. Additionally, subcellular fractionation for biomarker discovery enables the in-depth analysis of biomolecules by reducing the complexity of the protein mixture. In this chapter, four custom fractionation approaches and one commercial kit are compared for their efficacy and compatibility with subsequent proteomic analysis.

Key words Subcellular fractionation, Biomarker discovery, Protein biomarkers, Subcellular localization and translocation, Differential detergent fractionation

1 Introduction

The application of liquid chromatography and tandem mass spectrometry (LC-MS/MS) based proteomic analysis for biomarker discovery has been developed for the early detection, treatment, and prognosis of human diseases. Improvements to instrumentation in both LC and MS have made possible the detection and quantification of lower femtomoles of peptides. However, the complexity and dynamic ranges of proteins expressed in biological systems of cells, tissues, and organs of the human body still present challenges with current technology using the state-of-the-art LC-MS/MS.

Various chemical pre-fractionation methodologies for protein identification were developed using hydrophobic, hydrophilic, ion exchange, and size exclusion separations. These fractionation approaches prior to LC-MS/MS analysis have not provided enough fractionation power to overcome the 10^7–10^8 dynamic range of protein abundance in human tissues (1). Complementary methods

Ming Zhou and Timothy Veenstra (eds.), *Proteomics for Biomarker Discovery: Methods and Protocols*, Methods in Molecular Biology, vol. 1002, DOI 10.1007/978-1-62703-360-2_3, © Springer Science+Business Media, LLC 2013

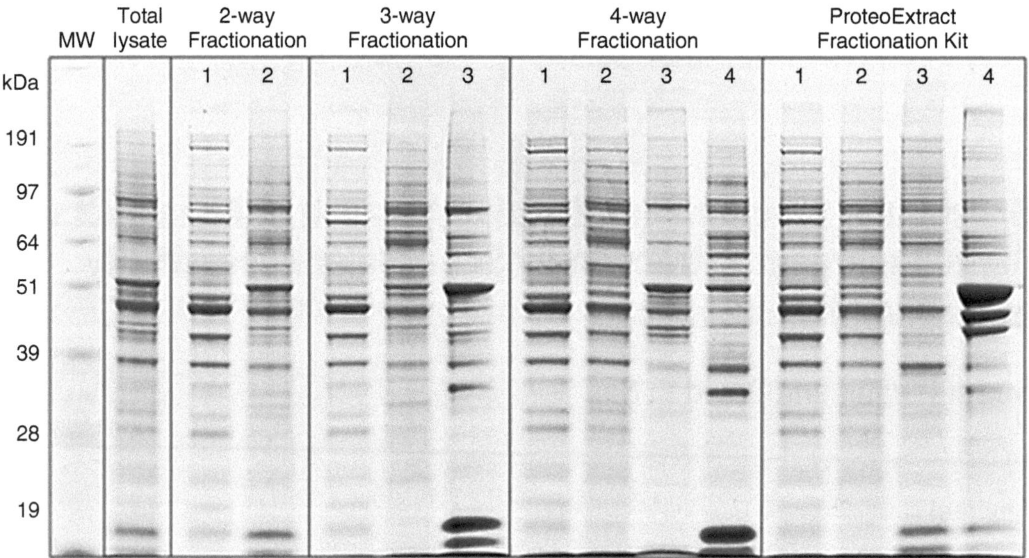

Fig. 1 The protein band patterns of various subcellular fractionation methods. *Left* to the *right*: Total cell lysate, 2-way subcellular fractionation, 3-way subcellular fractionation, 4-way subcellular fractionation, and ProteoExtract subcellular fractionation kit. Fractions 1, cytosol; 2, membrane; 3, nucleus; and 4, cytoskeleton

utilizing subcellular fractionation have been established using either density gradient centrifugation (DGC) (2–7) or differential detergent fractionation (DDF) (8, 9). The DGC method isolates highly pure subcellular organelles; however it is time consuming and requires fresh samples and a high degree of precision for reproducible results. A comparison of results from different institutes nationally or internationally would generate high variability due to different technical experimental variations. The alternative method using DDF simplifies fractionation of the subcellular structures based on protein solubility and/or hydrophobicity and employs more reproducible and universal procedures which are more compatible with previously frozen samples.

Proteins identified as biomarkers of disease are often associated with a specific function or biological alteration which manifests at a specific subcellular site. Frequently this molecular alteration in function requires the translocation of the protein from one site to another within the subcellular structure. To achieve organelle isolation by cellular structure, specimens need to be processed quickly before cellular proteases are activated. Samples should be stored at cold temperatures without freezing to preserve membrane integrity. In the clinical setting, it is difficult to maintain the tissue samples fresh in 4 °C and process them within a couple of hours. The current standard process for clinical tissue banks is snap-freezing after surgery, which disrupts the membrane structure dramatically. Therefore, we compared one commercial and four custom DDF techniques. Efficacy was judged visually by the pattern of protein bands upon electrophoretic separation and staining as shown in Fig. 1.

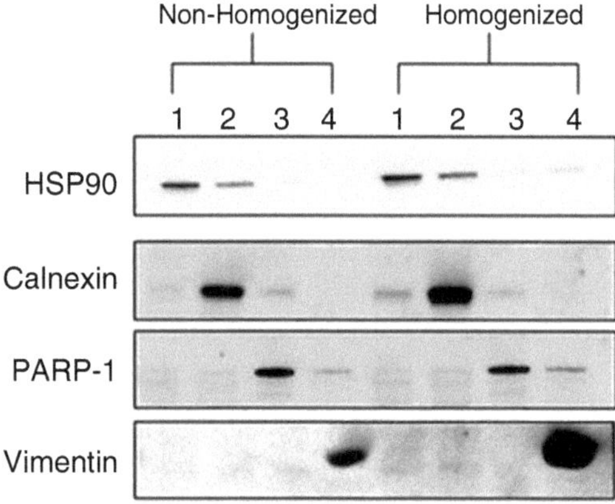

Fig. 2 Fractionation efficiency of the 4-way fractionation kit with/without homogenization using Omni homogenizer. Western blot for cytosolic fraction was demonstrated with Heat Shock Protein-90 (HSP-90), membrane fraction with calnexin, nuclear fraction with Poly [ADP-ribose] polymerase 1 (PARP-1), and cytoskeletal fraction with vimentin. Fractions 1, cytosol; 2, membrane; 3, nucleus; and 4, cytoskeleton

Differential banding patterns are evident in each fraction, which indicates effective fractionation resulting in a unique protein population for each fraction. In addition, the fractionation efficiency for cytosolic, membrane, nucleus, and cytoskeleton fractions was validated by Western blot (Fig. 2). These results demonstrate minimal carry-over between fractions, which indicates a relatively pure protein population for each fraction. We have found that these approaches provide efficient subcellular fractionation for LC-MS/MS analysis and generate data regarding subcellular localization for protein biomarker discovery in human disease (10–12).

2 Materials

2.1 Cells and Tissue Preparation

1. PANC-1 cell culture: Cells were grown in 10 % FBS and DMEM complete media.
2. Tissue homogenization: Omni homogenizer (Omni International, Kennesaw, GA, USA).

2.2 Chemicals

1. Differential detergent fractionation: EDTA, digitonin, Triton X-100, Tween-40, and deoxycholic acid were purchased from Sigma-Aldrich. The protease inhibitor cocktail tablets were purchased from Roche Applied Science (Indianapolis, IN, USA).

2.3 Subcellular Fractionation

2.3.1 Total Cell Lysate

1. 2× radioimmunoprecipitation assay (RIPA) buffer: 300 mM NaCl, 2 % (v/v) Triton X-100, 1 % (w/v) deoxycholate, 0.2 % SDS, 100 mM Tris–HCl (pH 8.0), 1× protease inhibitor cocktail (see Note 1).

2.3.2 Two-Way Subcellular Fractionation (2SF)

1. Digitonin extraction buffer: 10 mM HEPES, pH 7.4, at 4 °C, 0.015 % (w/v) digitonin, 300 mM sucrose, 100 mM NaCl, 3 mM $MgCl_2$, 5 mM EDTA, 1× protease inhibitor cocktail.
2. 2× RIPA buffer.

2.3.3 Three-Way Subcellular Fractionation (3SF)

1. Digitonin extraction buffer.
2. Modified-RIPA buffer (m-RIPA): 150 mM NaCl, 1 mM EDTA, 10 mM Tris–HCl (pH 7.5), 1 % (v/v) Triton X-100, 0.25 % (w/v) deoxycholate, 1× protease inhibitor cocktail.
3. 2× RIPA buffer.

2.3.4 Four-Way Subcellular Fractionation (4SF)

1. Digitonin extraction buffer. See Note 2 for modified compositions from ref. 8.
2. Triton X-100 extraction buffer: 10 mM HEPES, pH 7.4, at 4 °C, 0.5 % (v/v) Triton X-100, 300 mM sucrose, 100 mM NaCl, 3 mM $MgCl_2$, 5 mM EDTA, 1× protease inhibitor cocktail.
3. Tween-40/deoxycholate extraction buffer: 10 mM HEPES, pH 7.4, at 4 °C, 1 % (v/v) Tween-40, 0.5 % (v/v) deoxycholate, 1 mM $MgCl_2$, 1× protease inhibitor cocktail.
4. Cytoskeleton solubilization buffer: 5 % (w/v) SDS, 10 mM sodium phosphate, pH 7.4, 1× protease inhibitor cocktail.

2.3.5 ProteoExtract Subcellular Proteome Extraction Kit (S-PEK, Calbiochem)

1. All reagents are supplied in the kit.

2.4 SDS Polyacrylamide Gel Electrophoresis and Western Blot

1. NuPAGE® Bis–Tris gels: 10 % Bis–Tris SDS-PAGE gels are purchased from Life Technologies™ (Grand Island, NY, USA).
2. Invitrogen *XCell SureLock Mini-Cell* gel running apparatus.
3. 6× sample buffer: 0.35 M Tris–Cl, pH 6.8, 10 % SDS, 0.6 M DTT (w/v), 30 % glycerol (v/v), 0.12 % bromophenol blue (w/v).
4. Invitrogen *XCell II Blot Module* Western blotting apparatus.
5. Nitrocellulose membrane is purchased from Bio-Rad catalog# 162-0167.
6. Enhanced chemiluminescent (ECL Advance catalog# RPN-2135) reagent and blocking buffer are purchased from GE Healthcare.
7. Antibodies against HSP90 (cat# CA1023), vimentin (cat# IF01), and PARP-1 (cat# AM30) are from EMD Millipore; antibody against calnexin (cat# 610523) is from BD Biosciences.

3 Methods

3.1 Fresh/Frozen Cells and Tissue Preparation Prior to Fractionation

3.1.1 Adherent Fresh Cells

1. Trypsinize adherent cells and pellet in a 15 ml conical tube at 200–300 × *g*.
2. Remove media, suspend pellet with ice-cold PBS with gentle flicking, and transfer to preweighed clean tube to determine wet weight of pellet.
3. Remove supernatant after centrifugation at 200–300 × *g* and wash pellet by resuspension in ice-cold PBS followed by centrifugation.
4. Repeat two more times.

3.1.2 Suspension Cells

1. Suspension cells are collected in a 15 ml conical tube at 200–300 × *g*.
2. Remove media, suspend pellet with ice-cold PBS with gentle flicking, and transfer to preweighed clean tube to determine wet weight of pellet.
3. Remove supernatant after centrifugation at 200–300 × *g* and wash pellet by resuspension in ice-cold PBS followed by centrifugation.
4. Repeat two more times.

3.1.3 Frozen Tissue Specimen (Using Calbiochem S-PEK Kit)

1. Weigh out 23–50 mg tissue and dice it into pieces using a scalpel. Place pieces into a 2 ml centrifuge tube. Rinse pieces in 1 ml PBS on ice for 2 min, remove PBS, and repeat (centrifuge briefly at low speed if needed to move tissue pieces down to the bottom of tube). Make note of level of blood remaining in tissue, if applicable.
2. While washing tissue sample, warm up extraction buffer (EB) IV and protease inhibitor cocktail (PIC) to room temperature. All other buffers should remain on ice.
3. Remove all remaining PBS and add 1.0 ml EB I and 5.0 μl PIC (see Note 3). Using the Omni homogenizer, homogenize sample for 30–60 s on ice. Incubate at 4 °C for 10 min, rotating (see Note 4).
4. Centrifuge sample at 500–1,000 × *g* for 10 min at 4 °C. Avoid transfer of lipids which sometimes congeal around top of liquid layer.
5. Transfer supernatant to a clean tube and keep on ice (fraction 1).
6. Subsequent fractions are processed as seen in protocol for adherent cells below.
7. Protein bands of each fractionation are shown in Fig. 3.

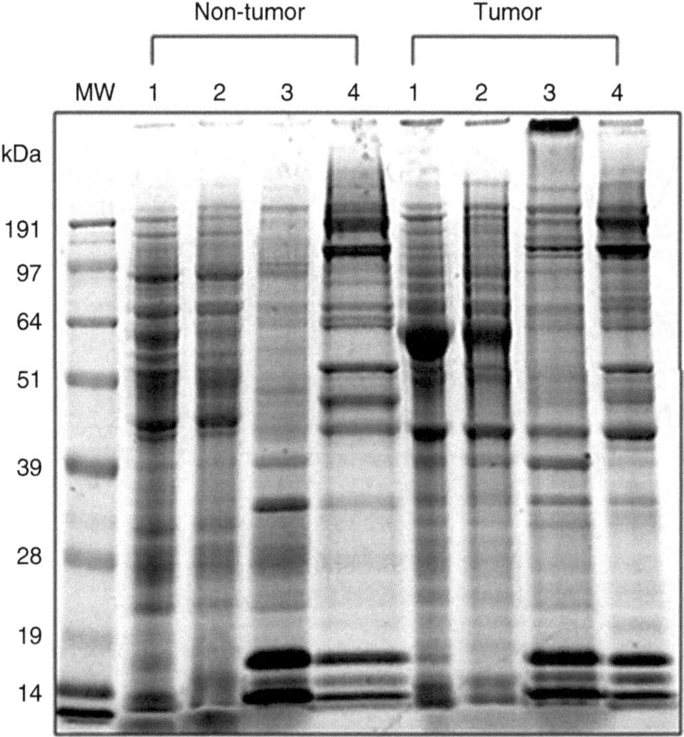

Fig. 3 The fraction of frozen pancreas tissues from non-tumor and tumor regions using ProteoExtract fraction kit. Fractions 1, cytosol; 2, membrane; 3, nucleus; and 4, cytoskeleton

3.2 Subcellular Fractionation Methods

3.2.1 Total Cell Lysate

1. Remove all remaining PBS and add 2× RIPA buffer (containing 1× protease inhibitor added shortly prior to use) to washed cell pellets (4–5 vol/g wet wt, resuspend by pipetting up and down).
2. Sonicate for 10 min followed by a centrifugation at 18,000 ×*g* for 10 min and transfer supernatant (total cell lysate) to a clean tube, aliquot, and store at −80 °C.
3. The protein band pattern of Coomassie staining is shown in Fig. 1.

3.2.2 Two-Way Subcellular Fractionation (2SF)

1. Remove all remaining PBS and add ice-cold digitonin extraction buffer to washed cell pellets (5 vol/g wet wt, gently resuspend by swirling). Incubate cells on ice with gentle agitation until 95–100 % of cells are permeabilized (5–10 min) as assessed by trypan blue exclusion.
2. Centrifuge at 1,000 ×*g* for 5 min and transfer supernatant (1, cytosolic fraction) to a clean tube, aliquot, and store at −80 °C. Wash pellet using 2 volumes of digitonin extraction buffer.
3. Resuspend digitonin insoluble pellets using ice-cold 2× RIPA buffer (containing 1× protease inhibitor added shortly prior

to use) in one-half volume to that used for high-salt PBS extraction, and let the pellets be solubilized by pipetting up and down.

4. Sonicate for 10 min followed by centrifugation at 18,000 ×g for 10 min and transfer supernatant (2, non-cytosolic fraction) to a clean tube, aliquot, and store at −80 °C.
5. The protein band patterns of Coomassie staining are shown in Fig. 1.

3.2.3 Three-Way Subcellular Fractionation (3SF)

1. Remove all remaining PBS and add ice-cold digitonin extraction buffer to washed cell pellets (5 vol/g wet wt, gently resuspend by swirling). Incubate cells on ice with gentle agitation until 95–100 % of cells are permeabilized (5–10 min) as assessed by trypan blue exclusion.
2. Centrifuge at 1,000×g for 5 min and transfer supernatant (1, cytosolic fraction) to a clean tube, aliquot, and store at −80 °C. Wash pellet using 2 volumes of digitonin extraction buffer.
3. Resuspend digitonin insoluble pellets in ice-cold PBS and incubation in an ice bath for 15–20 min. Remove supernatant after centrifugation at 1,000 ×g.
4. Resuspend pellets in ice-cold m-RIPA buffer (containing 1× protease inhibitor added shortly prior to use) in one-half volume to that used for high-salt PBS extraction. Incubate cells on ice with gentle agitation for 30 min.
5. Centrifuge at 5,000 ×g for 10 min and transfer the supernatant (2, membrane/organelle fraction) to a clean tube, aliquot, and store at −80 °C. Wash pellet using 2 volumes of m-RIPA buffer.
6. Resuspend m-RIPA insoluble pellets in ice-cold 2× RIPA buffer (containing 1× protease inhibitor added shortly prior to use) in an equal volume to that used for m-RIPA extraction, and let the pellets be solubilized by pipetting up and down.
7. Sonicate for 10 min followed by a centrifugation at 18,000 ×g for 10 min and transfer supernatant (3, nuclear component/cytoskeletal fraction) to a clean tube, aliquot, and store at −80 °C.
8. The protein band patterns of Coomassie staining are shown in Fig. 1.

3.2.4 Four-Way Subcellular Fractionation (4SF)

1. Remove all remaining PBS and add ice-cold digitonin extraction buffer to washed cell pellets (5 vol/g wet wt, gently resuspend by swirling). Incubate cells on ice with gentle agitation until 95–100 % of cells are permeabilized (5–10 min) as assessed by trypan blue exclusion (cell numbers: see Note 5). For further details and original protocols, see Ref. 8.

2. Centrifuge at 480 × *g* for 5 min and transfer supernatant (1, cytosolic fraction) to a clean tube, aliquot, and store at −80 °C. Wash pellet using an equal volume of digitonin extraction buffer.
3. Resuspend digitonin-insoluble pellets in ice-cold Triton X-100 extraction buffer in an equal volume of digitonin extraction buffer. Incubate cells on ice with gentle agitation for 30 min.
4. Centrifuge at 5,000 × *g* for 10 min and transfer supernatant (2, membrane/organelle fraction) to a clean tube, aliquot, and store at −80 °C. Wash pellet using an equal volume of Triton X-100 extraction buffer.
5. Resuspend Triton-insoluble pellets in ice-cold Tween/DOC extraction buffer at one-half volume of Triton buffer using dounce homogenizer (5 strokes, medium speed). Incubate cells on ice with gentle agitation for 30 min.
6. Centrifuge at 6,780 × *g* for 10 min and transfer supernatant (3, nuclear fraction) to a clean tube, aliquot, and store at −80 °C. Wash pellet using an equal volume of Tween/DOC extraction buffer.
7. Wash the detergent-resistant pellet in ice-cold PBS (pH 7.4) containing 1.2 mM PMSF by resuspension and centrifuge at 12,000 × *g*. Wash pellets once with −20 °C 90 % acetone. Then lyophilize the sample to dryness.
8. Resuspend pellets using cytoskeleton (CSK) solubilization buffer and store at −80 °C (4 °C for the cytoskeletal/nuclear matrix fraction).
9. The protein band patterns of Coomassie staining are shown in Fig. 1.

3.2.5 ProteoExtract Subcellular Proteome Extraction Kit (S-PEK, Calbiochem)

1. Trypsinize cells as usual and pellet in a 15 ml conical tube at 200–500 × *g*. Remove media. Wash pellet with S-PEK wash buffer, repeating for a total of two washes, pelleting at no more than 500 × *g* each time. While washing pellet, warm up extraction buffer (EB) IV and protease inhibitor cocktail (PIC) to room temperature. All other buffers should remain on ice (cell numbers: see Note 5). For further details and original protocols, see the manufacturer's manual.
2. Remove all remaining wash buffer and add 1.0 ml EB I and 5.0 μl PIC. Gently resuspend pellet and transfer entire sample to a 1.5 or 2.0 ml centrifuge tube. Incubate at 4 °C for 10 min, rotating.
3. Centrifuge sample at 500–1,000 × *g* for 10 min at 4 °C. Transfer supernatant to a clean tube and keep on ice (1, cytosolic fraction).
4. To pellet add a solution of 1.0 ml EB II plus 5.0 μl PIC. Resuspend by flicking tube.

5. Incubate at 4 °C for 30 min, rotating.
6. Centrifuge at 5,000–6,000 × *g* for 10 min at 4 °C.
7. Transfer sup to clean tube and keep on ice (2, membrane fraction).
8. To pellet add solution of 500 μl EB III plus 5 μl PIC plus 1.5 μl Benzonase. Resuspend pellet by flicking tube and incubate at 4 °C for 10 min, rotating.
9. Centrifuge at 7,000 × *g* for 10 min at 4 °C.
10. Transfer sup to clean tube and keep on ice (3, nuclear fraction).
11. To pellet add solution of EB IV plus 5 μl PIC and PHIC. Resuspend by pipetting up and down (4, cytoskeletal fraction).
12. If desired, desalt fractions 3 and 4 using Zeba desalting columns (Thermo/Pierce) according to manufacturer instructions (see Note 6).
13. Quantify a 1/10 dilution of each sample using the Thermo/Pierce BCA microplate assay.
14. Store samples at −80 °C.
15. The protein band patterns of Coomassie staining are shown in Fig. 1.

3.3 Protein Separation on SDS-PAGE and Fractionation Efficiency Test

1. Mix ~10 μg samples with an appropriate volume of sample buffer (sample buffer final concentration = 1×).
2. Incubate samples at 95 °C for 10 min.
3. Load onto NuPage Bis-Tris gel and run at 150 V constant for 1 h, or until samples have migrated to within 1 cm of the bottom of the gel.
4. While gel is running prepare for each gel: 4–6 sponges and 4 pieces of filter paper and one piece of nitrocellulose transfer membrane each cut to the size of the gel.
5. Soak all of these pieces in transfer buffer for at least 20 min before using for transfer.
6. After gel finishes running, carefully remove from plates, remove stacking gel, and trim bottom ridge away. Assemble blot as follows. Roll each layer to remove any air bubbles: 2–3 pieces sponge, 2 pieces filter paper, gel, membrane, 2 pieces filter paper, 2–3 pieces sponge. Add extra sponges as needed to ensure a tight fit inside transfer chamber.
7. Insert into transfer apparatus with transfer membrane towards positive side. Be sure fit is tight and that gel and membrane are forced completely together.
8. Place into electrophoresis chamber and fill inner chamber with NuPage transfer buffer to just over the top of sponges. Fill outer chamber with ~600 ml Millipore water.

9. Run at 25 V constant for 2 h.
10. Remove and disassemble sandwich carefully. Stain membrane with Ponceau S solution (0.1 % Ponceau S in 5 % acetic acid) for 5 min, then rinse with distilled water to visualize efficiency of transfer. Remove blot to 5 % blocking buffer for 1 h (room temp) while shaking.
11. Remove gel to gel code blue staining to ensure complete transfer (optional).
12. Remove blocking solution. Add primary antibody diluted in blocking buffer. Incubate overnight (4 °C) with gentle shaking.
13. Remove to TBST and wash, shaking, three times, 20 min each.
14. Add secondary antibody (HRP labeled). Incubate 1 h at RT shaking.
15. Wash 3–5 times, 5 min each with TBST, and proceed to ECL protocol.
16. Place membrane in reaction tray.
17. Combine 2.5 ml of each of the two ECL reagents and apply to the membrane and then let stand for 5 min.
18. Place membrane onto plastic wrap and seal completely, making sure to smooth out all wrinkles in the plastic wrap.
19. Visualize the blot using the LAS4000 with "precision" exposure and "standard" sensitivity settings.

4 Notes

1. Protease inhibitor must be added to each detergent extraction buffer shortly prior to use.
2. 10 mM PIPES, pH 6.8 and 7.4 are substituted with 10 mM HEPES, pH 7.4. The protease inhibitor, 1 mM PMSF, is replaced with 1× protease inhibitor cocktail.
3. Tissue homogenization is carried out in EB I instead of PBS to prevent loss of proteins into the wash buffer.
4. To evaluate mechanical homogenization effects, we used PANC-1 cells and compared the non-homogenized and homogenized fractionation efficiency by Western blotting (Fig. 2).
5. The number of cells we use in this protocol is 1–5 × 107 cells with 2–3 T75 flasks at 70–80 % confluence. For the ProteoExtract kit, the guidelines for optimal cell number in representative cell lines are provided in manufacturer's protocol. In addition, we have successfully reduced the volume of EB used in each step by 25–40 % to obtain optimal protein concentration.

6. When using the Calbiochem kit, fractions 3 and 4 are treated with Zeba Desalting columns (Thermo cat# 89882) to remove large amounts of salts and/or detergents which have been seen to interfere with migration of proteins on SDS-PAGE.

Acknowledgments

Ms. Kimberly McKinney and Dr. Jin-Gyun Lee are gratefully acknowledged for their technical assistance and comments. This work was supported by Carolinas HealthCare System, a start-up package and a research grant (CMC 08-026).

References

1. Corthals GL, Wasinger VC, Hochstrasser DF, Sanchez JC (2000) The dynamic range of protein expression: a challenge for proteomic research. Electrophoresis 21:1104–1115
2. Jung E, Heller M, Sanchez JC, Hochstrasser DF (2000) Proteomics meets cell biology: the establishment of subcellular proteomes. Electrophoresis 21:3369–3377
3. Bell AW, Ward MA, Blackstock WP et al (2001) Proteomics characterization of abundant Golgi membrane proteins. J Biol Chem 276:5152–5165
4. Andersen JS, Lyon CE, Fox AH et al (2002) Directed proteomic analysis of the human nucleolus. Curr Biol 12:1–11
5. Hanson BJ, Schulenberg B, Patton WF, Capaldi RA (2001) A novel subfractionation approach for mitochondrial proteins: a three-dimensional mitochondrial proteome map. Electrophoresis 22:950–959
6. Kikuchi M, Hatano N, Yokota S, Shimozawa N, Imanaka T, Taniguchi H (2004) Proteomic analysis of rat liver peroxisome: presence of peroxisome-specific isozyme of Lon protease. J Biol Chem 279:421–428
7. Fialka I, Pasquali C, Lottspeich F, Ahorn H, Huber LA (1997) Subcellular fractionation of polarized epithelial cells and identification of organelle-specific proteins by two-dimensional gel electrophoresis. Electrophoresis 18: 2582–2590
8. Ramsby ML, Makowski GS (1999) Differential detergent fractionation of eukaryotic cells. Analysis by two-dimensional gel electrophoresis. Methods Mol Biol 112:53–66
9. Ramsby ML, Makowski GS, Khairallah EA (1994) Differential detergent fractionation of isolated hepatocytes: biochemical, immunochemical and two- dimensional gel electrophoresis characterization of cytoskeletal and noncytoskeletal compartments. Electrophoresis 15:265–277
10. Lee YY, McKinney KQ, Ghosh S et al (2011) Subcellular tissue proteomics of hepatocellular carcinoma for molecular signature discovery. J Proteome Res 10:5070–5083
11. McKinney KQ, Lee YY, Choi HS et al (2011) Discovery of putative pancreatic cancer biomarkers using subcellular proteomics. J Proteomics 74:79–88
12. Hwang SI, Lundgren DH, Mayya V et al (2006) Systematic characterization of nuclear proteome during apoptosis: a quantitative proteomic study by differential extraction and stable isotope labeling. Mol Cell Proteomics 5:1131–1145

Chapter 4

Analysis of Secreted Proteins

Valeria Severino, Annarita Farina, and Angela Chambery

Abstract

Most biological processes including growth, proliferation, differentiation, and apoptosis are coordinated by tightly regulated signaling pathways, which also involve secreted proteins acting in an autocrine and/or paracrine manner. In addition, extracellular signaling molecules affect local niche biology and influence the cross-talking with the surrounding tissues. The understanding of this molecular language may provide an integrated and broader view of cellular regulatory networks under physiological and pathological conditions. In this context, the profiling at a global level of cell secretomes (i.e., the subpopulations of a proteome secreted by the cell) has become an active area of research. The current interest in secretome research also deals with its high potential for the biomarker discovery and the identification of new targets for therapeutic strategies. Several proteomic and mass spectrometry platforms and methodologies have been applied to secretome profiling of conditioned media of cultured cell lines and primary cells. Nevertheless, the analysis of secreted proteins is still a very challenging task, because of the technical difficulties that may hamper the subsequent mass spectrometry analysis. This chapter describes a typical workflow for the analysis of proteins secreted by cultured cells. Crucial issues related to cell culture conditions for the collection of conditioned media, secretome preparation, and mass spectrometry analysis are discussed. Furthermore, an overview of quantitative LC-MS-based approaches, computational tools for data analysis, and strategies for validation of potential secretome biomarkers is also presented.

Key words Secretome, Proteomics, Biomarker discovery, Mass spectrometry, Quantitative LC-MS

1 Introduction

The emergence of novel high-throughput technologies has enabled the comprehensive analysis of biological systems under physiological and pathological conditions. With the advent of the "omic" era, large-scale studies have been exploited to dissect molecular mechanisms regulating cellular processes at genome, transcriptome, and proteome level (1, 2). These strategies rely more on holistic approaches than on single-process studies, providing an integrated and broader view of cellular regulatory networks.

Over the last decades, the investigation of extracellular molecules involved in cell signaling has opened new avenues in proteomic research (3). In this context, the elucidation of the cell

Ming Zhou and Timothy Veenstra (eds.), *Proteomics for Biomarker Discovery: Methods and Protocols*, Methods in Molecular Biology, vol. 1002, DOI 10.1007/978-1-62703-360-2_4, © Springer Science+Business Media, LLC 2013

"secretome," i.e., the array of proteins released by cells, tissues, or organisms, is eliciting a growing interest for the potential implications for basic and applied research. In eukaryotic cells, proteins actively secreted through the well-characterized classical secretory pathway are synthesized as precursors carrying an N-terminal signal peptide and directed in the extracellular space via endoplasmic reticulum (ER)/Golgi-dependent routes (4). Several nonclassical protein secretion pathways have been also described, extending the definition of cell secretome to proteins deriving from the extracellular matrix or shed from the cell surface, as well as to proteins released through exosomes and membrane vesicles (5).

Under physiological conditions, secreted proteins (e.g., growth factors, cytokines, chemokines, enzymes, proteases, etc.) are part of a complex system of communication that coordinates cell functions through paracrine and/or autocrine mechanisms and modulates most basic cellular activities, including cell proliferation, differentiation and normal tissue development, and homeostasis. On the other hand, the alteration of the extracellular signaling strongly impairs the capability of cells to correctly respond to their microenvironment, often resulting in the onset of diseases such as cancer, autoimmunity, and diabetes (6).

As a consequence, the renewed interest in the large-scale analysis of secreted proteins is mainly related to their high potential for biomarker discovery of diagnostic and/or prognostic significance (7–10). Recently, this attractive strategy has taken advantage from the analysis of conditioned media of cultured cell lines and primary cells by applying proteomic and mass spectrometry methodologies (11, 12). Though such in vitro experimental systems only partially resemble the in vivo microenvironment, secretome studies offer a valid and complementary alternative to traditional approaches for biomarker mining based on the analysis of relevant biological fluids, including blood, serum, cerebrospinal fluid, urine, bile, etc. (13). The postulate supporting the feasibility of this approach is that candidate biomarkers identified in conditioned media of culture cells may be subsequently targeted to validate their presence and their prognostic value within biological fluids (14).

The main advantages of the secretome-based biomarker discovery platforms are related to (i) the easy accessibility to a large number of cell lines and primary cells whose culture conditions can be strictly controlled and (ii) the relatively low complexity of analytes associated with a reduced dynamic range with respect to biological fluids, which facilitates the detection of low abundance components. Nevertheless, the detection of secreted proteins still remains a challenging analytical task mainly due to technical difficulties that may hamper the subsequent mass spectrometry analysis.

This chapter describes a typical experimental workflow for the analysis of secreted proteins in conditioned media of cultured cells.

Critical points related to cell culture conditions, protein sample preparation, and LC-MS-based analysis are discussed. Finally, an overview of the different quantitative mass spectrometry methodologies, bioinformatic tools for data analysis, and approaches for secretome candidate biomarkers validation is also presented.

2 Materials

Prepare all solutions using ultrapure water and analytical grade reagents.

1. Dulbecco's Modified Eagle Medium (DMEM) or similar suitable cell culture media.
2. Protease inhibitors mixture (complete EDTA-free Protease Inhibitor Cocktail Tablets Roche, Mannheim, Germany).
3. 25 mM NH_4HCO_3 (197 mg of NH_4HCO_3 in 100 mL of H_2O).
4. Amicon Ultra-15, PLBC Ultracel-PL Membrane, 3 kDa (Millipore, Cat no: UFC900308).
5. 100 mM 1,4-Dithio-DL-threitol (DTT, 15.4 mg of DTT in 1 mL of H_2O).
6. 300 mM Iodoacetamide (IAM, 55.5 mg of IAM in 1 mL of H_2O).
7. Trypsin TPCK-treated from bovine pancreas.
8. Solvent A: H_2O containing 0.1 % formic acid (FA); Solvent B: CH_3CN containing 0.1 % FA.

3 Methods

3.1 Cell Culture for Secretome Preparation

An outline of the workflow for a typical secretome analysis is reported in Fig. 1. The first steps for cell culture and collection of conditioned media (CM) are of utmost importance for a reliable secretome analysis, which is strongly influenced by the quality of samples to be analyzed. This requirement mainly derives from the difficulties in detecting proteins secreted in low amounts within complex matrices such as CM. Appropriate procedures must be thus followed to prevent the masking effects derived by the presence, under common culture conditions, of the high-abundance serum proteins. In addition, particular care must be taken to avoid non-physiological cell lysis resulting in the unwanted release of intracellular proteins.

Although the culture conditions must be optimized for each cell line, depending on their type and specific requirements, a cell conditioning step is mandatory to remove or reduce the inter-

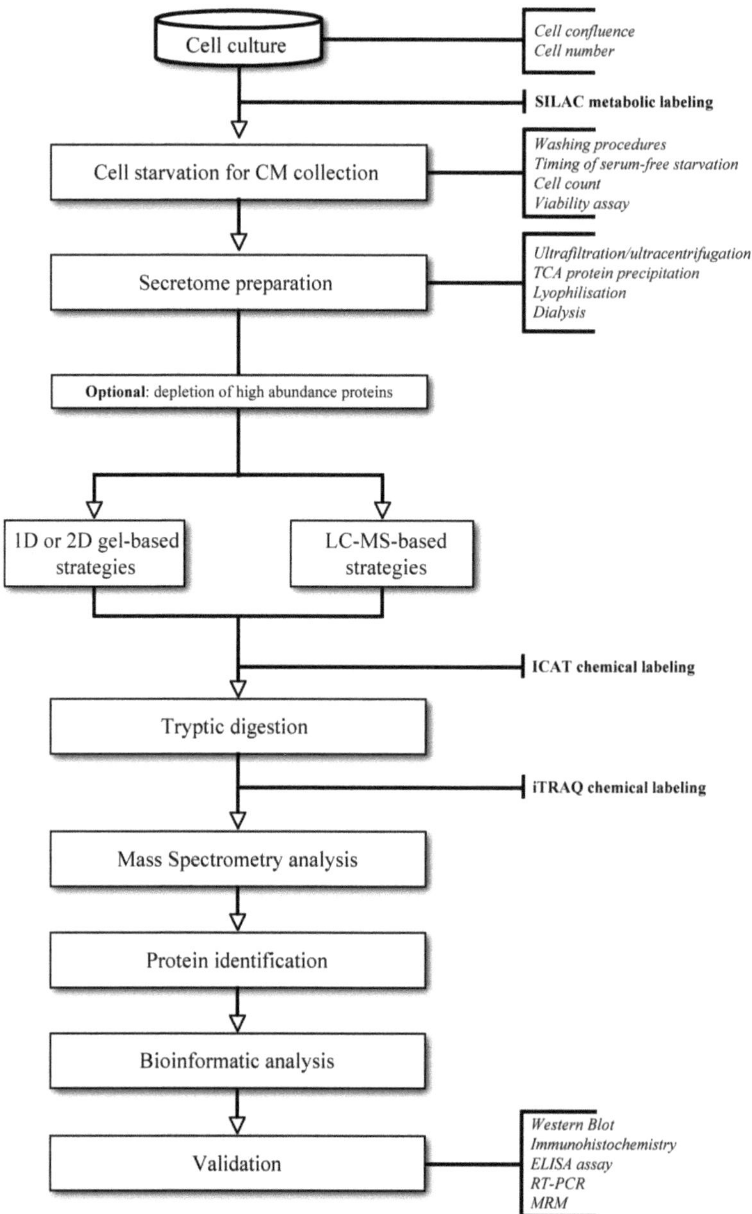

Fig. 1 Representative workflow for a typical secretome analysis

fering components of serum-supplemented media. In a basic workflow, individual secretome samples are prepared from cell lines or primary cultures in triplicate to address the biological variation (see Note 1), operating as follows:

1. Culture cells in DMEM or suitable medium containing 10–20 % (v/v) fetal bovine serum (FBS) or fetal calf serum (FCS) at 37 °C in humidified air containing 5–10 % CO_2.

2. Seed cells into 150 mm tissue culture dishes (see Note 2) and culture cells until a 60–70 % confluence is reached (see Note 3). It is highly recommended to seed the same number of cells or, for cell types growing with different replication rates, to seed the cells to have approximately the same number of cells for CM collection (see Note 4).
3. Wash gently the cells (see Note 5) with fresh serum-free DMEM or PBS. Alternatively, the percentages of serum within medium can be gradually lowered by performing washing steps in DMEM containing 2–0.5 % (v/v) FBS or FCS (see Note 6).
4. Add 20 mL of serum-free DMEM to each dish, and culture for 12–48 h (see Note 7).
5. Optionally, include at this stage in the experimental design a treatment of cells with stimulation factors (e.g., drugs, growth factors, small molecules). In this instance, a non-stimulated sample is analyzed in parallel as control.
6. It is strongly recommended at this stage to count cells (see Note 8) and perform viability assays (see Note 9).
7. Collect the CM from each dish and pool if necessary (see Note 1).
8. Centrifuge the CM at low speed (ranging from 160 to 1,500× *g*) at 4 °C for 5–10 min to pellet intact cells, remove cell debris, and transfer the supernatant to a new tube (see Note 10).
9. It is recommended to add the inhibitors mixture (complete EDTA-free Protease Inhibitor Cocktail Tablets, 1 tablet used in 50 mL).
10. If necessary, filter CM samples with a 0.2 μm low protein-binding membrane to further remove cell debris.

3.2 Secretome Preparation

At this stage, secreted proteins are diluted in CM at concentrations that may be very low for some cytokines and growth factors (ng/mL). In addition, most common CM contain salts, amino acids, antibiotics, and pH indicators, such as phenol red, that may hamper the subsequent MS analysis. Thus, procedures for protein concentration and desalting are needed in order to perform the secretome analysis. To this aim, several methodologies may be used in either a single or combined step(s) by means of different commercially available devices. The most commonly used are (i) ultrafiltration/ultracentrifugation (see Note 11), (ii) dialysis (see Note 12), and (iii) lyophilization and protein precipitation with trichloroacetic acid (TCA) or acetone (see Note 13).

Protein concentration and desalting may constitute a crucial point in secretome preparation since none of the above methodologies is 100 % efficient, thus potentially leading to massive sample losses. The advantages and disadvantages of each methodology should be considered for the selection of the more appropriate strategy for a given experimental system. Table 1 summarizes the experimental conditions and general procedures for secretome sample preparation reported so far for several cell lines. Similar data on primary culture isolates and stem cells, previously reported by Skalnikova and coworkers (12), have not been included in the table.

CM lyophilization has been used for secretome analysis mainly coupled with a desalting step, such as TCA precipitation (15–18) as reported below:

1. Lyophilize the CM, avoiding sample melting (see Note 14).
2. Resuspend the lyophilized CM in 1 mL of 25 mM NH_4HCO_3 or a similar suitable buffer.
3. Add 100 % TCA to reach a 10–20 % final concentration alone or in combination with specific carriers to precipitate the proteins (see Note 15).
4. Incubate 30 min on ice.
5. Centrifuge for 15 min at 4 °C at 15,800× *g*. Carefully remove and discard the supernatant (see Note 16).
6. Perform two successive washes of the protein pellet to remove TCA with 1 mL of cold diethyl ether and 1 mL of cold acetone.
7. Centrifuge for 10 min at 4 °C at 15,800× *g*. Discard the supernatant.
8. Air-dry the pellet at room temperature (see Note 17).
9. Resuspend the pellet in 25–100 μL of 25 mM NH_4HCO_3 and thoroughly sonicate the sample into an ultrasonic bath for 10–15 min to enhance protein solubilization.
10. Determine protein concentration by Bradford or an equivalent assay and store samples at −20 °C until use.

3.3 Thiol Alkylation and Tryptic Digestion for LC-MS-Based Analysis

The secretome samples prepared as described above may be analyzed by using different proteomic platforms. The traditional gel-based (2-DE or DIGE) techniques have been applied to the secretome analysis of several experimental systems (19–25). Detailed methods for gel-based methodologies and in-gel digestion protocols have been previously reported (26–28). However, gel-independent shotgun approaches, based on 1D or 2D LC-MS/MS techniques, are often preferred for secretome analysis, providing a significant advantage in terms of sensitivity which is crucial for the detection of low-abundance secreted molecules. A general protocol for processing secretome samples (17, 18) for the subsequent LC-MS-based analysis is described below:

1. For disulphide reduction, add 2.5 μL of 100 mM DTT (see Notes 18 and 19) to 100 μL of secretome samples in 25 mM NH_4HCO_3 (final concentration 2.5 mM DTT).

Table 1
Summary of the experimental conditions and general procedures for secretome sample preparation

Cell line	Description	Medium/ % serum	Washing procedure	Starvation (h)	Desalting/concentration	Ref.
ATCC TIB-152	Acute T-cell leukemia	RPMI 1640/10 % FBS	1 serum-free RPMI 1640	24	Ultrafiltration	(79)
Jurkat cells	Acute T-cell leukemia	RPMI 1640/10 % FBS	1 serum-free RPMI 1640	24	Ultrafiltration	(79)
ATCC JMV-13	B-cell leukemia	RPMI 1640/2 % FBS	5 PBS	12–24	Ultrafiltration	(80)
U1/U4	Bladder cancer	RPMI 1640/10 % FBS	1 serum-free RPMI 1640	24	Ultrafiltration	(79)
ATCC HTB-22/-129	Breast cancer	DMEM/10 % FBS	1 serum-free DMEM	24	Ultrafiltration	(79)
MCF-7	Breast cancer	DMEM/10 % FBS	1 serum-free DMEM	24	Ultrafiltration	(79)
MCF-7	Breast cancer	RPMI 1640/10 % FBS	4[a] serum-free RPMI 1640	24	Ultrafiltration	(66)
MDA-MB435S	Breast cancer	DMEM/10 % FBS	1 serum-free DMEM	24	Ultrafiltration	(79)
BEAS-2B	Bronchial epithelial cell	LHC-9	5 PBS	72	Dialysis/Ultraconcentration	(59)
ATCC HTB-31	Cervical carcinoma	DMEM/10 % FBS	1 serum-free DMEM	24	Ultrafiltration	(79)
C-33A	Cervical carcinoma	DMEM/10 % FBS	1 serum-free DMEM	24	Ultrafiltration	(79)
HuCCA-1	Cholangiocarcinoma	Ham F12/10 % FBS	2 serum-free Ham F12	24	Dialysis/SpeedVac	(20)
ATCC CCL-222/-227/-228	Colorectal carcinoma	RPMI 1640/10 % FBS	1 serum-free RPMI 1640	24	Ultrafiltration	(79)
CaCo2	Colorectal carcinoma	DMEM/ 10 % FCS	3 DMEM	nr	Ultrafiltration	(19)
Colo205	Colorectal carcinoma	RPMI 1640/10 % FBS	1 serum-free RPMI 1640	24	Ultrafiltration	(79)
HT29	Colorectal carcinoma	DMEM/10 % FCS	3 DMEM	nr	Ultrafiltration	(19)

(continued)

Table 1
(continued)

Cell line	Description	Medium/ % serum	Washing procedure	Starvation (h)	Desalting/concentration	Ref.
MICOL	Colorectal carcinoma	Ham F12/10 % FBS	5 PBS	12–24	Ultrafiltration	(80)
SW403	Colorectal carcinoma	RPMI/10 % FBS	5 PBS	12–24	Ultrafiltration	(80)
SW480/SW620	Colorectal carcinoma	RPMI 1640/10 % FBS	1 serum-free RPMI 1640	24	Ultrafiltration	(79)
SW948	Colorectal carcinoma	DMEM/10 % FCS	3 DMEM	nr	Ultrafiltration	(19)
A431	Epidermoid carcinoma	DMEM/10 % FBS	1 serum-free DMEM	24	Ultrafiltration	(79)
ATCC CRL-1555	Epidermoid carcinoma	DMEM/10 % FBS	1 serum-free DMEM	24	Ultrafiltration	(79)
HNGC2	Glioblastoma	DMEM/10 % FCS	3[a] serum-free DMEM	24	Ultrafiltration	(78)
LN229	Glioblastoma	DMEM/10 % FCS	3[a] serum-free DMEM	24	Ultrafiltration	(78)
U87MG	Glioblastoma	DMEM/10 % FCS	3[a] serum-free DMEM	24	Ultrafiltration	(78)
AMOSIII	Head/neck carcinoma	DMEM/10 % FBS	4 PBS + 1 serum-free DMEM	48	TCA–Sodium deoxycholate	(85)
HSC2	Head/neck carcinoma	DMEM/10 % FBS	4 PBS + 1 serum-free DMEM	24–48	TCA–Sodium deoxycholate	(85)
SCC4/SCC38	Head/neck carcinoma	DMEM/10 % FBS	4 PBS + 1 serum-free DMEM	24	TCA–Sodium deoxycholate	(85)
ATCC CCL-2	HeLa	DMEM/10 % FBS	1 serum-free DMEM	24	Ultrafiltration	(79)
Alexander HepG2	Hepatocellular carcinoma	DMEM/10 % FBS	2 serum-free DMEM	24	Dialysis/SpeedVac	(20)
ATCC HTB-52	Hepatocellular carcinoma	DMEM/10 % FBS	1 serum-free DMEM	24	Ultrafiltration	(79)
ATCC HB-8065/-8064	Hepatocellular carcinoma	DMEM/10 % FBS	1 serum-free DMEM	24	Ultrafiltration	(79)

HCC-S102	Hepatocellular carcinoma	RPMI 1640/10 % FBS	2 serum-free RPMI 1640	24	Dialysis/SpeedVac	(20)
Hep-G2	Hepatocellular carcinoma	DMEM/10 % FBS	1 serum-free DMEM	24	Ultrafiltration	(79)
Hep-3B	Hepatocellular carcinoma	DMEM/10 % FBS	2 serum-free DMEM	24	Dialysis/SpeedVac	(20)
SK-Hep-1	Hepatocellular carcinoma	DMEM/10 % FBS	1 serum-free DMEM	24	Ultrafiltration	(79)
SK-Hep-1	Hepatocellular carcinoma	DMEM/10 % FBS	2 serum-free DMEM	24	Dialysis/SpeedVac	(20)
A549	Lung cancer	RPMI/10 % FBS	5 PBS	12–24	Ultrafiltration	(80)
A549	Lung cancer	DMEM/10 % FBS	4 PBS	24	Ultrafiltration	(86)
CL_{1-0}/CL_{1-5}	Lung cancer	RPMI 1640/10 % FBS	1 serum-free RPMI 1640	24	Ultrafiltration	(79, 87)
CL_{1-0}/CL_{1-5}	Lung cancer	RPMI 1640/10 % FBS	3 serum-free RPMI 1640	24	Ultrafiltration	(46)
H358	Lung cancer	RPMI 1640/10 % FCS	4 PBS + 2 serum-free RPMI 1640	72	Ultrafiltration	(45)
LN18	Malignant astrocytoma	SILAC DMEM /10 % FBS	6 serum-free SILAC DMEM	24	Centrifugation	(39)
T98	Malignant astrocytoma	SILAC DMEM /10 % FBS	6 serum-free SILAC DMEM	24	Centrifugation	(39)
U118/U87	Malignant astrocytoma	SILAC DMEM /10 % FBS	6 serum-free SILAC DMEM	24	Centrifugation	(39)
CNE-2	Nasopharyngeal carcinoma	DMEM/10 % FBS	4 serum-free DMEM	24	Ultrafiltration	(21)
NPC-TW02/ -04/-BM1	Nasopharyngeal carcinoma	DMEM/10 % FBS	1 serum-free DMEM	24	Ultrafiltration	(79, 88)
HPDE6	Normal pancreatic ducts	RPMI 1640/5 % FBS	7 NaCl 0.9 %	24	Ultrafiltration	(67)
ATCC CRL-1624	Oral cancer	RPMI 1640/10 % FBS	1 serum-free RPMI 1640	24	Ultrafiltration	(79)

(continued)

Table 1 (continued)

Cell line	Description	Medium/ % serum	Washing procedure	Starvation (h)	Desalting/concentration	Ref.
OEC-M1	Oral cancer	RPMI 1640/10 % FBS	1 serum-free RPMI 1640	24	Ultrafiltration	(79)
SCC-4	Oral cancer	RPMI 1640/10 % FBS	1 serum-free RPMI 1640	24	Ultrafiltration	(79)
SKOV-3	Ovarian cancer	RPMI/10 % FBS	5 PBS	12–24	Ultrafiltration	(80)
AsPC1	Pancreatic carcinoma	RPMI 1640/5 % FBS	7 NaCl 0.9 %	24	Ultrafiltration	(67)
ATCC CRL-1469/-1420	Pancreatic carcinoma	DMEM/10 % FBS	1 serum-free DMEM	24	Ultrafiltration	(79)
MIA PaCa-2	Pancreatic carcinoma	DMEM/10 % FBS	1 serum-free DMEM	24	Ultrafiltration	(79)
MIA PaCa-2	Pancreatic carcinoma	RPMI 1640/5 % FBS	7 NaCl 0.9 %	24	Ultrafiltration	(67)
PANC-1	Pancreatic carcinoma	DMEM/10 % FBS	1 serum-free DMEM	24	Ultrafiltration	(79)
PANC-1	Pancreatic carcinoma	RPMI 1640/5 % FBS	7 NaCl 0.9 %	24	Ultrafiltration	(67)
PT45	Pancreatic carcinoma	RPMI 1640/5 % FBS	7 NaCl 0.9 %	24	Ultrafiltration	(67)
LNCaP	Prostate cancer	RPMI/10 % FBS	5 PBS	12–24	Ultrafiltration	(80)
PC3	Prostate cancer	RPMI/10 % FBS	5 PBS	12–24	Ultrafiltration	(80)
786-O/796-P	Renal cancer	RPMI/10 % FBS	5 PBS	12–24	Ultrafiltration	(80)
ACHN	Renal cancer	EMEM/10 % FBS	5 PBS	12–24	Ultrafiltration	(80)
CAL62	Thyroid cancer	DMEM/10 % FBS	3 PBS+1 serum-free DMEM	48	TCA–Sodium deoxycholate	(84)
TPC-1	Thyroid cancer	DMEM/10 % FBS	5 PBS	12–24	Ultrafiltration	(80)
TCP-1	Thyroid cancer	RPMI 1640/10 % FBS	3 PBS+1 serum-free RMPI 1640	48	TCA–Sodium deoxycholate	(84)

nr not reported

[a]One or more washing steps with prolonged incubation times ranging from 15 min to several hours

2. Vortex samples and incubate at 60 °C for 30 min.
3. Cool samples at room temperature and centrifuge briefly at low speed to collect the condensate to the bottom of tube.
4. For cysteine alkylation, add 2.5 μL of 300 mM IAM to yield a final concentration of 7.5 mM.
5. Incubate in darkness at room temperature for 30 min.
6. For trypsin digestion, add trypsin to a final enzyme:substrate ratio of 1:100 (w/w). For low amount of samples, add 5 μL of a 10 ng/μL trypsin stock solution (1 μg/μL in H_2O diluted 1:100).
7. Digest proteins at 37 °C overnight.
8. Centrifuge at 4 °C at 15,800× *g* for 10 min and transfer supernatants containing the tryptic peptides to a new tube.
9. Dry peptides in a vacuum centrifuge.
10. For the LC-MS analysis, dissolve samples in 5 % CH_3CN containing 0.1 % FA. Typically, 2 μg of tryptic digest is the maximum amount that can be injected on an LTQ instrument (see Subheading 3.4). Therefore, a sample concentration of 0.4 μg/μL is suitable for a 5 μL injection volume (see Note 20).

3.4 LC-MS Configuration for Secretome Analysis

Any tandem mass spectrometry platform may be used for the analysis of tryptic peptides obtained by the above procedures. Below, the method optimized for the LC-MS/MS analysis on an LTQ Orbitrap Velos instrument (Thermo Fisher Scientific, Waltham, MA, USA) equipped with a NanoAcquity system from Waters (Waters Corporation, Manchester, UK) is described:

1. Run the analytical separation for 65 min at a flow rate of 220 nL/min using a gradient of solvent A and solvent B. Set the gradient as follows: 5 % B for 1 min, from 5 to 35 % B in 54 min, and from 35 to 80 % B in 10 min. Any C_{18} column (0.75 × 150 mm, 5 μm 100 Å) is suitable for the analysis.
2. Perform the mass spectrometric analysis in the data-dependent mode which automatically switches between orbitrap-MS and LTQ-MS/MS (MS^2) modes of data acquisition.
3. Acquire full-scan MS spectra (m/z from 400 to 2,000) in the orbitrap with a resolution of $r=60{,}000$ and set the ion population to a target value of 500,000 charges in the linear ion trap.
4. Set the system to sequentially isolate the eight most intense ions for fragmentation in the linear ion trap using collision-induced dissociation (CID) at a target value of 7,000 charges (isolation width of 2 m/z). The resulting fragment ions are recorded in the LTQ.
5. Target ions already mass selected for CID are dynamically excluded for the duration of 45 s.

6. Further mass spectrometric settings are as follows: spray voltage 1.6 kV, temperature of the heated transfer capillary 300 °C, and relative normalized collision energy 35 % for CID. The minimal signal required for MS^2 is 500 counts. Set activation *q* to 0.25 and the activation time to 30 ms.
7. To clean the LC system, perform one blank injection between two consecutive analyses.

3.5 Protein Identification via Database Search

Several software packages are available for protein identification starting from the mass spectrometric data. The typical parameters used for a database search by using the MASCOT algorithm are reported below:

1. Convert the mass spectrometer .raw output file to the .mgf extension by using ReAdW 4.2.1 (http://sourceforge.net) or any similar software.
2. Analyze the obtained peak lists by using the MASCOT search engine (Matrix Science, Boston, MA, USA).
3. Set the specific search parameters in the MASCOT analyses as follows: fragment mass tolerance = 0.60 Da; peptide mass tolerance = 10 ppm; enzyme = trypsin; max missed cleavages = 1; peptide charge = 1^+, 2^+ and 3^+; data format = Mascot generic; and instrument = ESI-TRAP.
4. Set the following modifications: variable = oxidation of methionines; fixed = carbamidomethylation of cysteines.
5. Select an appropriate protein sequence database (e.g., UniProtSP database) and, eventually, a taxonomy restriction to limit the database search to entries from particular species or groups of species.
6. Optionally, the MASCOT output files may be analyzed for comparison by using appropriate tools such as the Scaffold software (Proteome Software Inc., Portland, OR).

3.6 Bioinformatics Analyses of Candidate Secreted Proteins

Proteins can be released to the extracellular space by different mechanisms, including classical and nonclassical secretion pathways. In the first instance, a signal peptide at the N-terminus is indicative of the protein secretion in the extracellular space. On the contrary, proteins released by nonclassical secretion pathways cannot be predicted as extracellular and could be categorized otherwise. Therefore, it may be useful to perform bioinformatics analyses of candidate secreted proteins by using classification tools and prediction servers. The classification of identified proteins may be performed according to the Gene Ontology (GO) hierarchy (http://www.geneontology.org/).

Specific algorithms based on a combination of several artificial neural networks are available to predict the presence of signal peptide cleavage sites within amino acid sequences such as SignalP (http://www.cbs.dtu.dk/services/SignalP/). This information may be eventually integrated by performing a prediction analysis for the nonclassical (i.e., not signal peptide-triggered) protein secretion pathways by using the SecretomeP server (http://www.cbs.dtu.dk/services/SecretomeP/). For these analyses a typical workflow is reported below:

1. Select the Retrieve tab of the UniProt toolbar (http://www.uniprot.org/).
2. Enter the list of UniProt identifiers into the search field or upload them from a saved file.
3. Download the sequence data in FASTA format.
4. Paste or upload the FASTA file into the SecretomeP search field and submit the analysis.
5. If the neural network (NN) parameter exceeds or is equal to a value of 0.5 (NN score ≥0.50), but no signal peptide is predicted, the protein is considered to be potentially secreted via a nonclassical pathway (29).
6. Proteins with a predicted N-terminal signal sequence can be confirmed using the SignalP software (30) and are considered to be secreted via a classical pathway (endoplasmic reticulum/Golgi-dependent pathway).

3.7 Quantitative LC-MS-Based Approaches

The reliable quantification of secreted proteins in CM is among the most challenging technical tasks in secretome research. Over the past years, mass spectrometry-based quantification approaches have gained a growing interest due to their capability of simultaneously identifying and quantifying proteins by using different LC-MS platforms and methodologies (Fig. 2). In most instances, these strategies rely on the principle of the stable isotope labeling by using specific mass tags that can be mass measured and concurrently provide quantitative information. Stable isotope incorporation can be achieved by metabolic, enzymatic, and chemical labeling (31).

A straightforward and accurate approach, widely used for the relative quantification of secreted proteins, is the stable isotope labeling by amino acids in cell culture (SILAC) (32). This metabolic labeling involves the addition to the growth media of isotopically tagged amino acids that are incorporated into proteins as they are synthesized by the growing cells. Basically, cells are cultured in a medium with normal essential amino acids (light label) and in a medium with isotopic modified forms of essential amino acids (heavy label). Lysine and arginine are the two most commonly used labeled amino acids, and ^{13}C, ^{15}N, ^{2}H, and ^{18}O are the mainly used isotopes for stable labeling. After some proliferation cycles, proteins are extracted, pooled, digested, and analyzed by LC-MS/MS. The heavy and light peptides elute as mass peak pairs separated by a defined mass difference. The ratios of the resulting relative

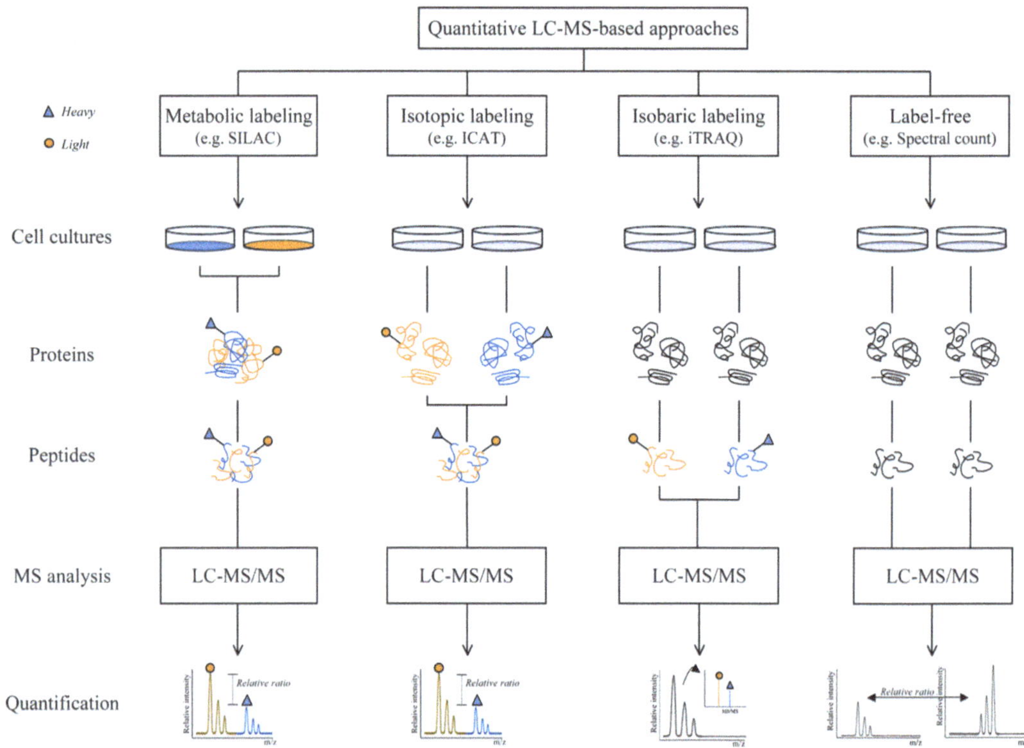

Fig. 2 Liquid chromatography-mass spectrometry-based quantification methodologies by metabolic and chemical stable isotope labeling and label-free approaches

peak intensities reflect the abundances of each measured peptide (33). In secretome analysis, the main advantage of SILAC with respect to chemical labeling is to allow the mixing of labeled and unlabeled cells before any subsequent fractionation and purification step, thus avoiding the introduction of errors in quantitation. However, several technical issues should be considered, including the incomplete incorporation of labels for certain cell lines and the loss of essential growth factors in the dialyzed serum required for SILAC (34, 35). Nevertheless, the SILAC approach has been applied for the secretome characterization of several cellular models such as the hepatoma cells (36), the esophageal squamous carcinoma (37), the pancreatic cancer (38), and the malignant astrocytoma cell lines (39). Moreover, secretomes of primary cell cultures including chondrocytes (40), astrocytes (41), and skeletal muscle cells undergoing myoblasts differentiation (42, 43) have been also analyzed by using the SILAC approach.

Another common approach used for quantifying secreted proteins is the isobaric tags for relative and absolute quantification (iTRAQ) (44). In this strategy, proteins are digested first and peptides labeled with different isobaric tags on Lys side chains and at N-termini. To date, four or eight different samples can be multiplexed (AbSciex, Framingham, MA, USA). After labeling, peptides are pooled and subjected to LC-MS/MS analysis. During MS scanning, peptides with different tags appear as a single peak due to the isobaric masses, while in MS^2 each iTRAQ tag is fragmented and releases a specific singly charged reporter group. These specific reporter ions are used for relative quantification (34). The iTRAQ approach may be easily utilized for all cell lines without the need to optimize the conditions for the incorporation of stable isotopes in cell cultures. On the other hand, secreted proteins have to be separately purified before labeling, leading to potential quantification errors. The iTRAQ tagging has been applied to the secretome analysis of lung cancer (45, 46) and primary cultures, such as macrophages (47) and adipocytes (48).

Chemical labeling of secretome proteins has been also performed using isotopic tags, by ICAT (isotope-coded affinity tag) or its cleavable version (cICAT) (49). This approach allows the comparison of two different cell states whose proteins are labeled on cysteine residues with light and heavy ICAT. Samples are then pooled, digested to peptides, and analyzed by LC-MS/MS. The light and heavy ICAT-modified peptides co-elute and can be easily distinguished from each other by a 8 or 9 Da mass shift, depending on the specific reagent used. The relative quantification is determined by the ratio of peptide pairs (33). However, a lack of quantitative data is possible for proteins not containing any cysteine residues (34). The ICAT methodology has been applied to secretome analysis of glioblastoma cells (50) and immortalized human epithelial cell line upon c-Myc induction (51). Other labeling approaches used for secretome protein quantification include the $^{18}O/^{16}O$-tagging reported for the study of activated Jurkat T-cells (52) and the comparison of isotope-labeled amino acid incorporation rates (CILAIR), an innovative strategy applied to the quantification of the insulin-induced changes in the rate of ^{13}C-lysine incorporation in human visceral adipose tissue cultures (53).

Isotopic labeling may provide very reliable quantitative results. However, the time-consuming tagging step and the relatively high costs for reagents often reduce the experimental flexibility by limiting the number of samples that can be compared. Label-free approaches represent a promising alternative for protein quantification by LC-MS. The quantification by label-free methods is performed by comparing the direct mass spectrometric signal intensity for any given peptide (31, 54–56) or by using the number of acquired spectra matching to a peptide/protein, a methodology named spectral counting (57, 58). Quantitative studies by means of label-free LC-MS using spectral counting have been recently applied to the secretome analysis of bronchial epithelial cell line (59) and the SW480 primary cell line compared to the SW620 lymph node metastatic cell line (60). A label-free quantitative algorithm combining measurements of spectral counting, ion intensity, and peak area on 1D PAGE has been also reported for the characterization of angiotensin II-stimulated smooth muscle cells secretome (61).

3.8 Validation of Selected Secretome Candidate Biomarkers

Secretome investigations by using proteomic approaches may provide a valuable tool in basic research. However, most interest in secretome analyses is related to their potential implications in applied research for the discovery of novel biomarkers (7–11). Nevertheless, the identification of disease-associated secreted proteins should be considered as a pilot study towards the development of clinically useful biomarkers. Indeed, in a conventional biomarker identification pipeline, additional steps are needed to go beyond the unbiased discovery phase to the targeted confirmation and validation stages (62). The high resolution, sensitivity, and accuracy of mass spectrometers allow the identification of thousand of proteins and peptides in a single LC-MS/MS run. However, the difficulty in handling these large datasets and often the lack of computational tools for their multiplexed analysis represent a serious limitation to the high-throughput potential of proteomic methodologies. These drawbacks are reflected in the comparison of a limited number of conditions within almost all proteomic studies (e.g., control/healthy vs. treated/disease samples) (62). As a consequence, each secretome research should include an array of validation analyses that can be performed at two different levels. At first, the presence and the expression levels of secreted proteins, selected on the basis of their relevance in a given cell model, should be validated on independent cell cultures (see Note 1). More importantly, a second level of validation should be performed on clinical samples, including biological fluids and tissues derived from biopsy. In this instance, a major advantage deals with the possibility to probe the presence of potential biomarkers within a large cohort of samples (62). On the other hand, this requirement represents a *condicio sine qua non,* a biomarker can be recognized as really informative, based on the rigorous criteria established by clinical research.

Several approaches can be used for secretome candidate biomarker validation, alone or in combination by using different methodologies. If specific antibodies are available, the traditional immunoassay-based techniques (i.e., ELISA assays, Western blotting analyses) allow a reliable quantitative screening on both CM and clinical samples. Complementary information may be obtained by immunohistochemistry (IHC) and by the analysis of mRNA expression profiles by quantitative RT-PCR. However, these approaches measure the intracellular expression levels of proteins, not necessarily related to their secretion levels. In addition, a linear correlation between mRNA and protein expression levels is not always expected (63).

More recently, there has been a renewed interest in validation methodologies based on LC-MS platforms, mainly triple quadrupole mass spectrometers, by exploiting multiple reaction monitoring (MRM). This technology allows the accurate and sensitive confirmation and quantification of specific analytes within complex biological matrices by the selective monitoring of their masses. The high specificity derives from the capability to monitoring, following the analyte collision-induced fragmentation, both parent and one or more product ions simultaneously. The selection of the appropriate parent/product ion pairs for the analyte of interest (transitions) is a precondition for the development of an MRM experiment. The enhanced selectivity and sensitivity of MRM offer powerful capabilities in detecting and quantifying secreted proteins within complex mixtures such as biological fluids. However, technological limitations related to the relatively low resolution of triple quadrupole instruments may hamper the correct selection of ions with overlapping isotopic distributions within Q1 and Q3 quadrupole analyzers (64). This drawback may be further amplified by the high dynamic range and the presence of high-abundance interfering peptides in biological matrices (64).

In conclusion, it is clear that many efforts should be devoted to the development of a rigorous validation workflow, also considering that the above methodologies do not provide a comprehensive solution for the discovery of novel biomarkers. Nonetheless, the analysis of secreted proteins by proteomic and mass spectrometry approaches holds great promise in providing valuable clues to correlate the results from the in vitro to the in vivo models, thus supporting the biomarkers discovery research.

4 Notes

1. It is advisable to prepare CM from at least three independent cell cultures to address biological variation. If the experimental setup requires a high number of data points, CM obtained under the same conditions may be pooled for secretome analysis. In this instance, it is suggested to perform data validation (see Subheading 3.8) on independent cultures.
2. Petri dishes are commonly used for cell culture of adherent cell lines. Cells growing in suspension can be cultured in flasks with a surface area of 175 cm^2.
3. Cell confluence strongly affects the quality of secretome samples since it may lead to a reduced mitotic index and, eventually, to cell death. The optimal confluence has been shown to be in the range 60–70 % (65).
4. Although the total yield of secreted proteins strictly depends on the specific cell type, usually a cell number comprised between 2.5×10^7 and 8×10^7 cells is suitable for secretome analysis of several normal and cancer cell lines (20, 24, 66, 67).

When a comparison of cell types with different proliferation rates is needed, an intrinsic bias in the amount of secreted proteins may be observed. To overcome this issue, it is suggested to collect CM deriving from a total number of cells as similar as possible when comparing two or more conditions (e.g., with/without stimulation, tumor/control cells).

5. The number and timing of cell washes should be optimized for each cell type, considering that a high number of washes results in a more efficient serum removal. Indeed, it has been reported that stringent and prolonged washing steps, even for several hours, strongly reduce BSA contamination (68, 69). However, two or three washing steps in serum-free medium before cell starvation often constitute a good compromise to ensure serum removal without negatively affecting cell viability.

6. Serum concentration can be gradually lowered by performing washes with CM containing decreasing concentrations of FBS. The use of media supplemented with low serum (1–2 %) has been also reported (70). In this instance, an additional cleanup for high-abundance protein depletion is generally required prior to downstream mass spectrometry analysis. These approaches are based on different strategies, including those based on immunodepletion, single or multidimensional chromatographic separation, and ligand-based enrichment (ProteoMiner beads). However, due to the high amount of serum proteins, whenever possible, it is suggested to perform the last conditioning step in serum-free medium.

7. It is frequently assumed that serum starvation reduces basal activity of cells (71). Although this procedure is widely used in cell biology, prolonged incubation in serum-deprived medium leads to cell detachment and death by activating apoptosis processes (72). For secretome analysis, the conditioning time and its effect on cells phenotype depend on cell type. Cell conditioning times are usually in the range 12–48 h. However, serum deprivation time should not exceed 30 h, being 24 h of incubation in serum-free media the ideal condition for most secretome analyses (65).

8. A reliable proteomic analysis usually requires that similar amounts of samples, determined by Bradford or an equivalent assay, are analyzed. However, for secretome analysis, significant differences in protein concentrations and composition, due to the alteration of protein release under both physiological and/or pathological conditions, may be masked or completely abolished by *a priori* analyzing the same protein amounts. As discussed in Note 4, a good normalization of the experimental system can be performed by analyzing CM collected from approximately the same number of cells (17, 18).

9. Assessment of cell viability and proliferation following the conditioning step is strongly recommended, because the presence of dead cells results in the artifactual release of intracellular proteins within CM. For cell viability evaluation, the trypan blue dye exclusion assay or standard flow cytometry analyses may be performed (17, 18, 73). Several additional spectrophotometric assays based on the use of tetrazolium salts, such as 3-(4,5-dimethylthiazol-2-yl)-2,5-diphenyltetrazolium bromide (MTT) assay and 2,3-bis(2-methoxy-4-nitro-5-sulfophenyl)-2H-tetrazolium-5-carboxanilide (XTT), are also commonly used (74, 75). For secretome collection, the decrease of viability must not exceed 3–5 %.
10. The detection of high-abundant cytosolic proteins such as β-actin and β-tubulin by Western blot analysis on collected CM has been used to point out the occurrence of unwanted cell lysis events (65, 76).
11. Ultrafiltration/ultracentrifugation techniques, allowing the simultaneous protein concentration and desalting, are widely employed for secretome sample preparation (19, 77–79). In these approaches, protein desalting and concentration are achieved by forcing samples to pass through semipermeable membranes with different cut-off. Choosing a membrane with appropriate retention characteristics is critical for ensuring the maximum protein recovery. Typical membrane cut-off of 3–5 kDa is used for secretome analysis (66, 67, 80). Depending on the specific device available, hydrostatic pressure is applied by using either a centrifuge or an ultrafiltration apparatus. Ultrafiltration techniques have proved to be more efficient than the carrier-assisted TCA precipitation (77). However, during ultrafiltration, some protein losses may occur because of their adsorption on filtration membranes.
12. Protein desalting can be performed by traditional dialysis techniques by using membrane tubing with cut-off of 3–5 kDa (20, 59). However, this procedure requires an additional step for protein concentration. The use of chromatographic columns, such as the silica-based resin with hydroxyl groups (StrataClean resin), has been also reported for secretome preparation (81, 82).
13. TCA precipitation has the advantage to combine protein concentration and efficient desalting in a single step. In addition, amongst the miscible organic solvents commonly used for protein precipitation, TCA is effective at lower concentrations compared to other solvents (i.e., about 15 % for TCA vs. about 75 % for acetone and 90 % for ethanol) permitting to not increase excessively the sample volume. In this way, protein concentration remains higher, thus improving the efficiency of the precipitation. The main disadvantage of TCA as well as

of other precipitation methods is associated with partial denaturation of proteins which may be difficult to redissolve in aqueous buffers (see Note 17).

14. It is recommended to keep the CM on ice throughout the duration of secretome concentration and preparation.
15. The enhanced efficiency of TCA-induced precipitation of proteins by using carriers has been widely demonstrated (83). In particular, carriers such as sodium lauroyl sarcosinate or sodium deoxycholate have been used with final percentages in the range of 0.01–0.1 % and 0.02–0.2 %, respectively (40, 84, 85).
16. A special attention has to be paid in supernatant withdrawal, since pellet deriving from TCA precipitation may be hardly visible because of the low protein amount.
17. Pellet overdrying should be avoided as it will be difficult to resuspend in aqueous buffers.
18. DTT and IAM solutions must be freshly prepared prior to use.
19. Prepare all solutions for sample preparation for LC-MS analysis by using exclusively LC-MS-grade reagents. Avoid contamination by external protein sources (e.g., keratin originating from dust, hairs, and finger skin). It is recommended to work in a clean and dust-free space and to wear a lab coat, gloves, and possibly a hair net. Ideally, all operations should be performed in a laminar flow hood.
20. Depending on the purity of samples and the LC-MS configuration, a preliminary analytical purification step on C_{18} columns may be performed before proceeding with LC-MS analysis. To this aim, several commercial devices are available, including the C_{18} microspin columns (Harvard Apparatus Ltd. Edenbridge, United Kingdom) and the ZipTip C_{18} pipette tips (Merck Millipore KGaA, Darmstadt, Germany).

Acknowledgments

We gratefully acknowledge Dr. Menotti Ruvo (IBB, CNR of Naples) and Dr. Carla Pasquarello Mosimann (Geneva Proteomics Core Facility) for their useful suggestions and support.

References

1. Pandey A, Mann M (2000) Proteomics to study genes and genomes. Nature 405: 837–846
2. Tyers M, Mann M (2003) From genomics to proteomics. Nature 422:193–197
3. May M (2009) From cells, secrets of the secretome leak out. Nat Med 15:828
4. Walter P, Gilmore R, Blobel G (1984) Protein translocation across the endoplasmic reticulum. Cell 38:5–8
5. Nickel W (2003) The mystery of nonclassical protein secretion. A current view on cargo proteins and potential export routes. Eur J Biochem 270:2109–2119

6. Karagiannis GS, Pavlou MP, Diamandis EP (2010) Cancer secretomics reveal pathophysiological pathways in cancer molecular oncology. Mol Oncol 4:496–510
7. Makridakis M, Vlahou A (2010) Secretome proteomics for discovery of cancer biomarkers. J Proteomics 73:2291–2305
8. Xue H, Lu B, Lai M (2008) The cancer secretome: a reservoir of biomarkers. J Transl Med 6:1–12
9. Pavlou MP, Diamandis EP (2010) The cancer cell secretome: a good source for discovering biomarkers? J Proteomics 73:1896–1906
10. Stastna M, Van Eyk JE (2012) Secreted proteins as a fundamental source for biomarker discovery. Proteomics 12:1–14
11. Dowling P, Clynes M (2011) Conditioned media from cell lines: a complementary model to clinical specimens for the discovery of disease-specific biomarkers. Proteomics 11:794–804
12. Skalnikova H, Motlik J, Gadher SJ, Kovarova H (2011) Mapping of the secretome of primary isolates of mammalian cells, stem cells and derived cell lines. Proteomics 11:691–708
13. Veenstra TD, Conrads TP, Hood BL, Avellino AM, Ellenbogen RG, Morrison RS (2005) Biomarkers: mining the biofluid proteome. Mol Cell Proteomics 4: 409–418
14. Mischak H, Allmaier G, Apweiler R et al (2010) Recommendations for biomarker identification and qualification in clinical proteomics. Sci Transl Med 2:46ps42
15. Lim JWE, Bodnar A (2002) Proteome analysis of conditioned medium from mouse embryonic fibroblast feeder layers which support the growth of human embryonic stem cells. Proteomics 2: 1187–1203
16. Lee MJ, Kim J, Kim MY et al (2010) Proteomic analysis of tumor necrosis factor-alpha-induced secretome of human adipose tissue-derived mesenchymal stem cells. J Proteome Res 9:1754–1762
17. Rocco M, Malorni L, Cozzolino R et al (2011) Proteomic profiling of human melanoma metastatic cell line secretomes. J Proteome Res 10: 4703–4714
18. Farina A, D'Aniello C, Severino V et al (2011) Temporal proteomic profiling of embryonic stem cell secretome during cardiac and neural differentiation. Proteomics 11:3972–3982
19. Klein-Scory S, Kubler S, Diehl H et al (2010) Immunoscreening of the extracellular proteome of colorectal cancer cells. BMC Cancer 70:1–19
20. Srisomsap C, Sawangareetrakul P, Subhasitanont P et al (2010) Proteomic studies of cholangiocarcinoma and hepatocellular carcinoma cell secretomes. J Biomed Biotechnol 2010:437143
21. Tang CE, Guan YJ, Yi B et al (2010) Identification of the amyloid beta-protein precursor and cystatin C as novel epidermal growth factor receptor regulated secretory proteins in nasopharyngeal carcinoma by proteomics. J Proteome Res 9:6101–6111
22. Del Galdo F, Shaw MA, Jimenez SA (2010) Proteomic analysis identification of a pattern of shared alterations in the secretome of dermal fibroblasts from systemic sclerosis and nephrogenic systemic fibrosis. Am J Pathol 177:1638–1646
23. Roca-Rivada A, Alonso J, Al-Massadi O et al (2011) Secretome analysis of rat adipose tissues shows location-specific roles for each depot type. J Proteomics 74: 1068–1079
24. Mathias RA, Wang B, Ji H et al (2009) Secretome-based proteomic profiling of Ras-transformed MDCK cells reveals extracellular modulators of epithelial-mesenchymal transition. J Proteome Res 8:2827–2837
25. De la Fuente A, Mateos J, Lesende-Rodriguez I et al (2012) Proteome analysis during chondrocyte differentiation in a new chondrogenesis model using human umbilical cord stroma mesenchymal stem cells. Mol Cell Proteomics 11:M111.010496
26. Weiss W, Gorg A (2009) High-resolution two-dimensional electrophoresis. Methods Mol Biol 564:13–32
27. Mathias RA, Ji H, Simpson RJ (2012) Proteomic profiling of the epithelial-mesenchymal transition using 2D DIGE. Methods Mol Biol 854:269–286
28. Shevchenko A, Tomas H, Havlis J, Olsen JV, Mann M (2006) In-gel digestion for mass spectrometric characterization of proteins and proteomes. Nat Protoc 1:2856–2860
29. Bendtsen JD, Jensen LJ, Blom N, Von Heijne G, Brunak S (2004) Feature-based prediction of non-classical and leaderless protein secretion. Protein Eng Des Sel 17:349–356
30. Bendtsen JD, Nielsen H, von Heijne G, Brunak S (2004) Improved prediction of signal peptides: SignalP 3.0. J Mol Biol 340:783–795
31. Bantscheff M, Schirle M, Sweetman G, Rick J, Kuster B (2007) Quantitative mass spectrometry in proteomics: a critical review. Anal Bioanal Chem 389:1017–1031
32. Ong SE, Blagoev B, Kratchmarova I et al (2002) Stable isotope labeling by amino acids in cell culture, SILAC, as a simple and accurate approach to expression proteomics. Mol Cell Proteomics 1:376–386
33. May C, Brosseron F, Chartowski P, Schumbrutzki C, Schoenebeck B, Marcus K (2011) Instruments and methods in proteomics. Methods Mol Biol 696:3–26
34. Brewis IA, Brennan P (2010) Proteomics technologies for the global identification and quantification of proteins. Adv Protein Chem Struct Biol 80:1–44

35. Mann M (2006) Functional and quantitative proteomics using SILAC. Nat Rev Mol Cell Biol 7:952–958
36. Chen CY, Chi LM, Chi HC et al (2012) Stable isotope labeling with amino acids in cell culture (SILAC)-based quantitative proteomics study of a thyroid hormone-regulated secretome in human hepatoma cells. Mol Cell Proteomics 11:M111.011270
37. Kashyap MK, Harsha HC, Renuse S et al (2010) SILAC-based quantitative proteomic approach to identify potential biomarkers from the esophageal squamous cell carcinoma secretome. Cancer Biol Ther 10:796–810
38. Gronborg M, Kristiansen TZ, Iwahori A et al (2006) Biomarker discovery from pancreatic cancer secretome using a differential proteomic approach. Mol Cell Proteomics 5:157–171
39. Formolo CA, Williams R, Gordish-Dressman H, MacDonald TJ, Lee NH, Hathout Y (2011) Secretome signature of invasive glioblastoma multiforme. J Proteome Res 10:3149–3159
40. Calamia V, Rocha B, Mateos J, Fernandez-Puente P, Ruiz-Romero C, Blanco FJ (2011) Metabolic labeling of chondrocytes for the quantitative analysis of the interleukin-1-beta-mediated modulation of their intracellular and extracellular proteomes. J Proteome Res 10:3701–3711
41. Greco TM, Seeholzer SH, Mak A, Spruce L, Ischiropoulos H (2010) Quantitative mass spectrometry-based proteomics reveals the dynamic range of primary mouse astrocyte protein secretion. J Proteome Res 9:2764–2774
42. Chan CY, Masui O, Krakovska O et al (2011) Identification of differentially regulated secretome components during skeletal myogenesis. Mol Cell Proteomics 10:M110.004804
43. Henningsen J, Rigbolt KT, Blagoev B, Pedersen BK, Kratchmarova I (2010) Dynamics of the skeletal muscle secretome during myoblast differentiation. Mol Cell Proteomics 9: 2482–2496
44. Ross PL, Huang YN, Marchese JN et al (2004) Multiplexed protein quantitation in *Saccharomyces cerevisiae* using amine-reactive isobaric tagging reagents. Mol Cell Proteomics 3:1154–1169
45. Chenau J, Michelland S, de Fraipont F et al (2009) The cell line secretome, a suitable tool for investigating proteins released in vivo by tumors: application to the study of p53-modulated proteins secreted in lung cancer cells. J Proteome Res 8:4579–4591
46. Chiu KH, Chang YH, Wu YS, Lee SH, Liao PC (2011) Quantitative secretome analysis reveals that COL6A1 is a metastasis-associated protein using stacking gel-aided purification combined with iTRAQ labeling. J Proteome Res 10:1110–1125
47. Lietzen N, Ohman T, Rintahaka J et al (2011) Quantitative subcellular proteome and secretome profiling of influenza A virus-infected human primary macrophages. PLoS Pathog 7:e1001340
48. Zhong J, Krawczyk SA, Chaerkady R et al (2010) Temporal profiling of the secretome during adipogenesis in humans. J Proteome Res 9:5228–5238
49. Gygi SP, Rist B, Gerber SA, Turecek F, Gelb MH, Aebersold R (1999) Quantitative analysis of complex protein mixtures using isotope-coded affinity tags. Nat Biotechnol 17:994–999
50. Khwaja FW, Svoboda P, Reed M, Pohl J, Pyrzynska B, Van Meir EG (2006) Proteomic identification of the wt-p53-regulated tumor cell secretome. Oncogene 25:7650–7661
51. Pocsfalvi G, Votta G, De Vincenzo A et al (2011) Analysis of secretome changes uncovers an autocrine/paracrine component in the modulation of cell proliferation and motility by c-Myc. J Proteome Res 10:5326–5337
52. Bonzon-Kulichenko E, Martinez-Martinez S, Trevisan-Herraz M, Navarro P, Redondo JM, Vazquez J (2011) Quantitative in-depth analysis of the dynamic secretome of activated Jurkat T-cells. J Proteomics 75:561–571
53. Roelofsen H, Dijkstra M, Weening D, de Vries MP, Hoek A, Vonk RJ (2009) Comparison of isotope-labeled amino acid incorporation rates (CILAIR) provides a quantitative method to study tissue secretomes. Mol Cell Proteomics 8:316–324
54. Bondarenko PV, Chelius D, Shaler TA (2002) Identification and relative quantitation of protein mixtures by enzymatic digestion followed by capillary reversed-phase liquid chromatography-tandem mass spectrometry. Anal Chem 74:4741–4749
55. Chelius D, Bondarenko PV (2002) Quantitative profiling of proteins in complex mixtures using liquid chromatography and mass spectrometry. J Proteome Res 1:317–323
56. Wang W, Zhou H, Lin H et al (2003) Quantification of proteins and metabolites by mass spectrometry without isotopic labeling or spiked standards. Anal Chem 75:4818–4826
57. Ishihama Y, Oda Y, Tabata T et al (2005) Exponentially modified protein abundance index (emPAI) for estimation of absolute protein amount in proteomics by the number of sequenced peptides per protein. Mol Cell Proteomics 4:1265–1272
58. Lu P, Vogel C, Wang R, Yao X, Marcotte EM (2007) Absolute protein expression profiling estimates the relative contributions of transcriptional and translational regulation. Nat Biotechnol 25:117–124
59. Malard V, Chardan L, Roussi S et al (2012) Analytical constraints for the analysis of human

cell line secretomes by shotgun proteomics. J Proteomics 75:1043–1054
60. Xue H, Lu B, Zhang J et al (2010) Identification of serum biomarkers for colorectal cancer metastasis using a differential secretome approach. J Proteome Res 9:545–555
61. Gao BB, Stuart L, Feener EP (2008) Label-free quantitative analysis of one-dimensional PAGE LC/MS/MS proteome: application on angiotensin II-stimulated smooth muscle cells secretome. Mol Cell Proteomics 7:2399–2409
62. Rifai N, Gillette MA, Carr SA (2006) Protein biomarker discovery and validation: the long and uncertain path to clinical utility. Nat Biotechnol 24:971–983
63. Gygi SP, Rochon Y, Franza BR, Aebersold R (1999) Correlation between protein and mRNA abundance in yeast. Mol Cell Biol 19:1720–1730
64. Kitteringham NR, Jenkins RE, Lane CS, Elliott VL, Park BK (2009) Multiple reaction monitoring for quantitative biomarker analysis in proteomics and metabolomics. J Chromatogr B Analyt Technol Biomed Life Sci 877: 1229–1239
65. Mbeunkui F, Fodstad O, Pannell LK (2006) Secretory protein enrichment and analysis: an optimized approach applied on cancer cell lines using 2D LC-MS/MS. J Proteome Res 5:899–906
66. Yao L, Zhang Y, Chen K, Hu X, Xu LX (2011) Discovery of IL-18 as a novel secreted protein contributing to doxorubicin resistance by comparative secretome analysis of MCF-7 and MCF-7/Dox. PLoS One 6:e24684
67. Schiarea S, Solinas G, Allavena P et al (2010) Secretome analysis of multiple pancreatic cancer cell lines reveals perturbations of key functional networks. J Proteome Res 9: 4376–4392
68. Pellitteri-Hahn MC, Warren MC, Didier DN et al (2006) Improved mass spectrometric proteomic profiling of the secretome of rat vascular endothelial cells. J Proteome Res 5: 2861–2864
69. Alvarez-Llamas G, Szalowska E, de Vries MP et al (2007) Characterization of the human visceral adipose tissue secretome. Mol Cell Proteomics 6:589–600
70. Stastna M, Chimenti I, Marban E, Van Eyk JE (2010) Identification and functionality of proteomes secreted by rat cardiac stem cells and neonatal cardiomyocytes. Proteomics 10:245–253
71. Pirkmajer S, Chibalin AV (2011) Serum starvation: caveat emptor. Am J Physiol Cell Physiol 301:C272–C279
72. Hasan NM, Adams GE, Joiner MC (1999) Effect of serum starvation on expression and phosphorylation of PKC-alpha and p53 in V79 cells: implications for cell death. Int J Cancer 80:400–405
73. Keene SD, Greco TM, Parastatidis I et al (2009) Mass spectrometric and computational analysis of cytokine-induced alterations in the astrocyte secretome. Proteomics 9:768–782
74. Mosmann T (1983) Rapid colorimetric assay for cellular growth and survival: application to proliferation and cytotoxicity assays. J Immunol Methods 65:55–63
75. Phillips HJ, Terryberry JE (1957) Counting actively metabolizing tissue cultured cells. Exp Cell Res 13:341–347
76. Weng LP, Wu CC, Hsu BL et al (2008) Secretome-based identification of Mac-2 binding protein as a potential oral cancer marker involved in cell growth and motility. J Proteome Res 7:3765–3775
77. Dowell JA, Johnson JA, Li L (2009) Identification of astrocyte secreted proteins with a combination of shotgun proteomics and bioinformatics. J Proteome Res 8: 4135–4143
78. Polisetty RV, Gupta MK, Nair SC et al (2011) Glioblastoma cell secretome: analysis of three glioblastoma cell lines reveal 148 non-redundant proteins. J Proteomics 74:1918–1925
79. Wu CC, Hsu CW, Chen CD et al (2010) Candidate serological biomarkers for cancer identified from the secretomes of 23 cancer cell lines and the human protein atlas. Mol Cell Proteomics 9:1100–1117
80. Caccia D, Zanetti Domingues L, Micciche F et al (2011) Secretome compartment is a valuable source of biomarkers for cancer-relevant pathways. J Proteome Res 10:4196–4207
81. Yoon JH, Yea K, Kim J et al (2009) Comparative proteomic analysis of the insulin-induced L6 myotube secretome. Proteomics 9:51–60
82. Canals F, Colome N, Ferrer C, Plaza-Calonge Mdel C, Rodriguez-Manzaneque JC (2006) Identification of substrates of the extracellular protease ADAMTS1 by DIGE proteomic analysis. Proteomics 6(Suppl 1):S28–S35
83. Chevallet M, Diemer H, Luche S, van Dorsselaer A, Rabilloud T, Leize-Wagner E (2006) Improved mass spectrometry compatibility is afforded by ammoniacal silver staining. Proteomics 6:2350–2354
84. Kashat L, So AK, Masui O et al (2010) Secretome-based identification and characterization of potential biomarkers in thyroid cancer. J Proteome Res 9:5757–5769
85. Ralhan R, Masui O, Desouza LV, Matta A, Macha M, Siu KW (2011) Identification of proteins secreted by head and neck cancer cell lines using LC-MS/MS: strategy for discovery of candidate serological biomarkers. Proteomics 11:2363–2376

86. Luo X, Liu Y, Wang R, Hu H, Zeng R, Chen H (2011) A high-quality secretome of A549 cells aided the discovery of C4b-binding protein as a novel serum biomarker for non-small cell lung cancer. J Proteomics 74: 528–538
87. Wang CL, Wang CI, Liao PC et al (2009) Discovery of retinoblastoma-associated binding protein 46 as a novel prognostic marker for distant metastasis in nonsmall cell lung cancer by combined analysis of cancer cell secretome and pleural effusion proteome. J Proteome Res 8:4428–4440
88. Chang KP, Wu CC, Chen HC et al (2010) Identification of candidate nasopharyngeal carcinoma serum biomarkers by cancer cell secretome and tissue transcriptome analysis: potential usage of cystatin A for predicting nodal stage and poor prognosis. Proteomics 10:2644–2660

Chapter 5

Preparation of Human Cerebrospinal Fluid for Proteomics Biomarker Analysis

Timothy J. Waybright

Abstract

The analysis of the cerebrospinal fluid (CSF) proteome in recent years has resulted in a valuable repository of data for targeting and diagnosing a variety of diseases, such as Parkinson's disease, Alzheimer's disease, traumatic brain injury, and amyotrophic lateral sclerosis. Human ventricular CSF contains numerous proteins that are unique to CSF due in part to the interaction of the biofluid with the brain. This allows researchers to obtain information from a region that would otherwise be inaccessible except through invasive surgery or during autopsy. Characterization of the CSF proteome requires that strict care be taken so that sample integrity and fidelity are maintained to ensure data reproducibility. Standardized methods in sample collection, storage, preparation, analysis, and data mining must be used for meaningful information to be obtained. The following method describes a simple and robust approach for preparing CSF samples for analysis via reversed-phase liquid chromatography (RPLC) and mass spectrometry (MS).

Key words Mass spectrometry, Strong cation exchange, Cerebrospinal fluid, Liquid chromatography

1 Introduction

Cerebrospinal fluid is produced in the choroid plexus of the third and fourth lateral ventricles via simple passive capillary diffusion from blood serum and by active choroidal epithelial secretion. It then circulates through the ventricles, subarachnoid space and parenchyma before being eventually reabsorbed by the blood. It acts as a liquid barrier between the brain and the skull, providing a buoyant shock-absorbing shield against sudden trauma. It also provides a transport and distribution mechanism for various physiological factors such as nutrients, metabolites, toxins, and neuroendocrine factors.

CSF volume in adults is approximately 150 mL, with approximately 25 mL in the ventricles and the remainder in the subarachnoid spaces. The total volume produced per day is around 500–700 mL, which represents a turnover of around 3–4x of the

Ming Zhou and Timothy Veenstra (eds.), *Proteomics for Biomarker Discovery: Methods and Protocols*, Methods in Molecular Biology, vol. 1002, DOI 10.1007/978-1-62703-360-2_5, © Springer Science+Business Media, LLC 2013

entire volume per day. Its uniqueness lies in its intimate contact with the extracellular spaces of the brain and central nervous system (CNS), allowing researchers to detect and study pathological conditions associated with neurodegenerative disorders such as Parkinson's and Alzheimer's diseases, tumors, infections, and hydrocephalus.

CSF is contained in the brain tissue compartment, separated from blood by a blood–brain barrier comprised of endothelial cells. This compartmentalization results in a unique and distinct set of proteins, many of which are associated with the brain and CNS. This is fortuitous for the field of proteomics, as the inability of larger molecules like proteins to cross the blood–brain barrier creates a pool of readily available material for analysis. The development of mass spectrometry-based proteomic technologies and methodologies offers broad opportunities for improving the biological understanding of diseases associated with the CNS. Early detection and monitoring of a disease for development into a viable and useful clinical method is the ultimate goal of many research projects today. Biomarker discovery through proteomic analysis of the CSF proteins has the potential for providing these future clinical tests to detect or monitor neurodegenerative diseases or traumas (1).

Much work has been done in the field of CSF proteomic biomarker analysis in recent years. Analysis of patients with pneumococcal meningitis using 2D-PAGE gel separation and mass spectrometry resulted in the identification of 34 protein spots that were related to non-survivors as compared to survivors or control subjects (2). These proteins were from varied groups, such as transporters, glycoproteins, kinases, translation proteins, and metabolic enzymes, to name a few. Analysis of the autoimmune response of IgG in Hashimoto's encephalitis (HE), a "poorly understood disease of the central nervous system," provided an insight into the mechanisms involved in the pathogenesis of the disease (3).

Neurodegenerative factors associated with the infection of the larvae of *Taenia solium* in developing countries are the cause of neurocysticercosis, which can result in epileptic episodes, mental disorders, and possible death. Proteomic studies have resulted in the identification of multiple proteins believed to be involved in the progression of the infection (4). Another disease of the CNS, neuromyelitis optica (NMO), also called Devic's syndrome, is an inflammatory demyelinating disease that affects the spinal cord and optic nerves. CSF samples from patients with NMO revealed differentially expressed proteins between diseased and control patients (5).

Various methods have been employed in CSF analysis (6, 7), resulting in the characterization of the CSF proteome (8) and the identification of potential biomarkers in Alzheimer's (9, 10), multiple sclerosis (11, 12), Parkinson's disease (13), and epilepsy and traumatic brain injuries (14, 15). The method described here using strong cation exchange chromatography is robust and has been employed in numerous studies (16–26) and can be applied in virtually all CSF studies.

2 Materials

2.1 Sample Collection and Pretreatment

2.1.1 Sample Collection

All samples for this study were collected according to the guidelines set forth in the special communication from the Biospecimens and Biomarkers Working Group (27).

2.1.2 Sample Pretreatment

1. Falcon Blue Max 15 mL Polypropylene conical tubes (Becton Dickinson, Franklin Lakes, NJ).
2. 100 mM Ammonium bicarbonate (NH_4HCO_3) at pH 8.4 (Sigma, St. Louis, MO).
3. Vortex.

2.2 Protein Quantitation by BCA Assay

1. Incubator.
2. Vortex mixer.
3. Albumin Standard (BSA) Ampule (2 mg/mL, 1 mL) (Pierce, Rockford, IL).
4. BCA™ Protein Assay Kit (Pierce, Rockford, IL).
5. BCA™ Protein Assay Reagent A (5 mL) (Pierce, Rockford, IL).
6. BCA™ Protein Assay Reagent B (0.1 mL) (Pierce, Rockford, IL).
7. 0.65 mL Safe-Lock Polypropylene tubes (Eppendorf, Hauppauge, NY).
8. NanoDrop ND-1000 spectrophotometer (Thermo Scientific, West Palm Beach, FL).
9. 100 mM Ammonium bicarbonate (NH_4HCO_3) at pH 8.4 (Sigma, St. Louis, MO).
10. Ultrapure water (double distilled, deionized >18 Ω, NANOPure Diamond water system, Barnstead International, Dubuque, IA).

2.3 Enzymatic Digestion and Desalting

1. 50 mM Ammonium bicarbonate (NH_4HCO_3) at pH 8.4 (Sigma, St. Louis, MO).
2. Porcine sequencing grade modified trypsin (Promega, Madison, WI).
3. Incubator.
4. Speed Vac.
5. Vortex.
6. C18 Extra-Clean SPE columns (Alltech, Raleigh, NC).

2.4 Strong Cation Exchange Fractionation

1. Polysulfoethyl A Ion Exchange column (PolyLC Inc., Columbia, MD).
2. 96-well Polypropylene round-bottomed plates (Agilent Technologies, Santa Clara, CA).

2.5 Final Sample Preparation for Proteomics Analysis

1. Autosampler vials (Wheaton, Millville, NJ).
2. Sonicator.
3. Vortex.

3 Methods

3.1 Sample Pretreatment

All samples were received frozen on dry ice. There were four tubes of each sample, with approximately 1.8 mL per vial. They were subsequently stored at −80 °C and thawed on wet ice immediately prior to the pretreatment (see Note 1).

1. After thawing the CSF samples on ice, remove a 1 mL aliquot from each and combine in a sterile 15 mL conical tube. Gently vortex to promote thorough mixing of the sample.
2. Add 2 mL of the 100 mM NH_4HCO_3 and mix briefly.

3.2 Protein Quantitation by BCA Assay

1. The albumin standard solution (BSA) is prepared per the recommended dilution scheme from the supplier's protocol (see Table 1) using ddH_2O/100 mM NH_4HCO_3 (pH 8.4) at a ratio of 2:1 for the diluent. Each standard dilution is prepared in a 0.65 mL tube (see Note 2).

Table 1
Dilution information for albumin (BSA) standards for a nominal working protein concentration range of 20–2,000 μg/mL

Vial	Volume of diluent[a]	Volume and source of BSA	Final BSA concentration (μg/mL)
A	0	300 μL of stock[b]	2,000
B	125 μL	375 μL of stock	1,500
C	325 μL	325 μL of stock	1,000
D	175 μL	175 μL of vial B dilution	750
E	325 μL	325 μL of vial C dilution	500
F	325 μL	325 μL of vial E dilution	250
G	325 μL	325 μL of vial F dilution	125
H	400 μL	100 μL of vial G dilution	25
I	400 μL	0	0 μg/mL = blank

[a]Prepare the albumin (BSA) standards in ddH_2O
[b]The albumin (BSA) stock solution concentration is 2 mg/mL

2. Aliquot 25 μL of each BSA standard solution into a 0.65 mL tube.
3. Aliquot 25 μL of each sample into a 0.65 mL tube.
4. Prepare the BCA working reagent by mixing 50 parts of BCA reagent A with 1 part of BCA reagent B in a 15 mL conical tube. Vortex gently for 5–10 s to ensure thorough mixing.
5. Add 200 μL of the working solution to each standard and sample tube and vortex gently for 5–10 s.
6. Place the tubes in an incubator for 30 min at 37 °C with gentle shaking.
7. Remove the tubes and cool to room temperature. Vortex gently again for 5–10 s and centrifuge for 1–2 s to ensure a thoroughly homogonous sample.
8. Prepare the NanoDrop spectrophotometer for use by gently cleaning the optical pedestal with water and a soft cloth or tissue.
9. Start the instrument computer program and select the BCA Assay option.
10. Place 2 μL of ddH$_2$O on the optical pedestal, ensuring that the instrument is set to record the absorbance at 562 nm, and blank the instrument (see Note 3).
11. Wipe off the water, place 2 μL of sample I on the pedestal, and re-blank the instrument.
12. Wipe off the sample, wipe stand with wet tissue, and place 2 μL if sample I on the optical pedestal again. Do this for each of the standard samples H thru A, taking two readings for each. Record the absorbance values for each standard concentration (see Note 4).
13. After all of the standards have been read, repeat the above step with the unknown samples.
14. Subtract the average absorbance of sample I (blank replicate) from all of the other standards and unknown replicate readings.
15. Prepare a standard curve by plotting the average absorbance at 562 nm of the standard sample readings (y-axis) versus the BSA concentration of the sample (x-axis). Using this standard curve, calculate the concentration of protein in your unknown samples.

3.3 Enzymatic Digestion and Desalting

Since the protein concentrations of potential biomarkers may be very low, care must be taken so that none of the sample is lost throughout the sample preparation stages. With this in mind, it was decided to perform in-solution digestion of proteins rather than the commonly used in-gel digestion.

1. Prepare a fresh solution of 1 M dithiothreitol (DTT) in ddH_2O (see Note 5).
2. Add the 1 mM DTT to the 6 mL samples for a final DTT concentration of 10 mM.
3. Boil the samples for 10 min followed by cooling to room temperature.
4. Prepare a fresh solution of sequencing grade trypsin by solubilizing it in 50 mM NH_4HCO_3 (pH 8.4) (see Note 6).
5. Based on the protein concentration in the unknown samples, add the trypsin solution to each sample so that the enzyme:protein ratio is 1:50 (see Note 7).
6. Digest for 16 h at 37 °C with gentle shaking.
7. Add a second aliquot of freshly prepared trypsin as above and incubate for an additional 8 h at 37 °C.
8. Lyophilize the samples at room temperature.
9. Resuspend the digested peptides in 1 mL of 0.1 % (v/v) TFA using sonication and gentle vortexing.
10. Prepare SPE cartridges by wetting the packing material with 2 mL of 60 % $MeOH/H_2O$ and flushing with 2 mL of 0.1 % (v/v) TFA.
11. Apply the digested CSF samples to the cartridge and remove the salts by flushing with 2 mL of 0.1 % (v/v) TFA.
12. Elute the digested peptides with 400 μL of 60 % acetonitrile/H_2O.
13. Lyophilize the samples at room temperature.
14. Re-solubilize the samples in 200 μL of 0.1 % (v/v) TFA.
15. Using the steps from the previous section, perform a BCA protein assay on the above samples.
16. Based on the calculated peptide concentration, aliquot a portion of the mixture equivalent to 170 μg into a separate 0.65 mL tube.
17. Bring the volume up to 400 μL with ddH_2O and lyophilize.

3.4 Strong Cation Exchange Fractionation

1. Prepare system solvent A (25 % acetonitrile in ddH_2O) and system solvent B (25 % acetonitrile in 0.5 M ammonium formate).
2. Attach the Polysulfoethyl A ion exchange column to the LC system and equilibrate the column with 3 % solvent B at 50 μL/min.
3. Solubilize the lyophilized samples from above in 100 μL of 25 % acetonitrile/0.1 % (v/v) FA.

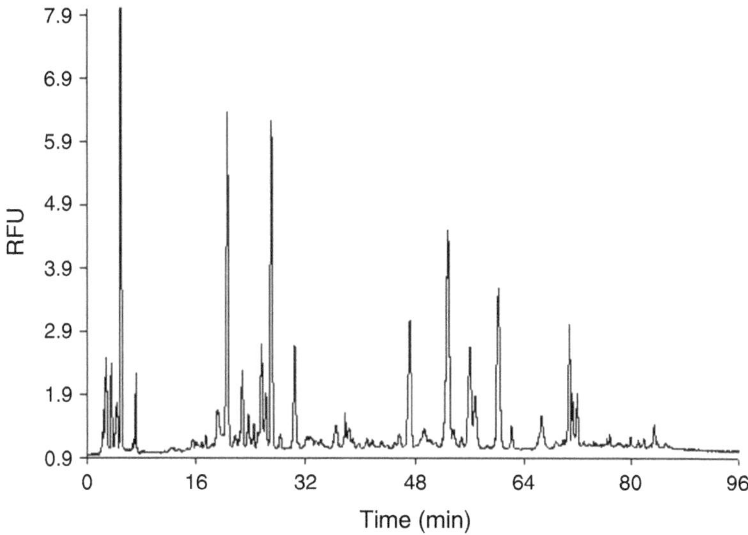

Fig. 1 A typical strong cation exchange chromatogram of cerebrospinal fluid

4. Inject the sample onto the column and elute the peptides into a 96-well round-bottomed plate, collection a total of 96 fractions (see Fig. 1), using the following step gradient:
   (a) 0–3 min—3 % solvent B.
   (b) 3–46 min—10 % solvent B.
   (c) 46–86 min—45 % solvent B.
   (d) 86–87 min—100 % solvent B.
   (e) 87–100 min—100 % solvent B.
5. Re-equilibrate the column as in step 2 above prior to injecting the next CSF sample.
6. Lyophilize the fractions at room temperature and store at −80 °C until ready for analysis via mass spectrometry.

3.5 Final Sample Preparation for Proteomics Analysis

1. Add 10 μL of 0.1 % (v/v) TFA to each well in the 96-well plate.
2. Place the plate in a sonicator for 1 min.
3. Place the plate on a vortex plate shaker for 1–2 min.
4. Transfer the solution to labeled autosampler vials and cap tightly.
5. Place on the LC autosampler for immediate analysis or store at −80 °C for future analysis.

4 Notes

1. For samples that have a larger volume, it is not necessary to put them on ice immediately. They may be placed in racks and allowed to partially thaw at room temperature until they are about halfway thawed. They should then be put on wet ice to finish thawing.
2. Additional tubes of 25 μL aliquot per tube may be prepared for long-term storage of the remaining standard solutions if desired.
3. Aliquots of 1 μL may be used on the spectrophotometer for absorbance readings, but we have found that 2 μL gives more consistent data.
4. The built-in standard curve generation function of the software works very well in generating the standard curve. However, it is limited to only five data points on the graph. If the protein content of the unknown samples can be estimated by visual comparison with the standards, then you can bracket the expected concentration and use only five of the standards. If the protein concentration ranges of your unknown samples are very broad, you have the choice of using all of the standards and plotting a standard curve manually, or breaking the samples into two groups and using the built-in components to generate two separate standard curves.
5. In lieu of using DTT, protein disulfide bonds may be reduced during the boiling step by using a commercially available solution such as Bond-Breaker TCEP Solution (Thermo Scientific, West Palm Beach, FL). This material is provided as a 0.5 M solution and provides a few advantages over DTT or β-mercaptoethanol in that it can be used in either acidic or basic conditions, has no pungent odor, is almost irreversible, and is more hydrophilic.
6. Excess trypsin solution prepared in 50 mM NH_4HCO_3 may be frozen and stored for short periods of time for future use.
7. This protein:enzyme ratio is not set in stone. It may be varied based on the researchers' preference or prior experience.

Acknowledgments

This project has been funded in whole or in part with federal funds from the National Cancer Institute, National Institutes of Health, under Contract HHSN261200800001E. The content of this publication does not necessarily reflect the views or policies of the Department of Health and Human Services, nor does mention of trade names, commercial products, or organizations imply endorsement by the United States Government.

References

1. Mischak H, Allmaier G, Apweiler R et al (2010) Recommendations for biomarker identification and qualification in clinical proteomics. Sci Transl Med 2:46ps42
2. Goonetilleke UR, Scarborough M, Ward SA et al (2010) Proteomic analysis of cerebrospinal fluid in pneumococcal meningitis reveals potential biomarkers associated with survival. J Infect Dis 202:542–550
3. Gini B, Lovato L, Cianti R et al (2008) Novel autoantigens recognized by CSF IgG from Hashimoto's encephalitis revealed by a proteomic approach. J Neuroimmunol 196: 153–158
4. Tian XJ, Li JY, Huang Y et al (2009) Preliminary analysis of cerebrospinal fluid proteome in patients with neurocysticercosis. Chin Med J (Engl) 122:1003–1008
5. Bai S, Liu S, Guo X et al (2009) Proteome analysis of biomarkers in the cerebrospinal fluid of neuromyelitis optica patients. Mol Vis 15:1638–1648
6. Hwang HJ, Quinn T, Zhang J (2009) Identification of glycoproteins in human cerebrospinal fluid. Methods Mol Biol 566: 263–276
7. Mouton-Barbosa E, Roux-Dalvai F, Bouyssie D et al (2010) In-depth exploration of cerebrospinal fluid by combining peptide ligand library treatment and label-free protein quantification. Mol Cell Proteomics 9:1006–1021
8. Waybright T, Avellino AM, Ellenbogen RG et al (2010) Characterization of the human ventricular cerebrospinal fluid proteome obtained from hydrocephalic patients. J Proteomics 73:1156–1162
9. Hu WT, Chen-Plotkin A, Arnold SE et al (2010) Novel CSF biomarkers for Alzheimer's disease and mild cognitive impairment. Acta Neuropathol 119:669–678
10. Maarouf CL, Andacht TM, Kokjohn TA et al (2009) Proteomic analysis of Alzheimer's disease cerebrospinal fluid from neuropathologically diagnosed subjects. Curr Alzheimer Res 6:399–406
11. Harris VK, Diamanduros A, Good P et al (2010) Bri2-23 is a potential cerebrospinal fluid biomarker in multiple sclerosis. Neurobiol Dis 40:331–339
12. Liu S, Bai S, Qin Z et al (2009) Quantitative proteomic analysis of the cerebrospinal fluid of patients with multiple sclerosis. J Cell Mol Med 13:1586–1603
13. Sinha A, Srivastava N, Singh S et al (2009) Identification of differentially displayed proteins in cerebrospinal fluid of Parkinson's disease patients: a proteomic approach. Clin Chim Acta 400:14–20
14. Cadosch D, Thyer M, Gautschi OP et al (2010) Functional and proteomic analysis of serum and cerebrospinal fluid derived from patients with traumatic brain injury: a pilot study. ANZ J Surg 80:542–547
15. Xiao F, Chen D, Lu Y et al (2009) Proteomic analysis of cerebrospinal fluid from patients with idiopathic temporal lobe epilepsy. Brain Res 1255:180–189
16. Barnea E, Sorkin R, Ziv T et al (2005) Evaluation of prefractionation methods as a preparatory step for multidimensional based chromatography of serum proteins. Proteomics 5:3367–3375
17. Barnidge DR, Tschumper RC, Jelinek DF et al (2005) Protein expression profiling of CLL B cells using replicate off-line strong cation exchange chromatography and LC-MS/MS. J Chromatogr B Analyt Technol Biomed Life Sci 819:33–39
18. Blonder J, Chan KC, Issaq HJ et al (2006) Identification of membrane proteins from mammalian cell/tissue using methanol-facilitated solubilization and tryptic digestion coupled with 2D-LC-MS/MS. Nat Protoc 1:2784–2790
19. Conrads KA, Yu LR, Lucas DA et al (2004) Quantitative proteomic analysis of inorganic phosphate-induced murine MC3T3-E1 osteoblast cells. Electrophoresis 25:1342–1352
20. Cutillas PR, Norden AG, Cramer R et al (2003) Detection and analysis of urinary peptides by on-line liquid chromatography and mass spectrometry: application to patients with renal Fanconi syndrome. Clin Sci (Lond) 104: 483–490
21. Dai J, Wang J, Zhang Y et al (2005) Enrichment and identification of cysteine-containing peptides from tryptic digests of performic oxidized proteins by strong cation exchange LC and MALDI-TOF/TOF MS. Anal Chem 77:7594–7604
22. Das S, Bosley AD, Ye X et al (2010) Comparison of strong cation exchange and SDS-PAGE fractionation for analysis of multiprotein complexes. J Proteome Res 9:6696–6704
23. Gao M, Deng C, Yu W et al (2008) Large scale depletion of the high-abundance proteins and analysis of middle- and low-abundance proteins in human liver proteome by multidimensional liquid chromatography. Proteomics 8:939–947
24. Le Bihan T, Goh T, Stewart II et al (2006) Differential analysis of membrane proteins in mouse fore- and hindbrain using a label-free approach. J Proteome Res 5:2701–2710
25. Liu X, Valentine SJ, Plasencia MD et al (2007) Mapping the human plasma proteome by SCX-LC-IMS-MS. J Am Soc Mass Spectrom 18:1249–1264

26. Zhang C, Liu P, Wang N et al (2007) Comparison of two tandem mass spectrometry-based methods for analyzing the proteome of healthy human lens fibers. Mol Vis 13:1873–1877

27. Manley GT, Diaz-Arrastia R, Brophy M et al (2010) Common data elements for traumatic brain injury: recommendations from the biospecimens and biomarkers working group. Arch Phys Med Rehabil 91:1667–1672

Chapter 6

Proteomic Analysis of Frozen Tissue Samples Using Laser Capture Microdissection

Sumana Mukherjee, Jaime Rodriguez-Canales, Jeffrey Hanson, Michael R. Emmert-Buck, Michael A. Tangrea, DaRue A. Prieto, Josip Blonder, and Donald J. Johann Jr.

Abstract

The discovery of effective cancer biomarkers is essential for the development of both advanced molecular diagnostics and new therapies/medications. Finding and exploiting useful clinical biomarkers for cancer patients is fundamentally linked to improving outcomes. Towards these aims, the heterogeneous nature of tumors represents a significant problem. Thus, methods establishing an effective functional linkage between laser capture microdissection (LCM) and mass spectrometry (MS) provides for an enhanced molecular profiling of homogenous, specifically targeted cell populations from solid tumors. Utilizing frozen tissue avoids molecular degradation and bias that can be induced by other preservation techniques. Since clinical samples are often of a small quantity, tissue losses must be minimized. Therefore, all steps are carried out in the same single tube. Proteins are identified through peptide sequencing and subsequent matching against a specific proteomic database. Using such an approach enhances clinical biomarker discovery in the following ways. First, LCM allows for the complexity of a solid tumor to be reduced. Second, MS provides for the profiling of proteins, which are the ultimate bio-effectors. Third, by selecting for tumor proper or microenvironment-specific cells from clinical samples, the heterogeneity of individual solid tumors is directly addressed. Finally, since proteins are the targets of most pharmaceuticals, the enriched protein data streams can then be further analyzed for potential biomarkers, drug targets, pathway elucidation, as well as an enhanced understanding of the various pathologic processes under study. Within this context, the following method illustrates in detail a synergy between LCM and MS for an enhanced molecular profiling of solid tumors and clinical biomarker discovery.

Key words Biomarker, Cancer, Laser capture microdissection (LCM), Liquid chromatography-mass spectrometry (LC-MS), Solid tumor heterogeneity, Frozen tissue

1 Introduction

At present, very few cancers can be cured and clinical outcomes for the majority of tumor types remain disappointing. What are sorely needed are novel molecular diagnostics and improved medications. However, new 'omics technologies (genomics, transcriptomics,

Ming Zhou and Timothy Veenstra (eds.), *Proteomics for Biomarker Discovery: Methods and Protocols*, Methods in Molecular Biology, vol. 1002, DOI 10.1007/978-1-62703-360-2_6, © Springer Science+Business Media, LLC 2013

proteomics, metabolomics) are transforming our understanding of disease and are helping to provide new therapeutic options (1). In this regard, future discovery of effective biomarkers through advanced molecular profiling is promising.

Recently, a new molecular diagnostic for multiple myeloma was developed and is now available commercially (2–4). Critical in the success of this assay development was the use and analysis of purified human subject cell populations, via flow cytometry. This approach permitted the heterogeneity and complexity of a hematologic/liquid malignancy to be reduced, thus enhancing effective scientific study, as well as assay development for clinical use. For solid tumors, laser capture microdissection (LCM) can be viewed as an analog to flow cytometry (5).

Solid tumors possess a heterogeneous cellular architecture. Significant heterogeneity exists among tumors of the same organ system as well as within individual tumor samples (6). Recently, intratumor heterogeneity has been effectively illustrated as a significant clinical issue (7). Thus, there is little surprise why the categorical results of large clinical trials are far from optimal when applied to individual patients. Additionally, new treatments based on devising a therapy regimen from a genetic test that is derived from an outpatient tumor biopsy appear to be a bit too simplistic (8).

Fundamentally, the salient elements of solid tumors include cancer cells proper along with cellular and structural stromal elements. The histology may be quite complex; for instance, an epithelial tumor may contain regions of carcinoma in situ, well to poorly differentiated carcinoma, inflammation, and neovascularity. The tumor microenvironment is composed of both normal and modified stromal cells that serve to nurture the malignant process. The tumor stroma is now recognized as an important area in cancer therapy and many new therapeutic strategies target aspects of this functional region (9).

Solid tumor heterogeneity is reflective of the diversity present at the molecular level and has profound biologic and therapeutic implications (10). For instance, breast cancer is actually many different diseases with the only common characteristic being the organ of origin. Hence, the ability to directly and effectively profile solid tumors at the proteome level is essential, since proteins are the final mediators of pathologic processes, and proteomics in particular can begin to characterize fundamental molecular events such as alternative protein splicing and post translational modifications. Additionally, although cell culture and model organism studies are quite important, they lack a true microenvironment, and invariably the cells utilize some different biochemical systems and clinical translation can be limited (11, 12). Therefore, methods to decompose solid tissue to better enable biological understanding and biomarker discovery are needed (13–15).

LCM and MS (16) are powerful independent analytical technologies. Both have been commonly used for molecular profiling of formalin-fixed paraffin-embedded tissue sections (17). Additionally, we have shown that the LCM-MS platform can be effectively used for proteomic profiling of thin fresh-frozen tissue sections obtained from a solid tumor in conjunction with a simple methanol-aided (18) solubilization and digestion process (19). Furthermore, we advocate that fresh frozen tissue can be particularly useful in some circumstances to avoid potential bias (20). In this chapter, we further illustrate this method for profiling the proteomes of a solid tumor using LCM coupled to biological MS for clinically relevant biomarker discovery.

2 Materials

2.1 LCM

1. TISSUE-Tek O.C.T. cryostat mounting medium (Sakura Finetek Inc., Torrance, CA).
2. Frozen tissue staining: Mayer's hematoxylin solution, eosin Y solution (alcohol-based), and Scott's tap water substitute bluing solution (magnesium sulfate buffered with sodium bicarbonate).
3. Frozen tissue dehydration: 100 % ethanol (ethyl alcohol, absolute, 200 proof for molecular biology). 70 % (v/v) and 95 % (v/v) ethanol baths were prepared using Milli-Q filter with purified water. Xylene is used in the final post dehydration step.
4. CapSure® Macro LCM Caps (Life Technologies, Applied Biosystems, Carlsbad, CA).
5. PixCell IIe, Veritas, or ArcturisXT (Life Technologies, Applied Biosystems, Carlsbad, CA).
6. Leica Cryostat CM 1850 UV, Leica Microsystems, Wetzlar, Germany.
7. Pre-cleaned glass microscope slides, 25 × 75 mm.
8. Membrane slide options include:
   (a) Pen-membrane glass slide.
   (b) Pen-membrane frame slide; both options available from Life Technologies, Applied Biosystems, Carlsbad, CA.

2.2 Protein Extraction and Digestion

1. Sequencing grade trypsin (Promega, Madison, WI, USA).
2. ZipTips packed with C_{18} reversed-phase resin (Millipore, Billerica, MA, USA).

2.3 Reverse-Phase LC-MS

1. LC-MS buffers: Buffer A consists of 0.1 % FA in purified H_2O. Buffer B consists of 0.1 % FA in HPLC grade acetonitrile (ACN).
2. MS sample rehydration with 0.1 % trifluoroacetic acid (TFA).

2.4 Computational Support for CID Spectra Analysis

1. Single computer workstation or a cluster computer that follows a Beowulf design model (see Note 1).
2. Software for protein database search and match to experimental mass spectrometry data (see Note 2).
3. Non-redundant human proteome database.
4. Software for reverse database creation for the assessment of a false-positive rate.
5. Software to analyze experimental data for biologic classification and implications (see Note 3).

3 Methods

Handling the tumor sample rapidly and effectively during the tissue acquisition step is critically important in order to obtain reliable downstream molecular results. Tissue degradation and possibly frank necrosis can begin once a solid tumor is ligated from its blood supply. Thus, a few simple but deliberate steps are recommended to minimize ischemic effects. As quickly as possible, the tissue should be snap frozen in liquid nitrogen and then placed in a freezer at −80 °C. Embedding the tumor tissue in cryostat-mounting medium (TISSUE-Tek O.C.T.) can then be performed.

Tissue sections, usually with a slice thickness range of 8–12 μm, are then serially cut from the frozen tissue block using a cryostat. As a convenient measurement, the majority of cells will have a diameter either larger than or within the range of the thickness of the section. Therefore, the recommended slice thickness will aid in the homogenization/lysis procedure. Clearly, a key aspect of this method is maximizing the effective liberation of proteins from captured cells.

MS is a critical component in any bottom-up proteomic analysis. Key sample handling factors such as effective lysis and digestion are requisites for effective large-scale protein identification. Optimal buffering conditions are required for successful digestion of small-size LCM specimens. Keeping proteins solubilized and denatured throughout the digestion process is essential; thus avoiding unnecessary manipulations and/or use of reagents that might interfere with LC-MS analysis is certainly advocated.

We deliberately chose to simplify and improve the analysis of LCM captured cells, and thus avoid the deficiencies associated with traditional approaches, which typically employ detergents or chaotropes. Hence, a simple two-step methanol-assisted solubilization and digestion protocol was developed. In the first step, 20 % buffered methanol is used to facilitate denaturation and solubilization of cytosolic proteins. In the second step, the digestion is carried out in a 60 % methanol buffer, targeting more hydrophobic proteins that are insoluble in 20 % buffered methanol. We have found

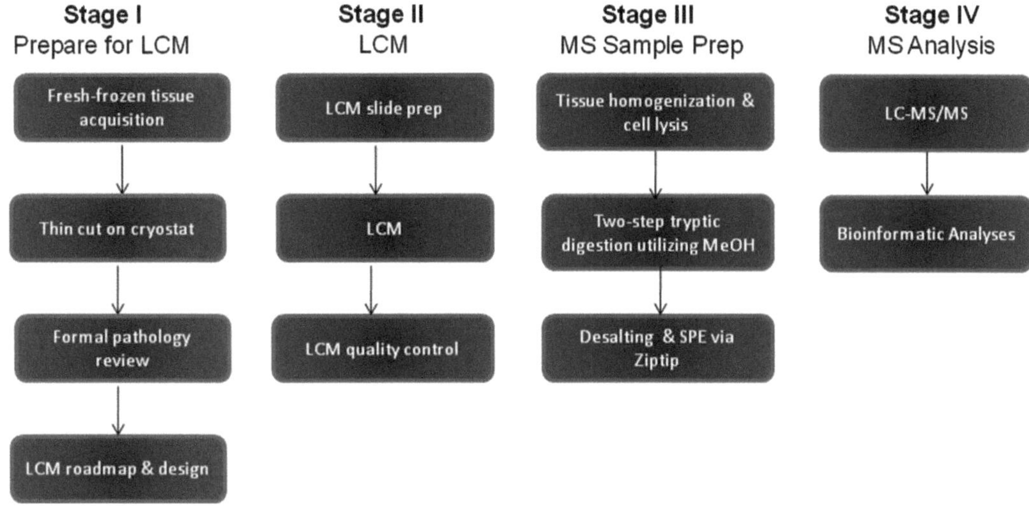

Fig. 1 Laser capture microdissection-mass spectrometry experimental design

this two-step approach results in enhanced proteome coverage. A comprehensive schematic of this experimental workflow, broken down into four stages, is illustrated in Fig. 1. Note: the use of organic solvents for micro and nanoscale proteomic sample prep methods has been very successfully pioneered by other groups (21).

3.1 Initial Pathologic Analysis (Prior to LCM)

A formal hematoxylin and eosin (H&E) staining procedure with cover slip should be performed using every tenth slide. Prior to LCM analysis, these slides should be reviewed with a pathologist to properly evaluate the histology, plan LCM sessions, and guard against potential bias in the *z*-dimension of the tumor tissue plane.

3.2 LCM Staining

The fresh frozen tissue slide must be fully defrosted before beginning the staining protocol. Placing the slide in the palm of your glove works well. As soon as condensate forms on the entire slide the protocol below may commence. To ensure good visualization and tissue capture, suggested times are provided for both membrane and glass slides (see Note 4).

Step	Solution	Comment	Time (membrane slide)	Time (glass slide)
1	70 % ethanol	Fix tissue section to slide	15 s	30 s
2	d.d. water	Remove OCT, rehydrate tissue	30 s	30 s
3	Hematoxylin	Stain nuclei	45 s	30 s
4	d.d. water	Remove excess hematoxylin	15 s	30 s
5	Bluing solution	Change hematoxylin hue	15 s	30 s

(continued)

Step	Solution	Comment	Time (membrane slide)	Time (glass slide)
6	70 % ethanol	Start dehydration	15 s	30 s
7	Eosin	Stain cytoplasm (1–2 quick dips)	1–2 s	2 s
8	95 % ethanol	Dehydration	30 s	1 min
9	95 % ethanol	Dehydration	30 s	1 min
10	100 % ethanol	Dehydration	30 s	2 min
11	100 % ethanol	Dehydration	30 s	2 min
12	Xylene	Ethanol removal	3 min	3 min

3.3 LCM Procedure

LCM analysis may begin on the slide(s) once they are air-dried. Typically, laser-based systems allow for dissections approaching 100 % purity. Staining with H&E allows microscopic visualization during microdissection and does not diminish protein recovery. Generally, we have found that depending on the type of tissue under study, approximately 5,000–50,000 cells are required to produce MS results with acceptable numbers of protein identifications, as well as protein class diversity (see Note 5). Figure 2 illustrates a stepwise approach for successful LCM tissue extraction, which is typical of either a PixCell IIe or Veritas system. LCM tissue extraction involves:

1. Establishing a histology area of interest (Fig. 2a).
2. Manual filling of the pattern to enable removal of cells (Fig. 2b).
3. LCM extraction of the cells from the selected region (Fig. 2c).

3.4 LCM Sample Prep Protocol

The sample preparation protocol for protein extraction and digestion from LCM samples captured on polymer cap is presented below.

3.4.1 Phase I: LCM Membrane-Based Tissue/Cell Extraction and Lysis

1. Prepare a hypotonic lysis buffer (ammonium bicarbonate to methanol: v/v = 80/20, pH ~ 8.0). For convenience, we recommend a 1 mL stock solution prepared by mixing 800 μL of 12.5 mM ammonium bicarbonate (final concentration) with 200 μL of 100 % MeOH and 2 μL of 0.5 mM TCEP (1 mM final concentration).
2. Carefully remove the LCM polymer membrane by peeling it off the CAP and then place it in a new siliconized tube (conical bottom).
3. Add 50 μL of hypotonic lysis buffer.
4. Incubate on dry ice for 30 min.

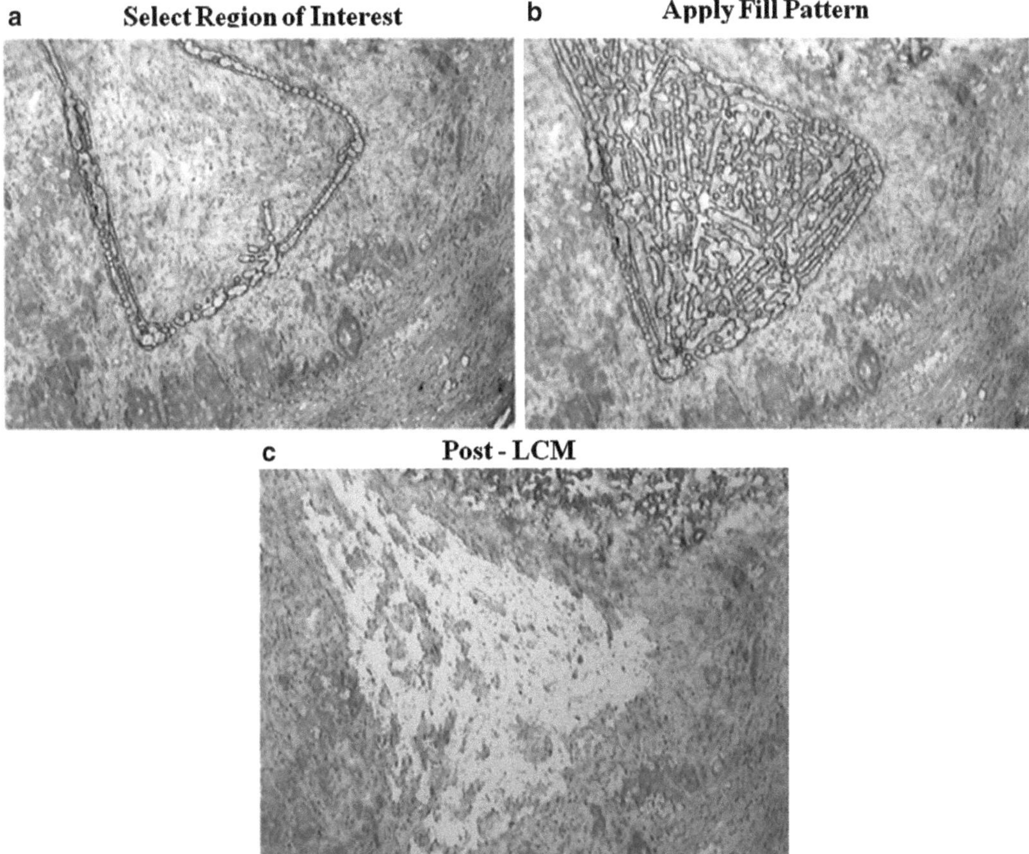

Fig. 2 Laser capture microdissection workflow

5. Thaw the sample in ice-cold water for 10 min.
6. Incubate the sample in a water bath for 2 h at 70 °C.
7. Cool the sample on ice for 20 min.
8. Adjust the buffer from 12.5 to 50 mM by adding 1.65 μL of 1 M ammonium bicarbonate.

3.4.2 Phase II: Initial Trypsin Digestion (Ratio of Trypsin:Protein = 1:50)

The background to the trypsin dilution protocol is based on single-cell protein content estimates in the range of 0.75 pg to 0.5 ng (see Note 6).

Prepare the dehydrated and frozen trypsin, e.g., Promega Trypsin Gold, 20 μg vial.

Mix with 20 μL of 50 mM ammonium bicarbonate, yielding a concentration of 1 μg/μL.

The availability of some tissue samples is quite limited in quantity. Therefore, this section attempts to accommodate these circumstances as well as situations with more abundant tumor tissue.

Recommended amount of trypsin as a function of sample cell count.

Cell count	Protein estimate (μg)	Trypsin:protein	Trypsin for sample (μg)
50	0.025	1:50	0.0005
500	0.25	1:50	0.005
5,000	2.5	1:50	0.05
15,000	7.5	1:50	0.15
50,000	25	1:50	0.5

1. Rehydrate trypsin by adding 20 μL of 50 mM ammonium bicarbonate, pH ~ 8.0.
2. Dilute trypsin in accordance with sample cell count per the trypsin dilution protocol.
3. Add the appropriate volume of trypsin solution and mix for 10 min.
4. Briefly vortex the sample.
5. Place the sample in the water bath sonicator for 5 min.
6. Transfer the sample tube to a small centrifuge and spin for ~15 s.
7. Incubate the sample digest for 6 h at 37 °C with good table motion.

3.4.3 Phase III: The Second Trypsin Digestion (Ratio of Trypsin:Protein = 1:20)

1. Make a 60 % methanol buffer (50 mM ammonium bicarbonate + 100 % methanol (v/v 40/60)). Add the appropriate volume of trypsin solution and mix for 10 min.
2. Briefly (~5 s) vortex the sample.
3. Place the sample in the water bath sonicator for 5 min.
4. Transfer the sample tube to a small centrifuge and spin for ~15 s.
5. Incubate the sample for 6 h at 37 °C with good table motion.
6. Lyophilize all samples to dryness.

3.4.4 Phase IV: Desalting Using ZipTip Columns (Solid Phase Extraction)

1. Rehydrate peptides in 20 μL 0.1 % TFA by sonication in a water bath for 2 min.
2. Prepare 10 μL aliquots of elution buffer consisting of 60 % ACN/0.1 % TFA (v/v) for each sample before beginning (to avoid contamination). Avoid drawing air through the tip during the procedure (from equilibration to elution). If you find that you make bubbles in the tip, try pulling the buffers in more slowly.
3. Set the Pipetman to 10 μL and attach the ZipTip.
4. Activate the ZipTip column by pipetting 20 μL of 60 % ACN and then discarding it to waste. Repeat this process three times.
5. Equilibrate the ZipTip column by pipetting 20 μL of 0.1 % TFA and then discarding it to waste. Repeat this process three

times. These steps act as a gradient for the mini-column, which activates the resin and conditions it to bind peptides.

6. Load the peptides by pipetting the sample up and down (discarding it back into its tube). Repeat this process ten times.
7. Wash the ZipTip column using the 0.1 % (v/v) TFA. Pipette up the solution and then discard it to waste. Repeat this process ten times.
8. Elute the sample by pipetting the ZipTip up and down in the elution buffer back into its tube in the 4 μL 60 % ACN/0.1 % TFA (already aliquoted). Repeat this process ten times. The organic phase elutes the peptides off the resin into the buffer. The sample is desalted as well as concentrated.
9. Lyophilize to dryness and dissolve the peptides in 10 μL of 0.1 % (v/v) TFA prior to LC-MS/MS analysis.

3.5 Guidelines for LC-MS Analysis of LCM Samples

Although there are a wide variety of mass spectrometer systems and liquid chromatography platforms, a linear ion trap mass spectrometer coupled with a reverse-phase liquid chromatography separation system is widely used in the proteomics community, and will be illustrated in this section. For our LCM-based proteomic studies, we utilize a reversed-phase column coupled directly on-line with a linear ion trap mass spectrometer (LTQ ThermoElectron, San Jose, CA). Certainly, newer instruments such as the Orbitrap have higher mass accuracy (1–2 ppm), resolving power, and dynamic range. This improves both the depth and integrity of the discovery process. This protocol is fully compatible with such instruments.

1. In our configuration, the solvent system is delivered using the HP 1100 pump (Agilent Technologies, Palo Alto, CA).
2. A nano-electrospray ionization source is employed applying a voltage of 1.7 kV, and a capillary temperature of 160 °C.
3. The LTQ is operated in a data-dependent mode in which the seven most abundant peptide molecular ions detected by each MS survey scan are dynamically selected. They are then passed for MS/MS (fragmentation) using collision-induced dissociation (CID) facilitated by a normalized collision energy of 35 %.
4. Dynamic exclusion is employed to avoid redundant acquisition of precursor ions previously selected for fragmentation.
5. Reversed-phase liquid chromatography separations are performed using a 75 μm i.d. × 10 cm long fused silica capillary column (Polymicro Technologies, Inc., Phoenix, AZ) with a flame-pulled tip (~5–7 μm orifice).
6. The column is slurry packed in-house with 5 μm, 300 Å pore size C-18 stationary phase (Vydac, Hercules, CA) using a

slurry-packing pump (model 1666, Alltech Associates, Deerfield, IL) (see Note 7).

7. Note: the total MS run time for each sample is 180 min.
   (a) After injecting 5 μL of sample, the column is washed for 30 min with 98 % mobile phase A (0.1 % FA in d.d. water).
   (b) Peptides are then eluted using a linear step gradient from 2 to 40 % mobile phase B (0.1 % FA in ACN) over 90 min.
   (c) Then, an elution gradient of 60–98 % for mobile phase B over 10 min at a constant flow rate of 0.25 μL/min is performed.
   (d) Next, the column is washed for 20 min with 98 % mobile phase B.
   (e) Finally, the column is re-equilibrated with 2 % mobile phase B for 30 min prior to subsequent loading of the next sample.

3.6 Data Processing Guidelines

As previously stated, the searching and matching of experimentally obtained spectra against a non-redundant protein database is computationally intensive, but highly parallelizable and therefore amendable to "divide and conquer" strategies employing cluster computers.

1. For our LTQ-derived data, the precursor ion tolerance is set to 1.5 Da, and the fragment ion tolerance to 0.5 Da. These two values effectively serve as binning parameters during data acquisition concerning parent and daughter (fragment) ions.
2. We require candidate peptides to possess tryptic terminus at both ends, and generally will allow for a maximum of two missed tryptic cleavages.
3. The following SEQUEST thresholds are routinely used to filter experimental peptides:
   (a) Delta-correlation score (dCn) ≥ 0.08.
   (b) Charge state cross correlation scores as follows:

       ≥2.1 for $[M+H]^{+}$ peptides.

       ≥2.3 for $[M+H]^{2+}$ peptides.

       ≥3.5 for $[M+H]^{3+}$ peptides.
4. The final list of protein identifications is created using a parsimony principle, reporting a minimal number of protein identifications from a pool of uniquely identified peptides.
5. Resultant raw data are routinely subjected to a false-positive rate assessment via decoy (reverse) database analysis (22).
6. In the final step the data are analyzed for biologic implications by Ingenuity Pathway Analysis and the Database for Annotation, Visualization, and Integrated Discovery (DAVID).

4 Notes

1. The processing of CID spectra is computationally intensive but highly parallelizable. Therefore, a cluster computer solution generally offers substantial time-savings. This approach follows a linear function that is dependent on the number of computational elements in the cluster configuration. Multicore computers may likely offer similar speed advantages as software tools become more adept in parallel task operations and interactions.
2. Commercial products include MASCOT (Matrix Science, http://www.matrixscience.com) and SEQUEST (Thermo Scientific, http://www.thermo.com). Open source solutions include (a) the X! Tandem database search engine (http://www.thegpm.org/tandem), (b) the Trans-Proteomic Pipeline (TPP, http://tools.proteomecenter.org/wiki/index.php?title=Software:TPP), and (c) the open mass spectrometry search algorithm (OMSSA) (pubchem.ncbi.nlm.nih.gov/omssa/).
3. Commercial products include Ingenuity Pathway Analysis (IPA, http://www.ingenuity.com). Public domain tools include the Database for Annotation, Visualization, and Integrated Discovery (DAVID, http://david.abcc.ncifcrf.gov). Innovative bioinformatic approaches towards the analysis of LCM-MS data are essential for progress. Along these lines, a quite interesting computational approach has been put forth by Karger and Sgroi (23).
4. For each step in the staining protocol, a different solution bath is recommended. Through experience this procedure has been found to make a significant difference to subsequent analyses. Additionally, when glass slides are used for LCM, the time for dehydration (steps 8–12) may need to be increased up to 1 min (occasionally up to 3 min) for each ethanol bath. The increased time may improve the pickup of captured cells from the glass slide onto the LCM Cap. Finally, enhanced dehydration is usually not required for membrane slides.
5. We have found that tissues with a compact cellular density provide greater protein yields and thus usually require a smaller quantity of cells. However, when encountering a new tumor tissue type, a few preliminary experiments are recommended to determine the general estimate of protein yield.
6. In addition to the provided reference, these estimates are also cited in the following text books: Molecular Biology of the Cell, third edition, by Alberts et al. and Molecular Cell Biology, fourth edition, by Lodish et al.
7. This protocol does not depend on custom columns. Adequate high-quality commercial columns are available (e.g., New

Objective, Inc.; http://www.newobjective.com). Additionally, some groups (Karger et al.) have developed custom high-pressure columns, which potentially result in an enhanced separation. Such innovations are vital for the improved analysis of minute samples, as is characteristic with LCM (24, 25).

References

1. Chen R, Mias GI, Li-Pook-Than J et al (2012) Personal omics profiling reveals dynamic molecular and medical phenotypes. Cell 148:1293–1307
2. Zhan F, Huang Y, Colla S et al (2006) The molecular classification of multiple myeloma. Blood 108:2020–2028
3. Zhan F, Hardin J, Kordsmeier B et al (2002) Global gene expression profiling of multiple myeloma, monoclonal gammopathy of undetermined significance, and normal bone marrow plasma cells. Blood 99:1745–1757
4. Zhou Y, Zhang Q, Stephens O et al (2012) Prediction of cytogenetic abnormalities with gene expression profiles. Blood 119:e148–e150
5. Emmert-Buck MR, Bonner RF, Smith PD et al (1996) Laser capture microdissection. Science 274:998–1001
6. Yachida S, Jones S, Bozic I et al (2010) Distant metastasis occurs late during the genetic evolution of pancreatic cancer. Nature 467: 1114–1117
7. Gerlinger M, Rowan AJ, Horswell S et al (2012) Intratumor heterogeneity and branched evolution revealed by multiregion sequencing. N Engl J Med 366:883–892
8. Longo DL (2012) Tumor heterogeneity and personalized medicine. N Engl J Med 366:956–957
9. Mbeunkui F, Johann DJ Jr (2009) Cancer and the tumor microenvironment: a review of an essential relationship. Cancer Chemother Pharmacol 63:571–582
10. Swanton C, Caldas C (2009) Molecular classification of solid tumours: towards pathway-driven therapeutics. Br J Cancer 100: 1517–1522
11. Page MJ, Amess B, Townsend RR et al (1999) Proteomic definition of normal human luminal and myoepithelial breast cells purified from reduction mammoplasties. Proc Natl Acad Sci USA 96:12589–12594
12. Ornstein DK, Gillespie JW, Paweletz CP et al (2000) Proteomic analysis of laser capture microdissected human prostate cancer and in vitro prostate cell lines. Electrophoresis 21:2235–2242
13. Johann DJ Jr, Blonder J (2007) Biomarker discovery: tissues versus fluids versus both. Expert Rev Mol Diagn 7:473–475
14. Johann DJ, Wei BR, Prieto DA et al (2010) Combined blood/tissue analysis for cancer biomarker discovery: application to renal cell carcinoma. Anal Chem 5:1584–1588
15. Wei BR, Simpson RM, Johann DJ et al (2012) Proteomic profiling of H-Ras-G12V induced hypertrophic cardiomyopathy in transgenic mice using comparative LC-MS analysis of thin fresh-frozen tissue sections. J Proteome Res 11:1561–1570
16. Aebersold R, Mann M (2003) Mass spectrometry-based proteomics. Nature 422: 198–207
17. Hwang SI, Thumar J, Lundgren DH et al (2007) Direct cancer tissue proteomics: a method to identify candidate cancer biomarkers from formalin-fixed paraffin-embedded archival tissues. Oncogene 26:65–76
18. Blonder J, Chan KC, Issaq HJ et al (2006) Identification of membrane proteins from mammalian cell/tissue using methanol-facilitated solubilization and tryptic digestion coupled with 2D-LC-MS/MS. Nat Protoc 1:2784–2790
19. Johann DJ, Rodriguez-Canales J, Mukherjee S et al (2009) Approaching solid tumor heterogeneity on a cellular basis by tissue proteomics using laser capture microdissection and biological mass spectrometry. J Proteome Res 8: 2310–2318
20. Hewitt SM, Badve SS, True LD (2012) Impact of preanalytic factors on the design and application of integral biomarkers for directing patient therapy. Clin Cancer Res 18: 1524–1530
21. Wang H, Qian WJ, Mottaz HM et al (2005) Development and evaluation of a micro- and nanoscale proteomic sample preparation method. J Proteome Res 4: 2397–2403
22. Peng J, Elias JE, Thoreen CC et al (2003) Evaluation of multidimensional chromatography coupled with tandem mass spectrometry (LC/LC-MS/MS) for large-scale protein analysis: the yeast proteome. J Proteome Res 2:43–50
23. Cha S, Imielinski MB, Rejtar T et al (2010) In situ proteomic analysis of human breast cancer epithelial cells using laser capture microdissection: annotation by protein set enrichment

analysis and gene ontology. Mol Cell Proteomics 9:2529–2544

24. Luo Q, Yue G, Valaskovic GA et al (2007) On-line 1D and 2D porous layer open tubular/LC-ESI-MS using 10-microm-i.d. poly(styrene-divinylbenzene) columns for ultrasensitive proteomic analysis. Anal Chem 79:6174–6181

25. Thakur D, Rejtar T, Wang D et al (2011) Microproteomic analysis of 10,000 laser captured microdissected breast tumor cells using short-range sodium dodecyl sulfate-polyacrylamide gel electrophoresis and porous layer open tubular liquid chromatography tandem mass spectrometry. J Chromatogr A 1218: 8168–8174

Chapter 7

Use of Formalin-Fixed, Paraffin-Embedded Tissue for Proteomic Biomarker Discovery

David B. Krizman and Jon Burrows

Abstract

Application of mass spectrometry to proteomic analysis of tissue is a highly desirable approach to discovery of disease biomarkers due to a direct correlation of findings to tissue/disease histology and in many respects obviating the need for model systems of disease. Both frozen and formalin-fixed, paraffin-embedded (FFPE) tissue can be interrogated; however, worldwide access to vastly larger numbers of highly characterized FFPE tissue collections derived from both human and model organisms makes this form of tissue more advantageous. Here, an approach to large-scale, global proteomic analysis of FFPE tissue is described that can be employed to discover differentially expressed proteins between different histological tissue types and thus discover novel protein biomarkers of disease.

Key words Formalin-fixed tissue, FFPE, Liquid Tissue®, Tissue microdissection, DIRECTOR® slides, Mass spectrometry, Global profiling

1 Introduction

A wealth of information that could dramatically expand the understanding of human disease is present in vast archived collections of formalin-fixed, paraffin-embedded (FFPE) tissue samples. These collections are valuable because they combine proteomic expression profiles with known patient clinical outcomes thus facilitating biomarker discovery and greatly reducing the time needed for validation of candidate biomarkers. Vast numbers of highly curated collections that define every human disease process and condition are in archives waiting to be exploited; however, proteins within FFPE tissue are heavily cross-linked due to the formalin fixation process and thus not easily amenable to standard proteomic analysis, especially for high-throughput in-depth analysis. The most common method for proteomic characterization of formalin-fixed tissue is immunohistochemistry (IHC) which lacks sensitivity, objective quantitation, and

Ming Zhou and Timothy Veenstra (eds.), *Proteomics for Biomarker Discovery: Methods and Protocols*, Methods in Molecular Biology, vol. 1002, DOI 10.1007/978-1-62703-360-2_7, © Springer Science+Business Media, LLC 2013

multiplexed analytical capabilities. Reports describing analysis of protein in FFPE tissue using standard methods other than IHC have met with limited degrees of success (1–3).

To address this issue, Oncoplex Diagnostics developed Liquid Tissue® reagents and protocol for complete solubilization of FFPE tissue with high efficiency and total representation of all proteins present in the tissue resulting in preparation of highly characterized tissue for proteomic analysis by mass spectrometry. Application of mass spectrometry to the field of proteomics has resulted in development of numerous strategies for large-scale, global proteomic analysis of complex cellular protein lysates making possible identification of thousands of expressed proteins from a single sample (4–11). Applying these strategies to Liquid Tissue lysates has been demonstrated across a wide variety of formalin-fixed tissue samples and across a variety of mass spectrometry platforms, and when combined with tissue microdissection the Liquid Tissue technology platform is capable of determining differentially expressed proteins between various histopathological states (12–18). Developing catalogs of differentially expressed proteins from biologically important and clinically relevant samples enhances our understanding of cellular processes and provides for increased biomarker discovery potential. This chapter will outline and describe the steps necessary to apply the Liquid Tissue technology platform, from tissue microdissection to identification of differentially expressed proteins, in order to discover protein biomarkers, directly from FFPE tissue, for use as diagnostic markers, prognostic markers, or therapeutic targets.

2 Materials

Collect all solutions, reagents, and materials prior to beginning the tissue dissection process. Prepare and store all reagents at room temperature unless otherwise indicated. Follow all waste disposal regulations when disposing of hazardous waste.

2.1 Tissue Dissection

1. Standard histopathology processing reagents, staining reagents, and materials including Coplin jars, xylene (or the xylene substitute, Surgipath® Sub-X), hematoxylin stain, ethyl alcohol, and deionized water.
2. Laser microdissection instrument for procuring specific cell populations from FFPE tissue section by laser microdissection (LMD) methodology: Leica LMD series laser microdissector (Leica Microsystems, Buffalo Grove, IL, USA); Zeiss PALM Microbeam series laser microdissector (Carl Zeiss Microimaging, Munich, Germany).
3. DIRECTOR® microdissection slides (Oncoplex Diagnostics, Rockville, MD, USA) for use on the Leica and Zeiss LMD instruments.

4. Collection tubes: Costar Thermowell 0.5 ml tubes with flat cap (Corning, Lowell, MA, USA).

2.2 Sample Preparation

1. Liquid Tissue® MS Protein Preparation kit (Oncoplex Diagnostics, Rockville, MD, USA). Store kit at −20 °C.
2. Collection tubes: Costar Thermowell 0.5 ml tubes with flat cap (Corning, Lowell, MA, USA).
3. Heating blocks or water baths set at 95 and 37 °C, or thermal cycler.
4. Micro BCA Protein Assay Reagent Kit (Pierce Biotechnology, Rockford, IL, USA).

2.3 Mass Spectrometry and Bioinformatics

1. High-resolution mass spectrometry instrument partnered with a high-performance liquid chromatography (LC) system capable of identifying individual peptides on a global scale from a complex proteome made up of tryptic digested peptide fragments, Orbitrap (Thermo Scientific, San Jose, CA, USA), LTQ FT Ultra (Thermo Scientific, San Jose, CA, USA), TOF/TOF 5800 system (AB Sciex, Framingham, MA, USA), and SYNAPT G2-S HDMS (Waters, Milford, MA, USA).
2. High-performance LC system (Ultimate 3000, Thermo Scientific, San Jose, CA, USA).
3. LC column consisting of 75 μm integrafrit packed with C18 resin coupled to 75 μm picofrit column packed with C18 resin (Phenomenex, Torrance, CA, USA).
4. Bioworks v3.2 software (Thermo Scientific, San Jose, CA, USA) for use in peptide identifications (see Note 1).
5. UniProt (http://www.uniprot.org/) is utilized for assignment of UniProt accession numbers to identified proteins.
6. Ingenuity Pathway Analysis (IPA) software (Ingenuity® Systems, www.ingenuity.com) for mapping significantly, differentially abundant proteins to HUGO gene symbols.
7. Scaffold software to develop spectral count information from global peptide identification (http://www.proteomesoftware.com).
8. ArrayTrack software (http://edkb.fda.gov/webstart/arraytrack/) for identifying differentially expressed proteins for consideration as candidate protein biomarkers.

3 Methods

Carry out all procedures at room temperature unless otherwise specified. There are multiple choices for successful application of these methods with respect to instruments and software; however, the following protocol outlines procedures utilizing: (1)

DIRECTOR microdissection technology in combination with a Leica LMD microdissection instrument for collecting approximately 30,000 cells from FFPE tissue samples, (2) Liquid Tissue reagents and protocol for preparing tissue for mass spectrometry analysis (Oncoplex Diagnostics Rockville, MD, USA), and (3) global profiling via nanoflow reverse-phase liquid chromatography system coupled to electrospray ionization resolved on a linear ion trap LTQ FT Ultra mass spectrometer (see Note 2).

3.1 Tissue Dissection

1. Cut 10 μm thick sections from a FFPE tissue block and place onto DIRECTOR microdissection slides. The number of slides required may vary depending on the tissue, but in general two sections should be more than sufficient to collect approximately 30,000 cells of interest.
2. Place the slides with cut tissue sections to be processed on an aluminum slide tray and bake in an oven at 58–61 °C for 30–60 min to melt the paraffin and to adhere tissue sections to slides.
3. Utilize standard histopathology practice to remove the paraffin with xylene (or Sub-X), wash in graded ethanol baths and deionized water, and stain with hematoxylin. Allow slides to completely air-dry prior to dissection.
4. Place one or more slides into a fully functional Leica LMD microdissection instrument. Refer to manufacturer's literature for optimal use of the instrument.
5. Place a collection tube into the tube holder according to manufacturer's instructions and pipet 20 μl of Liquid Tissue buffer into the cap. Load the tube into the tube holder of the microdissection instrument.
6. Identify the cells to be collected and utilize the laser function to tabulate and transfer approximately 30,000 cells into the buffer-filled cap. To collect approximately 30,000 cells, the tabulated area that needs to be transferred into the tube is 8 mm^2 area of a 10 μm thick tissue section. This calculation is based on an average width of a cell residing within tissue to be 15 μm.
7. Once the selected cells have been transferred into the collection tube, carefully remove the tube from the holder and cap the tube. Collect the buffer and tissue at the bottom of the tube by a brief spin in a microcentrifuge (see Note 3).
8. Store tube at −20 °C until sample processing.

3.2 Liquid Tissue Sample Preparation

1. Remove tube containing buffer/tissue from −20 °C storage and warm to room temperature.
2. Incubate the tube at 95 °C for 90 min in either a heating block or a thermal cycler with a heated lid. Every 20 min, remove the tube and shake down the buffer that has condensed so that it covers the tissue by flicking the tube in a downward motion with your wrist or by brief spin in the microcentrifuge.

3. After 90 min, remove the tube from the 95 °C heating block or thermal cycler, and microcentrifuge at 10,000 rcf for 1 min.
4. Cool the tube on ice for 2 min.
5. Resuspend the lyophilized trypsin in 20 μl of trypsin diluent to achieve a stock trypsin concentration of 1 μg/μl. Keep trypsin on ice and store at −20 °C.
6. Pipet a total of 1 μl (1 μg) of trypsin into sample tube and mix gently.
7. Incubate tube at 37 °C for 16–18 h in either a heating block, water bath, or thermal cycler to allow the trypsin to digest tissue in Liquid Tissue buffer.
8. Briefly microcentrifuge the tube to collect all fluid at the bottom of the tube and then incubate at 95 °C for 5 min to stop the reaction. Note the absence/presence of any pellet as all the tissue should be completely solubilized at this point.
9. Perform a Micro BCA assay and calculate the amount of digested protein present in the Liquid Tissue preparation according to manufacturer's recommendations.
10. The expected total protein yield from a Liquid Tissue preparation from approximately 30,000 cells is 4–8 μg.
11. Pipet 2 μl of 100 mM DTT for a final concentration of 10 mM and heat the sample at 95 °C for 5 min.
12. Store the Liquid Tissue preparation at −20 °C.

3.3 Mass Spectrometry and Bioinformatics

1. Once the amount of total protein in a Liquid Tissue preparation has been determined, suspend the preparation to a final concentration of 0.2 ng/μl in 0.1 % formic acid (FA).
2. Prepare the following LC column: vented column design using a 75 μm integrafrit packed with C18 resin coupled to 75 μm picofrit column packed with C18 resin.
3. Resolve Liquid Tissue preparation by triplicate injection of 1 μg of sample into the nanoflow reverse-phase liquid chromatography coupled online via electrospray ionization.
4. Elute peptides from the column using a linear gradient of 2 % mobile phase B (0.1 % formic acid (FA) in acetonitrile to 40 %) mobile phase B over 125 min at a constant flow rate of 250 nl/min followed by a column wash consisting of 95 % B for an additional 20 min at a constant flow rate of 400 nl/min.
5. Collect global broadband mass spectra (*m/z* 375–1,800) where the seven most abundant peptide molecular ions dynamically determined from the MS scan are selected for tandem MS using a relative CID energy of 35 %. Utilize dynamic exclusion to minimize redundant selection of peptides for CID.
6. Identify peptides by searching the LC-MS/MS data utilizing the SEQUEST function of BioWorks against the most updated ver-

sion of the UniProt-derived human proteome database using the following parameters: trypsin (KR), full enzymatic-cleavage, two missed cleavages sites, and variable modifications for methionine oxidation, pyroglutamic acid, and N-term Q for deamination.

7. Filter peptide identifications according to the following specific SEQUEST scoring criteria: delta correlation (ΔC_n) ≥ 0.08 and charge state-dependent cross correlation (Xcorr) ≥ 1.9 for $[M+H]^{1+}$, ≥ 2.2 for $[M+2H]^{2+}$, and ≥ 3.5 for $[M+3H]^{3+}$.
8. Derive inferred protein abundance from the Sequest data by spectral counting (SC) utilizing the Scaffold software program to tabulate total number of peptides identified per protein under strict filter criteria that derives <5 % false discovery rate (FDR) at the protein level and where each protein must be identified by a minimum of two unique peptides (see Note 4).
9. Utilizing the spectral count information, determine statistically significant changes in protein abundance across a collection of similarly analyzed FFPE tissues utilizing the hierarchical supervised cluster program ArrayTrack and where the variance in total spectral count peptides identified between two different tissue types is determined utilizing the Mann–Whitney rank-sum test (significance level $p \leq 0.05$, Fisher's exact test). Utilize filter criteria in the cluster analysis requiring that 60 % of the samples in a supervised group have a minimum peptide count of two or greater for a given protein.
10. From the cluster analysis, develop a list of candidate protein biomarkers by utilizing ArrayTrack to map significantly, differentially abundant proteins found between different populations of tissue (see Note 5). UniProt accessions that fail to map should be converted to HGNC identifiers by manual inspection at www.uniprot.org and remapped to IPA to maximize protein identifications available for downstream bioinformatic analyses. Peptides identified by UniProt accession numbers can be translated to HUGO (HGNC) gene symbols utilizing IPA.
11. Functional analysis of candidate protein biomarkers can be performed utilizing the "core analysis" function in IPA and inferred biofunctions exhibiting a $p < 0.05$ and a minimum of two associated proteins are considered significant.

4 Notes

1. While software programs are denoted throughout this protocol, in reality other software programs exist to perform many of the same bioinformatic tasks and thus can be substituted at the discretion of the user.
2. Just as multiple software programs can be utilized to achieve the same result, multiple mass spectrometry and tissue

microdissection instrument platforms can be used in substitution for those listed in this protocol.

3. If the microdissected tissue does not appear at the bottom of the tube (remains in the cap) after the quick spin step, the tissue can be transferred to the bottom of the tube using the following. Add 20 μl of 100 % acetonitrile (ACN) to the cap followed by closing the cap onto the tube and perform a quick centrifuge spin. This will result in a small pellet at the bottom of the tube. The ACN is removed by speedvac centrifugation at 35 °C for 6 min resulting in a dried pellet which can then be resuspended in 20 μl of Liquid Tissue buffer. At this point the Liquid Tissue sample preparation can then be carried out as described. Alternatively, all dissection can be performed into a dry cap followed by utilization of the ACN method to transfer all dissected tissue to bottom of the tube for sample preparation.
4. Spectral counting is only an approximation of the actual amount of a protein in a given Liquid Tissue sample, and thus proteins identified as differentially expressed between two distinct tissue groups based on either histology or disease constitute "candidate" protein biomarkers of the intended indication.
5. Candidates must further be validated in the context of the disease or condition and validation can be performed on the same tissues using either immunohistochemical methodology or by mass spectrometry-based selected reaction monitoring (SRM) of Liquid Tissue preparations.

Acknowledgments

The author wishes to thank Dr. Sheeno Thyparambil and Marlene Darfler for critical review. This work was supported in part by NIH grant 1R43CA132081-01.

References

1. Ikeda K, Monden T, Kanoh T et al (1998) Extraction and analysis of diagnostically useful proteins from formalin-fixed, paraffin-embedded tissue sections. J Histochem Cytochem 46:397–403
2. Brooks SA, Dwek MV, Leathem AJ et al (1998) Release and analysis of polypeptides and glycopolypeptides from formalin-fixed, paraffin wax-embedded tissue. Histochem J 30:609–615
3. Becker KF, Schott C, Hipp S et al (2007) Quantitative protein analysis from formalin-fixed tissues: implications for translational clinical research and nanoscale molecular diagnosis. J Pathol 211:370–378
4. Gygi SP, Rist B, Gerber SA et al (1999) Quantitative analysis of complex protein mixtures using isotope-coded affinity tags. Nat Biotechnol 17:994–999
5. Washburn MP, Wolters D, Yates JR 3rd (2001) Large-scale analysis of the yeast proteome by multidimensional protein identification technology. Nat Biotechnol 19:242–247
6. Gavin AC, Bösche M, Krause R et al (2002) Functional organization of the yeast proteome by systematic analysis of protein complexes. Nature 415:141–147
7. Ong SE, Blagoev B, Kratchmarova I et al (2002) Stable isotope labeling by amino acids

in cell culture, SILAC, as a simple and accurate approach to expression proteomics. Mol Cell Proteomics 1:376–386

8. Ross PL, Huang YN, Marchese JN et al (2004) Multiplexed protein quantitation in Saccharomyces cerevisiae using amine-reactive isobaric tagging reagents. Mol Cell Proteomics 3:1154–1169
9. Ishihama Y, Oda Y, Tabata T et al (2005) Exponentially modified protein abundance index (emPAI) for estimation of absolute protein amount in proteomics by the number of sequenced peptides per protein. Mol Cell Proteomics 4:1265–1272
10. Phillips GR, Florens L, Tanaka H et al (2005) Proteomic comparison of two fractions derived from the transsynaptic scaffold. J Neurosci Res 81:762–775
11. Hawkridge AM, Heublein DM, Bergen HR 3rd et al (2005) Quantitative mass spectral evidence for the absence of circulating brain natriuretic peptide (BNP-32) in severe human heart failure. Proc Natl Acad Sci USA 102:17442–17447
12. Hood BL, Darfler MM, Guiel TG et al (2005) Proteomic analysis of formalin fixed prostate cancer tissue. Mol Cell Proteomics 4:1741–1753
13. Patel V, Hood BL, Molinolo AA et al (2008) Proteomic analysis of laser-captured paraffin-embedded tissues: a molecular portrait of head and neck cancer progression. Clin Cancer Res 14:1002–1014
14. Jain MR, Liu T, Hu J et al (2008) Quantitative proteomic analysis of formalin fixed paraffin embedded oral HPV lesions from HIV patients. Open Proteomics J 1:40–45
15. Cheung W, Darfler MM, Alvarez H et al (2008) Application of a global proteomic approach to archival precursor lesions: deleted in malignant brain tumors 1 and tissue transglutaminase 2 are upregulated in pancreatic cancer precursors. Pancreatology 8:608–616
16. Huang SK, Darfler MM, Nicholl MB et al (2009) LC/MS-based quantitative proteomic analysis of paraffin-embedded archival melanomas reveals potential proteomic biomarkers associated with metastasis. PLoS One 4:4430
17. DeSouza LV, Krakovska O, Darfler MM et al (2010) mTRAQ-based quantification of potential endometrial carcinoma biomarkers from archived formalin-fixed paraffin-embedded tissues. Proteomics 10:3108–3116
18. Bateman NW, Sun M, Bhargava R et al (2011) Differential proteomic analysis of late-stage and recurrent breast cancer from formalin-fixed paraffin-embedded tissues. J Proteome Res 10:1323–1332

Chapter 8

Phosphopeptide Enrichment Using Offline Titanium Dioxide Columns for Phosphoproteomics

Li-Rong Yu and Timothy Veenstra

Abstract

Identification of phosphoproteins or phosphopeptides as cancer biomarkers is an emerging field in phosphoproteomics. Owing to the low stoichiometric nature of protein phosphorylation, phosphoproteins or phosphopeptides must be enriched prior to downstream mass spectrometry analysis. Titanium dioxide (TiO_2) has been prevalently used to enrich phosphopeptides from complex proteome samples due to its high affinity for phosphopeptides, and the method is straightforward. In this protocol, an offline phosphopeptide enrichment procedure using TiO_2 columns is described. Peptides from a proteome lysate are loaded onto a TiO_2 column in an acidic environment, followed by column washing with aqueous, organic, and ammonium glutamate (NH_4Glu) buffers at acidic conditions. Phosphopeptides are eluted using an ammonia solution at high pH. Use of NH_4Glu significantly reduces nonspecific bindings while a high recovery rate (84 %) of phosphopeptides is retained. The method is optimized for large-scale phosphoproteomic analysis and phosphoprotein biomarker discovery starting from sub-milligram or milligrams of proteome samples.

Key words Phosphoproteomics, Phosphopeptide enrichment, Titanium dioxide, Biomarker, Proteomics, Ammonium glutamate, Mass spectrometry, MS/MS, Liquid chromatography, Cancer

1 Introduction

Mass spectrometry (MS)-based phosphoproteomic technologies have been increasingly applied to the understanding of basic biological processes and mechanisms of disease development. An emerging field in phosphoproteomics has been the identification of phosphoproteins or phosphopeptides as cancer biomarkers. Specific phosphoprotein signatures would facilitate disease diagnosis and patient stratification, predict, and monitor therapeutic outcomes for personalized cancer therapy. Recent development in phosphoproteomic technologies (1) provides great opportunities in applying such tools to the discovery of phosphoprotein

Ming Zhou and Timothy Veenstra (eds.), *Proteomics for Biomarker Discovery: Methods and Protocols*, Methods in Molecular Biology, vol. 1002, DOI 10.1007/978-1-62703-360-2_8, © Springer Science+Business Media, LLC 2013

biomarkers. Several phosphoproteins/kinases have been approved by regulatory agencies as cancer biomarkers (2). Initial endeavors have been made to identify phosphoproteins from body fluids, and hundreds of phosphopeptides have been identified (3–5).

Owing to the low stoichiometric nature of protein phosphorylation, phosphoproteins or phosphopeptides must be enriched prior to downstream MS characterization and quantitative analysis. Several approaches have been developed and compared for the enrichment of phosphopeptides, including immobilized metal affinity chromatography (IMAC), phosphoramidate chemistry (PAC), and titanium dioxide (TiO_2) chromatography (2, 6). Each approach has its own advantages and disadvantages. Among them, TiO_2 has been prevalently used in phosphoproteomics because it is compatible with a wide range of reagents used in biological sample preparation, including salts, detergents, and other small molecules (7). In addition, the procedure for phosphopeptide enrichment by TiO_2 is relatively simple and straightforward compared to IMAC. TiO_2 chromatography has been incorporated into sequential elution from IMAC (i.e., SIMAC) to purify mono-phosphorylated peptides (8). Both online and offline TiO_2 chromatographic approaches have been employed in phosphoproteomics (9–11).

While TiO_2 has high affinity for phosphopeptides (10, 12–14), significant nonspecific bindings from non-phosphopeptides become an issue when digested proteome samples are applied to TiO_2 columns (11). Recent studies revealed that these nonspecific bindings can be minimized by conditioning and/or washing the TiO_2 column before and/or after sample loading with a buffer containing 2,5-dihydroxybenzoic acid (DHB) (10), aliphatic hydroxy acids (15), or ammonium glutamate (NH_4Glu) (11), etc. While several of these reagents might result in the loss of phosphopeptides, TiO_2 column wash with NH_4Glu retains a high recovery rate (e.g., 84 %) of phosphopeptides in terms of amounts (11).

In this chapter, an offline phosphopeptide enrichment procedure using a TiO_2 column is described. The method incorporates a column wash with an NH_4Glu buffer. This offline approach is flexible for both analytical and preparative scale isolation of phosphoproteomes by using different size of TiO_2 columns. A large-scale enrichment of phosphopeptides from 1 to 2 mg of digested proteome sample is suitable for current multidimensional liquid chromatography-tandem mass spectrometry (LC-MS/MS) approaches. As a demonstration of this protocol, 0.5 mg of peptides from a HeLa cell lysate is loaded onto the TiO_2 column in an acidic environment. The column is then washed with aqueous, organic, and NH_4Glu buffers at acidic conditions. Phosphopeptides are eluted with a high recovery rate using an ammonia solution at high pH (~11.7).

2 Materials

2.1 Cell Culture of HeLa Cells

1. Human cervical epithelial cancer cell line HeLa (ATCC).
2. DMEM-10 medium: Dulbecco's modified Eagle's medium supplemented with 10 % cosmic calf serum (CCS), 100 U/mL penicillin, 100 μg/mL streptomycin, and 2 mM L-glutamine.
3. T75 flasks: sterilized cell culture flasks (75 cm^2).

2.2 Proteome Sample Preparation

1. Lysis buffer: 8 M urea in 50 mM Tris–HCl, pH 8.3, with 100-fold dilution of phosphatase inhibitor cocktail 1 (Sigma-Aldrich) and 100-fold dilution of phosphatase inhibitor cocktail 2 (Sigma-Aldrich) (see Note 1).
2. Digital Sonifier cell disruptor with 1/8″ microtip (Branson Ultrasonics Corporation, Model 250).
3. Bicinchoninic acid (BCA) protein assay kit (Pierce).
4. 50 mM Tris–HCl, pH 8.3.
5. Sequencing grade modified trypsin (Promega).
6. Water purified from a Barnstead Nanopure system (ThermoFisher Scientific) (ultrapure water) (see Note 2).

2.3 Solid Phase Extraction

1. Extract-Clean™ high capacity 200 mg C18 cartridges (Alltech Associates Inc.).
2. One vacuum manifold for solid phase extraction (SPE).
3. HPLC grade acetonitrile (CH_3CN).
4. SPE washing buffer: 0.1 % trifluoroacetic acid (TFA).
5. SPE elution buffer: 80 % acetonitrile, 0.1 % TFA.

2.4 TiO_2 Chromatography

1. One TiO_2 (5 μm particle size) column and one cartridge holder: column size is 4 mm inner diameter (ID) and 10 mm long (GL Sciences Inc., Part No. 5020-08690).
2. One syringe pump: Harvard Pump 11 Plus single or dual syringe pump (Harvard Apparatus, Part No. 702208 or 702209).
3. Syringes: 1–2 syringes of 5 mL (22-gauge needle) capacity for column washing, 2–3 syringes of 2.5 mL (22-gauge needle) capacity for sample loading, column washing, and phosphopeptide elution (Hamilton Company, Part No. 81516 and 81416).
4. PEEK (polyetheretherketone) polymer tubing: 1/16″ OD (outside diameter) × 0.015″ ID × 5′ (IDEX Health & Science, Part No. 1565).
5. Two-piece fingertight fittings for 1/16″ OD tubing (IDEX Health & Science, Part No. F300x).

6. One Rheodyne® needle port adapter: the needle port adapter has a 10–32 threaded nut, a PEEK ferrule, and a Teflon liner (IDEX Health & Science, Part No. 9013).
7. One polymer tubing cutter for 1/16″ and 1/8 OD″ tubing (IDEX Health & Science, Part No. A-327).
8. Sample loading buffer: 50 % CH_3CN/0.1 % TFA.
9. TiO_2 washing buffer 1: 0.1 % TFA.
10. TiO_2 washing buffer 2: 200 mM L-glutamic acid ammonium salt (NH_4Glu), pH 2.0 (see Note 3).
11. TiO_2 washing buffer 3: 80 % CH_3CN/0.1 % TFA.
12. TiO_2 elution buffer: 1.5 % ammonium hydroxide (NH_4OH), pH 11.7 (see Note 4).

3 Methods

The protocol described in this chapter is aimed to enrich phosphopeptides from complex proteome samples as illustrated in the use of the tryptic digest from a HeLa cell lysate. It is highly efficient in isolating phosphopeptides with a high recovery rate and a significant reduction of non-phosphopeptides. The method yields enough phosphopeptides for subsequent multidimensional liquid chromatographic separation and MS analysis, which is usually employed for phosphoprotein biomarker discovery and phosphoproteome analysis, as demonstrated by using sub-milligram or milligrams of protein. The experimental setup and principle of TiO_2 phosphopeptide enrichment are illustrated in Fig. 1. Instead of isolating phosphopeptides from each fraction after liquid chromatographic fractionation, the current method is time efficient since phosphopeptide purification is performed for a single un-fractionated sample from the total cell lysate. The protocol is simple and straightforward, and can be applied to any mixtures of peptide digests of complex proteome samples. Since the flow rate of sample loading and column wash is strictly controlled, the procedure for phosphopeptide enrichment and experimental results is highly reproducible.

3.1 HeLa Cell Culture and Lysis

1. Grow human cervical epithelial cancer cell line HeLa in T75 flasks in DMEM-10 media at 37 °C in a humidified 5 % CO_2 atmosphere to reach 90 % confluence.
2. Wash the cells with ice-cold PBS buffer three times. Add 3 mL of ice-cold lysis buffer per flask of cells to directly disrupt the cells in the flasks. Harvest the cell lysate using a cell scraper and transfer the lysate to a 15-mL Falcon tube (see Note 5).

3.2 Proteome Sample Preparation and Digestion

1. Sonicate the cell lysate at ~20 % output five times 10 s on ice with 1-min intervals to further disrupt the cells using a digital sonifier cell disruptor (see Notes 6 and 7).

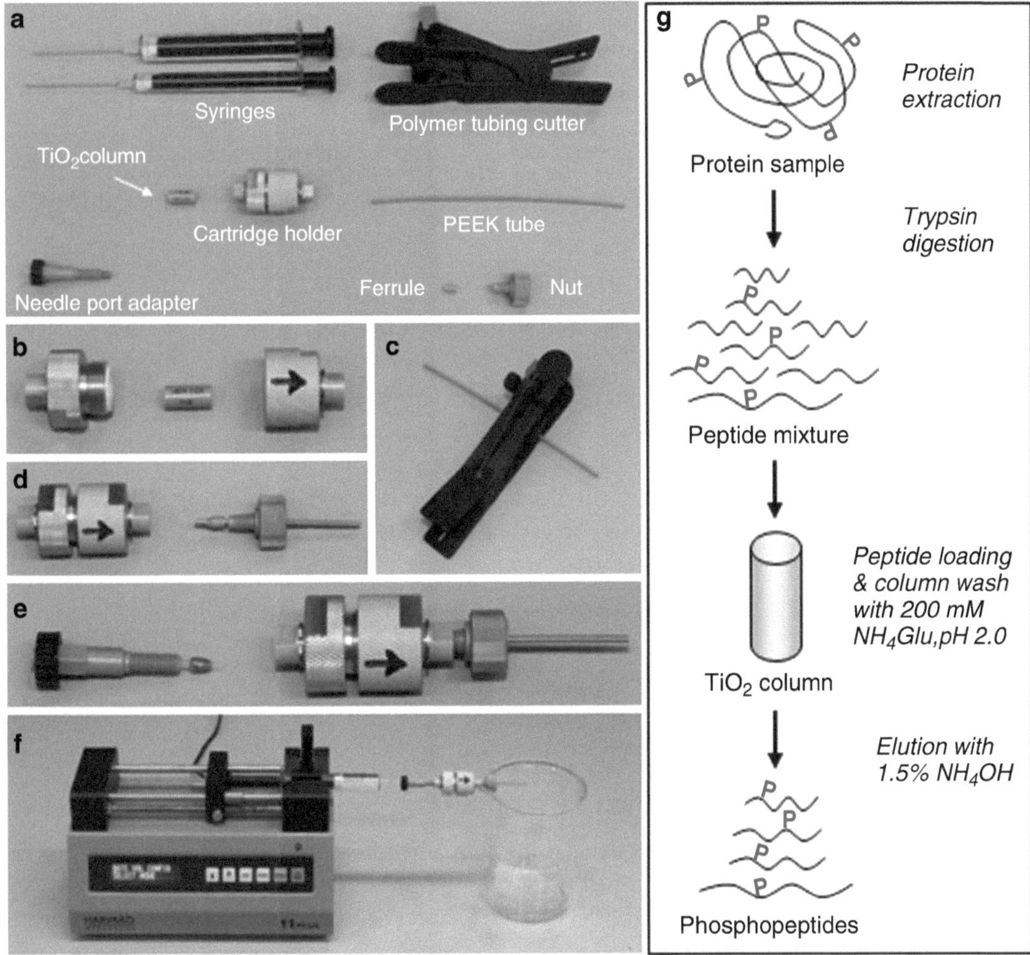

Fig. 1 The principle of assembling the TiO_2 cartridge column, setup of the syringe pump, and phosphopeptide enrichment. (**a**) A TiO_2 column, a cartridge holder, 1/16″ OD PEEK tubing, 1/16″ OD fittings, a Rheodyne® needle port adapter, a polymer tubing cutter, and 2–3 syringes are needed to assemble the TiO_2 cartridge column device for phosphopeptide enrichment. (**b**) The TiO_2 column is assembled in a cartridge holder by finger-tightening. (**c**) A ~2 in. of 1/16″ OD PEEK tube is cut using a polymer tubing cutter. (**d**) Insert the tube into a 1/16″ ferrule, and connect the tube to the outlet of the TiO_2 column using a 1/16″ nut by finger-tightening. (**e**) Connect the Rheodyne® needle port adapter to the inlet the TiO_2 column by finger-tightening. (**f**) Insert the needle of a syringe into the needle port adapter of the assembled TiO_2 cartridge column and place them onto the syringe pump. The device is ready for phosphopeptide enrichment. (**g**) Schematic diagram showing the principle of TiO_2 enrichment of phosphopeptides with a column wash of 200 mM NH_4Glu, pH 2.0

2. Centrifuge the lysate at 18,000 ×*g* for 15 min at 4 °C, and collect the supernatant.
3. Determine the protein concentration using a BCA assay kit according to the manufacture's instruction (see Note 8).
4. Dilute the protein sample fourfold with 50 mM Tris–HCl, pH 8.3 (see Note 9), and add sequencing grade modified trypsin at a ratio of 40:1 (w/w, protein/trypsin).
5. Incubate the sample for 8 h at 37 °C.

3.3 Reversed-Phase Solid Phase Extraction

1. Condition the high capacity C18 columns with CH_3CN. Use 3 mL of CH_3CN for each column (see Note 10).
2. Wash each column with 5 mL of SPE washing buffer (0.1 % TFA).
3. Dilute the digested protein sample with equal volume of ultra-pure water and load the samples onto the columns slowly (~1 drop/s).
4. Repeat step 2.
5. Elute the peptides with 2 mL of SPE elution buffer (80 % CH_3CN/0.1 % TFA) for each column and collect the eluates.
6. Lyophilize the eluates in a speed vacuum concentrator to dryness.

3.4 Assemble the TiO_2 Cartridge Column and Set Up the Syringe Pump

1. Confirm the flow direction of the TiO_2 column. Insert the column into the cartridge holder and strongly tighten the holder using hands.
2. Confirm the flow direction of the assembled cartridge column. Cut 2 in. of PEEK tube (1/16″ OD × 0.015″ ID) using a polymer tubing cutter. Insert the tube into a 1/16″ ferrule and connect the tube to the outlet of the TiO_2 column using a 1/16″ nut (see Note 11) by finger-tightening. Ensure the PEEK tube is bottomed in the outlet port of the cartridge.
3. Connect the Rheodyne® needle port adapter to the inlet of the TiO_2 column by finger-tightening (see Note 12). Ensure the Teflon tube of the adapter is bottomed in the inlet port of the cartridge.
4. Insert the needle of a 5-mL syringe into the needle port adapter of the assembled TiO_2 cartridge column, and place them onto the syringe pump.

3.5 TiO_2 Phosphopeptide Enrichment from Digested HeLa Cell Lysate

1. Configure the syringe pump flow rate unit as mL/min. Set the syringe inner diameter as 10.3 mm for the 5-mL syringe, and set the flow rate as 0.2 mL/min and volume as 3 mL (see Note 13). Condition the TiO_2 column with TiO_2 washing buffer 1 (0.1 % TFA) using a 5-mL syringe (see Notes 14–16).
2. Change the syringe inner diameter to 7.28 mm for the 2.5-mL syringe, and set the flow rate as 0.05 mL/min and volume as 1 mL for the syringe pump (see Note 17). Dissolve 0.5 mg of peptides from the HeLa cell lysate in 1 mL of the sample loading buffer (50 % CH_3CN/0.1 % TFA) (see Note 18). Load the sample onto the column using a 2.5-mL syringe (see Note 19).
3. Change the syringe pump flow rate to 0.2 mL/min and volume to 0.75 mL. Wash the column with TiO_2 washing buffer 1 using a 2.5-mL syringe.

4. Set the syringe pump parameters the same as in step 1. Wash the column with 3 mL of TiO_2 washing buffer 2 (200 mM NH_4Glu, pH 2.0) at 0.2 mL/min using a 5-mL syringe.
5. Change the syringe inner diameter to 7.28 mm for the 2.5-mL syringe, and set the flow rate as 0.2 mL/min and volume as 0.75 mL for the syringe pump. Wash the column with TiO_2 washing buffer 1 using a 2.5-mL syringe.
6. Change the syringe pump volume to 1.5 mL. Wash the column with TiO_2 washing buffer 3 (80 % CH_3CN/0.1 % TFA) at 0.2 mL/min using a 2.5-mL syringe.
7. Elute the phosphopeptides with 1.5 mL of TiO_2 elution buffer (1.5 % NH_4OH) at 0.2 mL/min using a 2.5-mL syringe.
8. Lyophilize the sample in a speed vacuum concentrator.
9. Redissolve the dried phosphopeptides in 150 μL of 0.2 % phosphoric acid (H_3PO_4). After a quick spin, the sample is ready for capillary reversed-phase LC-MS/MS analysis.

An example of nanoflow RPLC-MS^2-MS^3 analysis, the phosphopeptides enriched from a HeLa cell lysate using this protocol, is shown in Fig. 2. The analysis is for 5 μL injection of phosphopeptides onto a 75 μm ID capillary column coupled online with a linear ion trap MS (LTQ, Thermo Electron, San Jose, CA). MS/MS (MS^2) and phosphate neutral loss-dependent MS/MS/MS (MS^3) data are acquired. The lower panel is the base peak chromatogram of the analysis. The upper panel shows the MS^3 or MS^2 spectra of the phosphopeptides eluted from the chromatographic peaks at 13.8 min, 15.2 min, and 24.8 min, respectively. Phosphopeptides were effectively enriched as demonstrated from the relatively high abundance levels of the LC chromatographic peaks from which these phosphopeptides were eluted.

4 Notes

1. Phosphatase inhibitors should be added in the lysis buffer to prevent phosphate groups from being removed from the proteins during cell lysis. If the phosphatase inhibitors cocktails are prepared in-house, the inhibitors should cover a broad range of phosphatases. Highest purity of chemicals should be used throughout the whole phosphopeptide enrichment experiment.
2. All the solutions should be prepared using Type 1 ultrapure water with resistivity of ~18.2 MΩ cm such as those obtained from Barnstead Nanopure™ or Milli-Q systems.
3. Attention should be paid to avoid precipitation during the preparation of 200 mM L-glutamic acid ammonium salt (NH_4Glu), pH 2.0. To prepare 50 mL of this buffer, weigh 1.64 g NH_4Glu and transfer to a beaker. Add ultrapure water

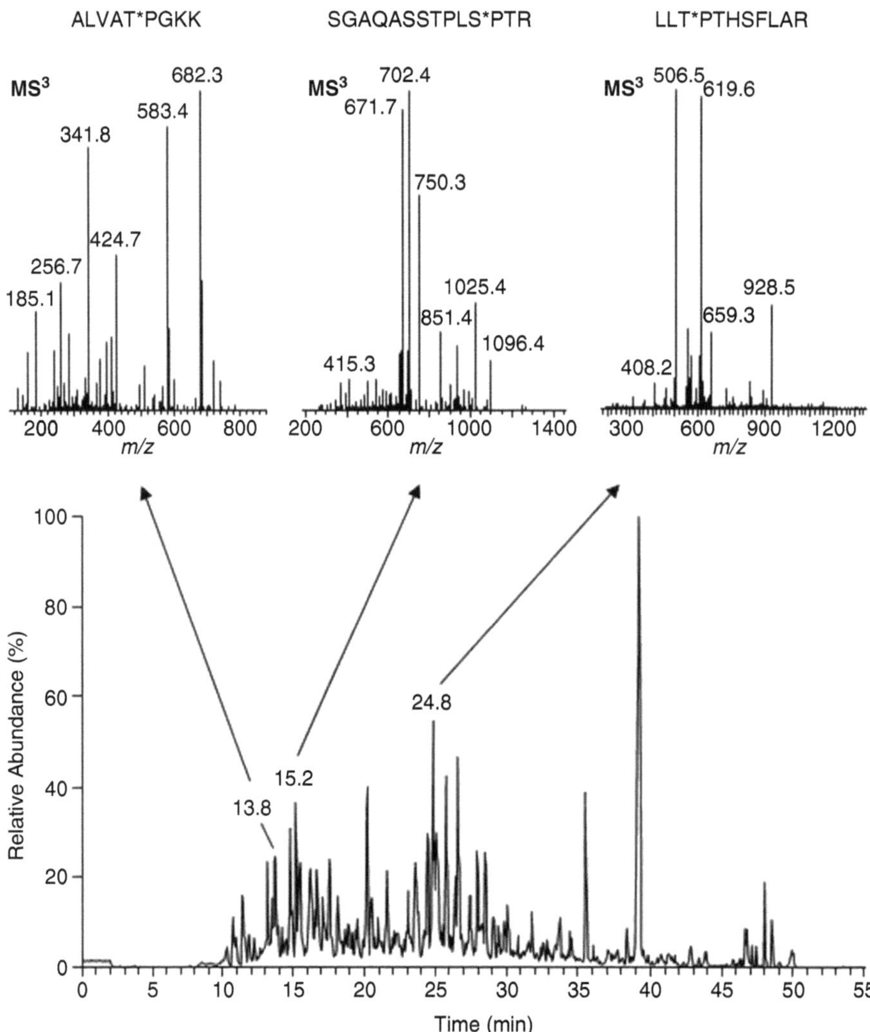

Fig. 2 Nanoflow reversed-phase liquid chromatography coupled online with a linear ion trap mass spectrometer for the analysis of TiO_2-enriched phosphopeptides from a HeLa cell lysate using collision-induced dissociation (CID) for tandem MS and phosphate neutral loss-dependent MS/MS/MS (RPLC-MS^2-MS^3). The analysis is for 5 μL injection of phosphopeptides onto a 75 μm ID capillary column. The *lower panel* is the base peak chromatogram of the analysis. The *upper panel* shows the MS^3 or MS^2 spectra of the phosphopeptides ALVAT*PGKK, SGAQASSTPLS*PTR, and LLT*PTHSFLAR identified, respectively, from nucleolin, lamin-A/C, and microtubule-associated protein 7. The relatively high abundance levels of the LC chromatographic peaks, from which these phosphopeptides were eluted, demonstrate that phosphopeptides were effectively enriched from the HeLa cell lysate. The *asterisk* marks the phosphorylated Ser or Thr residues

to a volume of ~25 mL. Adjust pH to 2.0 by adding ~12 % HCl drop-wise. Make up to 50 mL with ultrapure water using a cylinder.

4. The TiO_2 elution buffer is prepared by diluting ~28 % ammonium hydroxide (NH_4OH), which is commercially available. A syringe may be used to measure the 28 % NH_4OH solution.

5. Procedures to detach and collect the cells first and then lyse the cells could affect protein phosphorylation status since phosphorylation changes were observed within a minute of treatment (16). Therefore, in situ cell lysis is preferred. After the cell lysate is collected, the sample may be stored at −80 °C if the procedure needs to be interrupted.
6. The time interval between each sonication can be extended to avoid the sample getting warm or hot.
7. DNA is broken down during sonication, so the sample is no longer glutinous after this step.
8. Sample dilution may be needed so that the protein concentration is within the measurable standard range. Other protein assays such as the Bradford assay can be used as long as the lysis buffer is compatible with the assay.
9. Alternatively, dilution can be performed using 25–50 mM ammonium bicarbonate (NH_4HCO_3). Fourfold dilution of the sample is to bring down the urea concentration to 2 M or less in which trypsin can work effectively without significant loss of activity.
10. SPE should be performed using a vacuum manifold and is suggested to be in a chemical hood.
11. The orientation of ferrule to nut on tubing should be correct. The blunt end of the ferrule must face nut threads.
12. The tightness of the needle port should be adjusted using the syringe needle. Insert the syringe needle into the Teflon tube and tighten the fittings into the port until an increased resistance is felt during the insertion of the syringe needle. Do not over tighten the fittings.
13. Each time when a syringe of different size is used, the parameter of syringe inner diameter should be changed for the syringe pump, and the flow rate and volume settings need to be adjusted to meet the desired values.
14. The TiO_2 column is designed for use in the pH range of 2–12. Be cautious that all the solvents used do not exceed this pH range.
15. Make sure the TiO_2 cartridge column is properly tightened and connected without leaking. This can be done by checking the connections at the column inlet and outlet, and by comparing the solvent volume delivered by the syringe to that collected at the outlet of the column.
16. The TiO_2 column can be reusable after reconditioning. If this step is for reconditioning, it is important to ensure the column is fully re-equilibrated by 0.1 % TFA by checking the eluate pH.

17. The flow rate for sample loading is much slower than that for column washing to ensure all the phosphopeptides are retained on the column.
18. Fifty percent of CH_3CN facilitate dissolving of complex proteome samples. Up to 80 % of CH_3CN can be used.
19. The capacity of the TiO_2 column for phosphopeptides is very high. However, it is suggested to use multiple such columns for milligrams of peptide samples, and each is loaded, for example, with 0.5 mg of peptides. The phosphopeptides enriched from each column are then pooled and analyzed by multidimensional LC-MS/MS.

Acknowledgments

This study was supported in part with funds from the National Center for Toxicological Research, US Food and Drug Administration (NCTR/FDA) and in part with federal funds from the National Cancer Institute, National Institutes of Health, under contract N01-CO-12400. The content of this publication does not necessarily reflect the views or policies of the US Food and Drug Administration, the Department of Health and Human Services, nor does mention of trade names, commercial products, or organizations imply endorsement by the United States Government.

References

1. Salih E (2005) Phosphoproteomics by mass spectrometry and classical protein chemistry approaches. Mass Spectrom Rev 24:828–846
2. Yu LR, Issaq HJ, Veenstra TD (2007) Phosphoproteomics for the discovery of kinases as cancer biomarkers and drug targets. Proteomics Clin Appl 1:1042–1057
3. Bahl JM, Jensen SS, Larsen MR, Heegaard NH (2008) Characterization of the human cerebrospinal fluid phosphoproteome by titanium dioxide affinity chromatography and mass spectrometry. Anal Chem 80:6308–6316
4. Zhou W, Ross MM, Tessitore A et al (2009) An initial characterization of the serum phosphoproteome. J Proteome Res 8:5523–5531
5. Carrascal M, Gay M, Ovelleiro D, Casas V, Gelpi E, Abian J (2010) Characterization of the human plasma phosphoproteome using linear ion trap mass spectrometry and multiple search engines. J Proteome Res 9:876–884
6. Bodenmiller B, Mueller LN, Mueller M, Domon B, Aebersold R (2007) Reproducible isolation of distinct, overlapping segments of the phosphoproteome. Nat Methods 4:231–237
7. Jensen SS, Larsen MR (2007) Evaluation of the impact of some experimental procedures on different phosphopeptide enrichment techniques. Rapid Commun Mass Spectrom 21:3635–3645
8. Thingholm TE, Jensen ON, Robinson PJ, Larsen MR (2008) SIMAC (sequential elution from IMAC), a phosphoproteomics strategy for the rapid separation of monophosphorylated from multiply phosphorylated peptides. Mol Cell Proteomics 7:661–671
9. Pinkse MW, Mohammed S, Gouw JW, van Breukelen B, Vos HR, Heck AJ (2008) Highly robust, automated, and sensitive online TiO_2-based phosphoproteomics applied to study endogenous phosphorylation in Drosophila melanogaster. J Proteome Res 7:687–697
10. Larsen MR, Thingholm TE, Jensen ON, Roepstorff P, Jorgensen TJ (2005) Highly selective enrichment of phosphorylated peptides from peptide mixtures using titanium dioxide microcolumns. Mol Cell Proteomics 4:873–886
11. Yu LR, Zhu Z, Chan KC, Issaq HJ, Dimitrov DS, Veenstra TD (2007) Improved titanium dioxide

enrichment of phosphopeptides from HeLa cells and high confident phosphopeptide identification by cross-validation of MS/MS and MS/MS/MS spectra. J Proteome Res 6:4150–4162

12. Pinkse MW, Uitto PM, Hilhorst MJ, Ooms B, Heck AJ (2004) Selective isolation at the femtomole level of phosphopeptides from proteolytic digests using 2D-NanoLC-ESI-MS/MS and titanium oxide precolumns. Anal Chem 76:3935–3943
13. Kuroda I, Shintani Y, Motokawa M, Abe S, Furuno M (2004) Phosphopeptide-selective column-switching RP-HPLC with a titania precolumn. Anal Sci 20:1313–1319
14. Sano A, Nakamura H (2004) Titania as a chemo-affinity support for the column-switching HPLC analysis of phosphopeptides: application to the characterization of phosphorylation sites in proteins by combination with protease digestion and electrospray ionization mass spectrometry. Anal Sci 20:861–864
15. Sugiyama N, Masuda T, Shinoda K, Nakamura A, Tomita M, Ishihama Y (2007) Phosphopeptide enrichment by aliphatic hydroxy acid-modified metal oxide chromatography for nano-LC-MS/MS in proteomics applications. Mol Cell Proteomics 6:1103–1109
16. Kruger M, Kratchmarova I, Blagoev B, Tseng YH, Kahn CR, Mann M (2008) Dissection of the insulin signaling pathway via quantitative phosphoproteomics. Proc Natl Acad Sci USA 105:2451–2456

Chapter 9

iTRAQ-Labeling for Biomarker Discovery

Leroi V. DeSouza, Sébastien N. Voisin, and K.W. Michael Siu

Abstract

Various mass-tagging approaches have been developed over the last few years that have enabled mass spectrometry-based relative and absolute quantification of proteins from complex samples. This, in turn, has facilitated proteomics research to address issues ranging from alterations in the proteome of various model systems in response to various stimuli to biomarker discovery studies. Here we describe the use of one such mass-tagging approach, viz., iTRAQ labeling, as applied to cancer biomarker discovery. When applied to a cohort of tens of clinical samples, this technology can provide useful leads that serve as a basis for a more targeted validation-scale study.

Key words iTRAQ-labeling, LC-MS/MS analysis, Trypsin digestion, Relative quantification, Strong cation exchange, Reversed phase, Mass-tagging, Biomarker discovery

1 Introduction

No sooner did the identification of proteins by mass spectrometry (MS) become routine than researchers began to desire a means of quantifying the various proteins in their samples. This was in recognition of the fact that alterations in the proteome of a cell are often not reflected in terms of absolute presence or absence of specific proteins but in terms of more subtle changes in relative abundances. Traditional molecular biological based methods, such as Western analysis, could detect changes in expressions of targeted proteins, provided suitable antibodies were available or could be raised in-house. Alternately, traditional two-dimensional polyacrylamide gel electrophoresis (2D-PAGE) or the more recent differential gel electrophoresis (DIGE) could also provide a means of quantification, with the added benefit of being able to do this for a considerably larger number of proteins than possible by Western analysis. Spots of interest could then be excised and analyzed by MS to provide identification. However, this was a tedious approach wherein success depended in no small amount on the skill of the individual performing the analysis. Thus, there was considerable room for improvement which incubated the impetus for developing

Ming Zhou and Timothy Veenstra (eds.), *Proteomics for Biomarker Discovery: Methods and Protocols*, Methods in Molecular Biology, vol. 1002, DOI 10.1007/978-1-62703-360-2_9, © Springer Science+Business Media, LLC 2013

an MS-based approach for quantification. This need was answered in the form of a number of mass-tagging approaches that were developed to label either proteins or peptides at various stages of sample preparation. Tagging could be achieved by a variety of strategies, including metabolic labeling (of which the most frequently applied is stable isotope labeling by amino acids in cell culture, SILAC) where cells are grown in media enriched with isotopically labeled amino acids (1); enzymatic labeling of peptides by digestion of proteins in the presence of ^{18}O-labeled water (2); or chemical labeling of peptides through the use of mass tags by either a cysteine-labeling strategy, e.g., isotope-coded affinity tag (ICAT) (3), or an amine-labeling one as adopted by the tandem mass tag (TMT) (4) or isobaric tags for relative or absolute quantitation (iTRAQ) reagents (5). More recently label-free strategies involving various kinds of spectral or ion counting methods have been developed to estimate changes in expression levels, perhaps in a semiquantitative fashion. A detailed review on the various strategies for quantification can be found in Elliot et al. (6). Our laboratory has had considerable success with the iTRAQ reagents, which we have used to discover putative cancer biomarkers from tissue homogenates of several different kinds of cancer. While iTRAQ reagents are available as 4-plex or 8-plex versions, our work was performed exclusively on the 4-plex version and it is, therefore, the version that will form the basis of this chapter. The 8-plex reagent uses the same principle as the 4-plex but allows a comparison among eight samples.

The iTRAQ reagent comprises three chemical groups: (1) a terminal "peptide reactive group (PRG)," an *N*-hydroxysuccinimide (NHS), that binds primary amines of the peptide substrate; (2) the middle "balancer region" that decreases in the extent of its isotopic labeling in synchronization with increasing labeling of the third group; and (3) the reporter ion that is isotopically labeled and cleaves under tandem mass spectrometry (MS/MS). Isotopic labeling in the balancer and the reporter ion are matched such that the four tags have an identical mass, i.e., they and the labeled peptides are isobaric. iTRAQ modifies primary amines, which are present at the N-terminus and the side chain of lysine residues on peptides and proteins. Unfortunately, trypsinization is ineffective on iTRAQ-labeled lysine residues; this means proteins must first be cleaved into tryptic peptides prior to iTRAQ-labeling. Thus labeling will result in a lysine-terminating tryptic peptide bearing two iTRAQ tags, while an arginine-terminating tryptic peptide containing one tag. The cleaved reporter ion has a mass of 114.1, 115.1, 116.1, or 117.1 Da in the 4-plex version, while that in the 8-plex version contains an additional four options, viz., 113.1, 118.1, 119.1, and 121.1 Da. As stated earlier, the balancer region in each version compensates for the differences in the reporter ion masses, thereby ensuring that the total mass of the tag is the same across all the tags in the set. Any particular peptide from the different samples in a set remains isobaric after tagging, and the four differentially tagged peptides are detected as a single peak in the MS spectrum. The ions in this peak, when fragmented under MS/MS, yield typically an information-rich spectrum that yields qualitative information from the characteristic *b* and *y* series of ions that identify the peptide and quantitative information from the relative abundances of the reporter ions that cleave from the labeled peptide ions. The abundance of the reporter ion in each of these instances is a reflection of the abundance of that peptide in the initial sample. Therefore, the relative abundance of the peptide (and consequently, the originating protein) can be determined as a ratio of the peak area of the reporter ion for that peptide from a particular sample relative to that of the same peptide from the reference sample in the set.

Our studies typically begin with homogenization of biopsied tissues in phosphate-buffered saline (PBS) with proteases, followed by clarification of the homogenate by centrifugation. The protein content of the soluble fraction of each sample is then determined using a Bradford analysis and a 100 μg of total protein aliquoted for each analysis. The entire procedure involves four main stages. In the first, the samples are digested, labeled individually with the appropriate iTRAQ tag, and then pooled. In the second, labeled peptides in the pooled sample are fractionated using strong cation exchange (SCX) liquid chromatography (LC) in a first dimension of separation into 30 fractions. In the third, the contents of each SCX fraction is further resolved in a second dimension of reversed-phase (RP) LC, the eluent from which is directly analyzed by electrospray ionization mass spectrometry. Lastly, each data file acquired is processed by searching against a nonredundant protein database to identify and quantify the individual proteins in the sample.

2 Materials

The iTRAQ reagents kit is commercially available from AB SCIEX with all the necessary buffers and reagents necessary for the procedure. The only exception is the trypsin which is available also from AB SCIEX or from a number of other commercial sources. Water, organic solvents, and acids should all be LC-MS grade.

2.1 Trypsin Digestion and Labeling

1. Dissolution buffer: 0.5 M triethylammonium bicarbonate (TEAB).
2. Denaturant: 2 % SDS.
3. Reducing agent: 50 mM TCEP (tris-(2-carboxyethyl) phosphine).
4. Cysteine blocking reagent: 200 mM methyl methanethiosulfonate (MMTS).
5. Trypsin (sequencing grade): 10 μg per 100 μg of total sample protein.

6. Dissolution buffer, denaturant, reducing agent, and cysteine blocking reagent are all provided as part of the iTRAQ-labeling kit.

2.2 Strong Cation Exchange Separation

1. Buffer A: 15 mM KH_2PO_4 in 25 % acetonitrile, pH 3.0.
2. Buffer B: 15 mM KH_2PO_4 in 25 % acetonitrile, 350 mM KCl, pH 3.0.
3. Buffer C: 15 mM KH_2PO_4 in 25 % acetonitrile, 1 M KCl, pH 3.0.

2.3 Reversed-Phase Liquid Chromatography Separation

1. Buffer A: 0.1 % formic acid, 5 % acetonitrile.
2. Buffer B: 0.1 % formic acid, 95 % acetonitrile.

2.4 Columns

1. SCX Column: 2.1 mm internal diameter × 100 mm length PolyLC Polysulfoethyl A column packed with 5 μm beads with 300 Å pores (The Nest Group, Southborough, MA).
2. Guard column: 2.1 mm internal diameter × 10 mm length PolyLC Polysulfoethyl A column packed with 5 μm beads with 300 Å pores (The Nest Group, Southborough, MA).
3. C_{18} reversed-phase columns: Can be purchased from various vendors or packed in-house. The Pepmap column from LC Packings is an example of a commercial column we have used previously. We are currently using in-house-packed equivalents for which we use integra frit columns (New Objective, Woburn, MA) 75 μm i.d., packed to 150 mm length with 3 μm C_{18} beads (Kromasil), 100 Å pores.
4. C_{18} trap: 300 μm i.d. × 5 mm length C18 RP cartridge (LC Packings, Sunnyvale, CA).

3 Methods

3.1 Denaturing, Alkylating, Digesting, and Labeling

1. To each of the four samples containing up to 100 μg of total protein, add dissolution buffer from the kit to make up the volume to 20 μL (see Note 1 for planning of iTRAQ sets).
2. Add 1 μL of denaturant from the kit and vortex to mix.
3. Add 2 μL of reducing reagent, vortex to mix, then spin down in a picofuge.
4. Incubate tubes at 60 °C for 1 h, then spin down in a picofuge.
5. Add 1 μL cysteine blocking reagent from the kit, vortex to mix, and spin down again.
6. Incubate tubes at room temperature for 10 min.
7. While incubating the samples at step 6, dissolve a vial of trypsin in cold milli-Q water to a working concentration of 1 μg/μL.

8. After completion of the 10 min incubation, add 10 μL of the trypsin solution from step 7 into each sample tube. Vortex to mix, then spin down in a picofuge.
9. Incubate at 37 °C overnight.
10. Spin down in a picofuge.
11. Allow the iTRAQ reagent vials to reach room temperature, then spin down and add 70 μL of ethanol (supplied in iTRAQ kit) to each vial. Vortex to mix and spin again.
12. Transfer entire content of the iTRAQ reagent vial to a sample tube (see Notes 2–4 for factors affecting labeling). Vortex briefly to mix, then spin down.
13. Incubate tubes at room temperature for 1 h.
14. Combine the contents of each of the four samples for each set into one tube, vortex, and spin down.

3.2 Strong Cation Exchange Fractionation

1. Prior to SCX separation, after combining the individual samples as mentioned in step 14 in the section above, the sample is vacuum centrifuged to reduce the volume to a tenth of the volume of the sample injection loop on the HPLC system to be used for SCX separation.
2. Dilute the pooled sample at least tenfold with SCX buffer A (this is done in order to dilute the salt concentration in the labeled sample prior to loading onto the SCX column) and adjust the pH to 3.0 with phosphoric acid.
3. Transfer sample to autosampler vial. Set up HPLC to inject the entire sample onto an SCX column that is preconditioned and equilibrated with SCX buffer A. To prolong the life of the column, a guard column (2.1 mm i.d. × 10 mm length) of the same material fitted immediately upstream of the analytical column is recommended.
4. The SCX LC flow rate is set at 200 μL/min and the gradient is as shown in Table 1.
5. Program the fraction collector to collect a fraction every 2 min after an initial wait of 2 min to accommodate the void volume. This will result in a total of 30 fractions over the first 62 min. The gradient steps after 60 min are designed to strip and re-equilibrate the column.
6. After the fractions are collected and the LC method, including stripping and re-equilibration completed, remove the column, cap the ends and store under manufacturer recommended conditions.
7. Speed vacuum the SCX fractions to dryness (this can take 3–4 h) and store at −20 °C until ready for LC-MS analysis.

Table 1
Gradient used for strong cation exchange fractionation of iTRAQ-labeled samples

Time (min)	0	2	58	60	65	75	80	100
Buffer A (%)	100	100	0	0	0	0	100	100
Buffer B (%)	0	0	100	100	0	0	0	0
Buffer C (%)	0	0	0	0	100	100	0	0

3.3 Reversed-Phase Liquid Chromatography-Mass Spectrometry Analysis

The second dimension separation is usually performed online with mass spectrometric analysis. We have a nano-LC pump delivering the eluent at a flow rate of 200 nL/min to an in-house packed reversed-phase C18 nano-LC column of 75 μm i.d. × 150 mm long. We also use a C18 trapping column for online desalting. The trap is mounted on a switching valve so as to enable desalting with the high aqueous buffer (LC MS Buffer A) at a higher flow rate. During this desalting stage, the eluent from the trapping column is directed to waste. Following desalting, the valve switches to bring the trap in-line with the nano-LC analytical column (see Fig. 1 for an example of the valve connections), the eluent from which is directly analyzed by the mass spectrometer (see Note 5 for order in which SCX fractions should be run).

1. Redissolve each SCX fraction to be analyzed by LC-MS in 20 μL of 0.1 % formic acid (SCX fractions from later in the SCX gradient will contain more salt and, therefore, will require larger volume to ensure complete dissolution).
2. Depending on the autosampler configuration, the sensitivity of the mass spectrometer used and the capacity of the trap column being used load up to 25 % of the dissolved fraction onto the trap using an autosampler.
3. Desalt the sample using Buffer A at a flow rate of 20 μL/min for 5 min (the valve would be in the loading mode as shown in Fig 1a), then switch the trap to bring it in line with the analytical column (i.e., the valve would now be in the injection mode as shown in Fig. 1b).
4. Run the RP LC gradient as shown in Table 2.
5. The mass spectrometer is usually set up to run in an automated acquisition mode and acquires spectra of peptides eluting directly off the column. Depending on the particular instrument, the mass spectrometer can be set up to acquire a survey (MS) scan followed by multiple tandem MS scans. Our set up typically uses an MS scan (400–1,500 Th for 1 s) followed by up to 5 MS/MS scans (80–2,000 Th for 1 s each) in each cycle (see Notes 6 and 7 for mass spectrometer method set up).
6. If the MS software permits it, program the MS acquisition method to use a dynamic exclusion list that ignores a particular

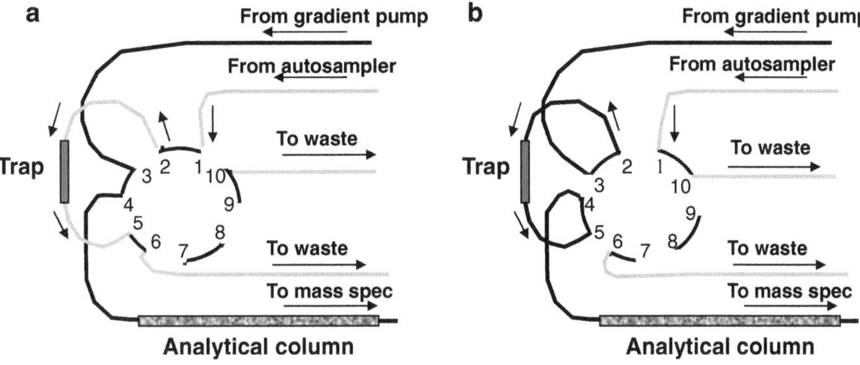

Fig. 1 Switching valve connections with valve in the (**a**) loading position and (**b**) injection position. Flow path of solvent from the gradient pump is in *black*, while the flow path of solvent from the autosampler pump is in *grey*

Table 2
Gradient used for reversed-phase liquid chromatography separation of iTRAQ-labeled samples

Time (min)	0	5	10	70	85	95	98	135
% B	5	5	15	35	80	80	5	5

m/z ratio for a short period after one MS/MS scan. This will minimize redundancy and aid in the identification and quantification of a larger number of peptides. The width of this exclusion window would depend on the LC peak width which can range from 30 to 60 s. Some groups prefer to use an exclusion window whose width is half of the LC peak width. This ensures that a spectrum is acquired for the peptide a second time when the peptide ion is likely to be at or near its maximum intensity, thereby giving a more robust fragmentation spectrum on which to base the identification and quantification.

7. To ensure that the lower abundance peptides are also detected, another feature used in our MS method is to target the least intense peaks (above a defined cutoff) on every third or fourth cycle.

3.4 Data Analysis

The data resulting from the MS analysis will need to be analyzed using an appropriate search algorithm and protein database. While there are a number of options available for performing the peptide identification aspect, currently, the quantification is most conveniently performed by AB SCIEX's Proteinpilot software. This software provides both identification and quantification as well as performs a grouping of redundant proteins. The particular details and nuances of the software are beyond the scope of this discussion, but in short, some of the more salient features are (1) the quantification data provided for each protein is based on ratios for peptides that are not shared with any other protein or group of proteins, and (2) the ratios are weighted in accordance with the confidence of the peptides identified and abundances of the iTRAQ signature ions. In addition, an adjustment factor, the applied bias, is also used to correct for variations in the initial total protein used in each sample in the set, and lastly, in the newest versions of the software, a background subtraction option is available that helps to alleviate a tendency of ratio compression that has been reported to be associated with lower abundance proteins in complex samples. The output from such an analysis is a list of proteins each with three ratios corresponding to its expression level in each of the three test samples relative to the designated reference (see Notes 8 and 9 for considerations during interpretation of data).

4 Notes

1. When planning the sample sets, ensure that there is sufficient amount of the sample that will be used as the reference in each set. As the iTRAQ technique provides relative quantification, the protein abundances are expressed as ratios relative to their expression in the reference sample. Currently, the preferred approach is to create a reference sample that is a pool of all the available samples. This pooled sample is then used as one of the four samples in each 4-plex iTRAQ set, thereby allowing for a meaningful cross comparison between the ratios from multiple iTRAQ sets that are analyzed as part of the study, e.g., if there are six samples for each of two states being studied, use 50 μg of each of the 12 samples to form a pool. Then use 100 μg of the pool as one of the samples in each iTRAQ set with the remaining three labels used to label the individual samples. This will result in four iTRAQ sets with a common reference sample in each set. Thus ratios of any given protein in each of the individual samples expressed relative to the reference sample will be comparable across the four sets. The excess aliquot of the reference sample can be saved and used in any subsequent study, thereby providing a means for a cross-study comparison.
2. While there are a number of factors that affect proper quantification, one that is a major factor is the possible presence of any primary amine-containing reagents. The presence of primary amines in your sample will affect labeling efficiency. Incomplete labeling of the peptides will, in turn, lead to unreliable quantification. One possible source of primary amines would be the buffer. If possible, avoid Tris-based buffers in favor of PBS or Hepes-based buffers during sample preparation.

3. Moisture and water will cause quenching of the label and also result in inefficient labeling. The iTRAQ reagents should, therefore, not be diluted or mixed in any aqueous solution other than when added to the actual sample at the time of labeling.
4. Labeling is also more efficient if the volume of the sample digest prior to addition of the label reagent is less than 50 μL. If the volume is larger, speed vacuum the sample and dissolve in 30 μL dissolution buffer. As an option, to confirm labeling, remove a small aliquot from each individual sample prior to pooling with the other samples from the iTRAQ set and analyze by MS/MS (these aliquots will need to be desalted, e.g., by ZipTipping with C18 tips (Millipore), prior to MS analysis). Confirm the presence of the expected iTRAQ reporter ion in the MS/MS spectra.
5. When analyzing SCX fractions, we typically load and run the samples on the RP nano-LC in reverse order. This is because the iTRAQ label might affect binding to the reversed phase, and the earlier fractions have higher amounts of unincorporated label. Because of this, we avoid injecting the earliest SCX fractions as they contain the flow through with the highest amount of unincorporated labels and reagents from the labeling step.
6. Collision energy required to optimally dissociate iTRAQ-labeled peptides is higher than that of unlabeled peptides. Therefore, the rolling collision energy used during data acquisition should be raised accordingly to ensure adequate fragmentation of the iTRAQ signature ions. The amount by which the collision energy needs to be raised should be determined empirically for the instrument in use.
7. Depending on the mass spectrometer being used, it is recommended that each fraction be injected a minimum of three times with an exclusion list generated after each injection to increase the overall numbers of peptides and proteins identified.
8. When iTRAQ analysis is used on a cohort of clinical samples, it is useful to incorporate individual non-disease controls as part of the iTRAQ sets as a safeguard against the selection of a protein that is unrelated to the disease state as a potential marker. Other than the reference pool, these individual controls provide a basis for assessing the range of interindividual expression levels for each of the proteins. Proteins that show consistent differential expression across the majority of test samples with no such differential expression in the individual control samples can then be regarded as potential biomarkers for that disease/sample state.
9. On a cautionary note and particularly in the context of biomarker discovery, iTRAQ analysis is typically used as a hypothesis-generating tool or a tool to generate leads. Confirmation of biomarker status usually requires a much larger-scale analysis

involving hundreds of samples instead of the dozen(s) that are used for iTRAQ analysis. Therefore, once a potential marker has been discovered using iTRAQ analysis, it is usually necessary to invest in the development of antibodies for use in confirmatory immunohistochemical analyses. Such a complementary approach not only provides independent verification of the iTRAQ results, but can also be performed on the larger scale using tissue microarray analysis (TMA) or ELISA, the results of which can then be subjected to much more stringent statistical analyses than possible with the iTRAQ results themselves.

Acknowledgments

We thank the Canadian Cancer Society Research Institute and the Canadian Institutes of Health Research for funding research that enabled the development and application of the iTRAQ technology. We also acknowledge the collaboration of our biomedical and clinical colleagues, especially Terence J. Colgan, Ranju Ralhan, Ian J. Witterick, Christina M. MacMillan, and Iona T. Leong (Mount Sinai Hospital, Toronto); Alexander D. Romaschin and George M. Yousef (St. Michael's Hospital, Toronto); and the late Abhijit Guha and Gelareh Zadeh (Hospital for Sick Children and Western Hospital, Toronto).

References

1. Ong S-E, Blagoev B, Kratchmarova I et al (2002) Stable isotope labeling by amino acids in cell culture, SILAC, as a simple and accurate approach to expression proteomics. Mol Cell Proteomics 1:376–386
2. Yao X, Freas A, Ramirez J, Demirev PA, Fenselau C (2001) Proteolytic 18O labeling for comparative proteomics: model studies with two serotypes of adenovirus. Anal Chem 73:2836–2842
3. Gygi SP, Rist B, Gerber SA, Turecek F, Gelb MH, Aebersold R (1999) Quantitative analysis of complex protein mixtures using isotope coded affinity tags. Nat Biotechnol 17:994–999
4. Tandem mass tag (TMT) (2008) http://www.piercenet.com/products/browse.cfm?fldID=C5A5F2FF-5056-8A76-4E4C-971BF496D176
5. Ross PL, Huang YN, Marchese JN et al (2004) Multiplexed protein quantitation in saccharomyces cerevisiae using amine-reactive isobaric tagging reagents. Mol Cell Proteomics 3:1154–1169
6. Elliot MH, Smith DS, Parker CE, Borchers C (2009) Current trends in quantitative proteomics. J Mass Spectrom 44:1637–1660

Chapter 10

Analysis of Glycoproteins for Biomarker Discovery

Jintang He, Yashu Liu, Jing Wu, and David M. Lubman

Abstract

Glycoproteins play an important role in cell signaling and cell–cell interaction. The alterations of glycoproteins are often relevant to progression of diseases, and these changed glycoproteins can be important biomarkers. The lectin-based glycoproteomic technology has extensively been used for high-throughput screening of potential glycoprotein biomarkers. Here we describe a multi-lectin affinity chromatography and label-free quantitative glycoproteomic approach for discovery of glycoprotein biomarkers relevant to differentiation of glioblastoma stem cells.

Key words Glycoprotein, Biomarker, Lectin affinity chromatography, Mass spectrometry, Proteomics

1 Introduction

Glycoproteins are a group of proteins with oligosaccharide chains covalently attached to them. These proteins are ubiquitously distributed, and play a critical role in various biological processes, such as cell signaling, cell–cell interaction, immune recognition, cell proliferation, and differentiation (1). According to recent findings, the alterations of glycoproteins are often associated with the progression of diseases including cancer (2–4). These altered glycoproteins may be important biomarkers and therapeutic targets.

The proteomic technology provides a powerful tool for discovery of potential biomarkers. The quantitative comparison of the proteome of different samples can result in identification of differentially expressed proteins which are potential biomarkers. Glycoproteomics is a branch of proteomics that studies the entire complement of glycoproteins. Glycoproteomic studies have greatly benefited from lectin affinity chromatography (5–10) that can be used to enrich glycoproteins, as well as from mass spectrometry which plays a vital role in the characterization of glycoproteins (11–14). Lectins are a class of proteins which can specifically bind to the glycan moieties of glycoproteins (15). Currently, hundreds of lectins have been purified and characterized. Concanavalin A (Con A) and

Ming Zhou and Timothy Veenstra (eds.), *Proteomics for Biomarker Discovery: Methods and Protocols*, Methods in Molecular Biology, vol. 1002, DOI 10.1007/978-1-62703-360-2_10,

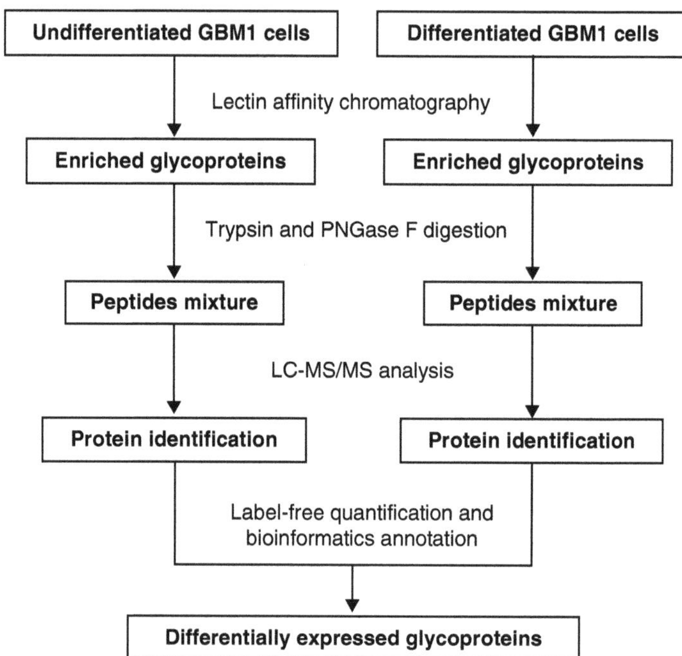

Fig. 1 Workflow of lectin-based glycoproteomic technology for biomarker discovery

wheat germ agglutinin (WGA) are the most commonly used lectins for glycoprotein enrichment where they have high affinity to most of the N-linked glycoproteins.

In this chapter, we describe a lectin-based glycoproteomic approach for discovery of glycoprotein markers relevant to differentiation of a human glioblastoma-derived stem-like neurosphere HSR-GBM1. A diagram of the workflow of this approach is shown in Fig. 1.

2 Materials

2.1 Protein Extraction

1. Protein extraction buffer: 20 mM Tris–HCl, pH 7.4, 150 mM NaCl, and 1 % (w/v) octyl-β-D-glucopyranoside. Store at 4 °C. Add 1 % protease inhibitor cocktail (Sigma-Aldrich) mixture immediately before use.
2. Dounce glass homogenizer with a tight-fitting pestle (Wheaton Science).
3. Temperature controlled centrifuge (Thermo Fisher Scientific).
4. 1.5 mL Eppendorf tubes.

2.2 Lectin Affinity Chromatography

1. Agarose-bounded Con A, WGA, and peanut agglutinin (PNA) (Vector Laboratories) (see Note 1).
2. Binding buffer: 20 mM Tris–HCl, pH 7.4, 150 mM NaCl, 1 mM $MgCl_2$, 1 mM $CaCl_2$, and 1 mM $MnCl_2$.

3. Elution buffer: 300 mM methyl-R-D-mannopyroside (Sigma-Aldrich), 300 mM N-acetylglucosamine (Sigma-Aldrich) and 200 mM D-galactose (Sigma-Aldrich) in binding buffer.
4. 5 mL disposable screw endcap spin column.
5. Microcon YM-10 centrifugal filter.
6. Tris(2-carboxyethyl)phosphine hydrochloride (TCEP), iodo-acetamide (IAA), and ammonium bicarbonate.

2.3 Digestion with Trypsin and PNGase F

1. TPCK modified sequencing grade porcine trypsin (Promega).
2. PNGase F (Sigma).
3. Speedvac concentrator system.

2.4 LC-MS/MS

1. Paradigm MG4 micropump system (Michrom Biosciences, Inc.).
2. Paradigm Platinum Peptide Nanotrap (Michrom Biosciences, Inc.).
3. Magic C18AQ column (0.1 mm × 150 mm, C18 AQ particles, 5 μm, 200 Å, Michrom Biosciences, Inc.).
4. Solvent A: 0.3 % formic acid in H_2O (HPLC grade).
5. Solvent B: 0.3 % formic acid in acetonitrile (HPLC grade).
6. LTQ linear ion trap mass spectrometer (Thermo Electron Corporation).

2.5 Mass Spectrometry Data Analysis

1. Proteome Discoverer software version 1.1.0.263 (Thermo Fisher Scientific).
2. Swiss-Prot database.

2.6 Western Blotting

1. 0.05 % Tween-20 in PBS (PBST).
2. Blocking buffer: 1 % BSA in PBST.
3. Mini-PROTEAN 3 system (Bio-Rad).
4. Polyvinylidene difluoride (PVDF) membrane.
5. Trans-Blot SD Semi-Dry System (Bio-Rad).
6. Peroxidase-conjugated goat anti-rabbit and anti-mouse IgG (H + L) (Abcam).
7. Supersignal West Pico Chemiluminescent HRP Substrate (Thermo Fisher Scientific).

3 Methods

3.1 Protein Extraction

Carry out all procedures at 4 °C unless otherwise specified.

1. Collect ~20 million of undifferentiated and differentiated HSR-GBM1 cells, wash the cells twice with PBS, and centrifuge at 500 × *g* for 5 min at 4 °C. Store the cell pellets in −80 °C until use.

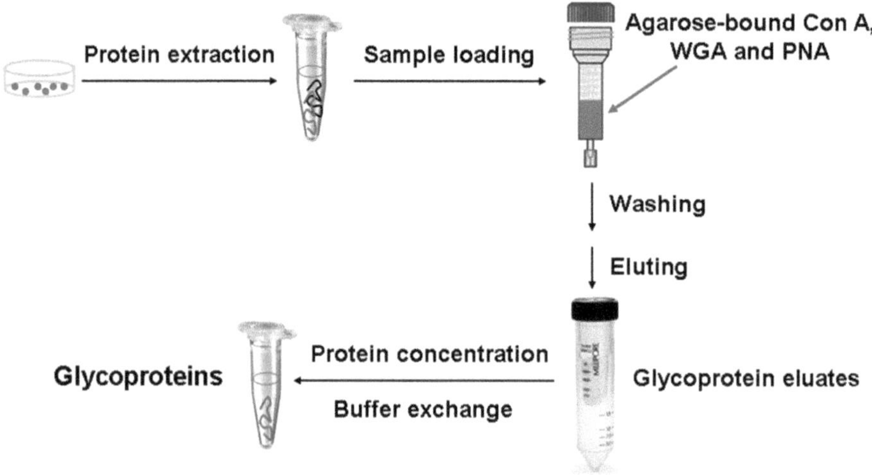

Fig. 2 Illustration of the lectin affinity chromatography method for glycoprotein enrichment

2. Resuspend the cell pellets in 1 mL of protein extraction buffer, and then transfer the suspension to a Dounce glass homogenizer and homogenize the cells with 25 strokes using a tight-fitting pestle. After 10-min incubation on ice, repeat the process.
3. Transfer the cell lysates to 1.5 mL eppendorf tubes (see Note 2) and centrifuge at 40,000 × *g* for 30 min at 4 °C. Collect the supernatants to new eppendorf tubes and determine protein concentrations of each sample by the Bradford method (16).

3.2 Glycoprotein Enrichment by Lectin Affinity Chromatography

Glycoproteins were enriched using multi-lectin affinity chromatography essentially as previously described (17). The workflow of this method is shown in Fig. 2.

1. Pack 2 Pierce centrifuge columns with agarose-bound lectin mixtures. Each column contains 0.8 mL of Con A, 0.8 mL of WGA, and 0.8 mL of PNA. Wash the columns with 7.2 mL of binding buffer.
2. Dilute cell lysates four times with ice-cold binding buffer, where each sample contains ~2 mg of proteins. Load the diluted samples to lectin columns, suspend the beads, and incubate for 15 min. Collect the flow-through and repeat the step (see Note 3).
3. Wash each lectin column with 4 × 2.4 mL of binding buffer to remove the nonspecific binding proteins (see Note 4).
4. Add 2.4 mL of elution buffer to each column, incubate for 15 min, and collect the eluate. Repeat this step twice, and pool the eluates of the same sample.
5. Transfer the eluates to Microcon YM-10 centrifugal filter devices, centrifuge at 5,000 × *g* for 45 min at 20 °C, and wash

each sample with 6 mL of 25 mM ammonium bicarbonate (see Note 5).

6. Add 200 μl of 10 mM TCEP to each sample, and incubate for 20 min at room temperature. And then add 200 μl of 30 mM IAA to each sample, and incubate in the dark for 20 min at room temperature (see Note 6).
7. Wash each sample with 6 mL of 25 mM ammonium bicarbonate twice. Determine protein concentration and store the glycoprotein samples at −80 °C until further use.

3.3 Digestion with Trypsin and PNGase F

1. Add trypsin to each glycoprotein sample (1:50 w/w) and incubate for 12–16 h at 37 °C.
2. Add 1 μL of formic acid to each sample to stop the enzymatic reaction, and dry the sample with a speedvac concentrator.
3. Suspend the dried sample with 25 mM ammonium bicarbonate, and then add 1 U of PNGase F, mix by vortex and incubate for 16–20 h at 37 °C.
4. Add 1 μL formic acid to the digest to stop the enzymatic reaction, and dry the sample with a speedvac concentrator. Keep the dried sample at −80 °C before analysis by mass spectrometry.

3.4 LC-MS/MS

1. Dissolve the peptide mixtures in 70 μL of 0.1 % formic acid, centrifuge samples at 15,000 × *g* for 10 min, and transfer the supernatants to new eppendorf tubes.
2. Inject 10 μL of each sample using an autosampler, transfer peptides to a Nanotrap column using a Paradigm MG4 micropump system, and separate peptides using a C18 separation column. A 90-min linear gradient of acetonitrile (solvent B)/water (solvent A) containing 0.3 % formic acid is used (see Note 7). The flow rate is 300 nL/min. The gradient is as follows:

   0–10 % B for 10 min

   10–40 % B for 90 min

   40–100 % B for 8 min

   100 % B for 1 min

   100–0 % B for 1 min

   0 % B for 10 min
3. Analyze the peptides using an LTQ linear ion trap mass spectrometer. Operate the mass spectrometric analysis in positive ion mode and acquire the data with an Xcalibur software. For each cycle of one full mass scan (range of *m*/*z* 400–2,000), select the five most intense ions in the spectrum for MS/MS analysis unless they appear in the dynamic or mass exclusion lists. Set the ESI spray voltage at 1.5 kV, and the capillary voltage at 30 V. The ion activation was achieved by utilizing helium with a normalized collision energy of 35 %.

3.5 Mass Spectrometry Data Analysis

1. Search all MS/MS spectra against the Swiss-Prot database using SEQUEST algorithm incorporated in Proteome Discoverer software version 1.1.0.263. The search parameters are as follows: (1) fixed modification, carbamidomethyl of cysteine; (2) variable modification, oxidation of M and asparagine to aspartate; (3) allowing two missed cleavages; (4) peptide ion mass tolerance 1.50 Da; (5) fragement ion mass tolerance 1.4 Da; and (6) peptide charges +1, +2, and +3.
2. Set false discovery rate (FDR) to be <0.01 for each database search. The Xcorr values for all the charge states (+1, +2, and +3) are automatically adjusted to achieve the predetermined FDR value. If multiple proteins share the same peptide sequences, report them as a protein group.
3. Export the database search results to excel files, and quantify proteins using label-free approaches such as spectral counting (18) and spectral index (19) (see Note 8). This will result in a list of differentially expressed proteins.
4. Bioinformatically interrogate the proteins according to the Swiss-Prot annotation to identify a list of differentially expressed glycoproteins which are potential glycoprotein biomarkers.

3.6 Validation of Potential Biomarkers by Western Blotting

1. Separate 12 μg of proteins from undifferentiated and differentiated HSR-GBM1 cells by SDS-PAGE using a Mini-PROTEAN 3 system.
2. Transfer proteins to a PVDF membrane using a Trans-Blot SD Semi-Dry System according to the manufacturer's instruction manual (see Note 9).
3. Wash the PVDF membrane with PBST and block the membrane with blocking buffer for 1 h at room temperature.
4. Incubate the membrane with various primary antibodies overnight at 4 °C.
5. Wash the PVDF membrane 3 × 10 min with PBST.
6. Incubate the membrane with peroxidase-conjugated goat anti-rabbit or anti-mouse IgG (H + L) for 1 h at RT.
7. Wash the PVDF membrane 3 × 10 min with PBST.
8. Visualize the blots with Supersignal West Pico Chemiluminescent HRP Substrate.

4 Notes

1. Con A, WGA, and PNA have different carbohydrate specificities. Con A is specific to α-linked mannose; WGA is specific to *N*-acetyl-D-glucosamine; and PNA is specific to β-galactose.

2. All the microcentrifuge tubes used in this work are Eppendorf Protein LoBind tubes which can reduce sample-to-tube binding and thus improve sample recovery.
3. Do not discard the flow-through until confirming that the enrichment is successful.
4. Monitor the protein concentration of the flow-through. If there is still a significant amount of protein in the flow-through buffer after this 4×2.4 mL washing step, wash the column using another 2.4 mL of binding buffer.
5. After this step, the glycoproteins are concentrated, the glycans and salts in the eluates are removed, and meanwhile, the buffer system is changed to ammonium bicarbonate which is suitable for the following tryptic digestion.
6. IAA is light sensitive. Prepare the IAA buffer freshly before use and keep it away from light.
7. To identify more proteins by the following mass spectrometry analysis, a longer gradient (for example, a 4-h gradient) may be used to try to separate the peptides as much as possible.
8. Quantification is based on comparison of spectral counts or spectral indexes of each protein in different samples.
9. PVDF membrane is hydrophobic and should be pre-wet in methanol before use.

Acknowledgments

This work was funded under the National Institute of Health Grant No. R01 49500 (D.M.L.) and the National Cancer Institute Grant R21 CA124441 (D.M.L.) and R01 CA154455 (D.M.L.).

References

1. Bertozzi CR, Kiessling LL (2001) Chemical glycobiology. Science 291:2357–2364
2. Reis CA, Osorio H, Silva L, Gomes C, David L (2010) Alterations in glycosylation as biomarkers for cancer detection. J Clin Pathol 63: 322–329
3. Hakomori S (1989) Aberrant glycosylation in tumors and tumor-associated carbohydrate antigens. Adv Cancer Res 52:257–331
4. Fuster MM, Esko JD (2005) The sweet and sour of cancer: glycans as novel therapeutic targets. Nat Rev Cancer 5:526–542
5. Qiu R, Regnier FE (2005) Use of multidimensional lectin affinity chromatography in differential glycoproteomics. Anal Chem 77:2802–2809
6. Kaji H, Saito H, Yamauchi Y, Shinkawa T, Taoka M, Hirabayashi J et al (2003) Lectin affinity capture, isotope-coded tagging and mass spectrometry to identify N-linked glycoproteins. Nat Biotechnol 21:667–672
7. Yang Z, Harris LE, Palmer-Toy DE, Hancock WS (2006) Multilectin affinity chromatography for characterization of multiple glycoprotein biomarker candidates in serum from breast cancer patients. Clin Chem 52:1897–1905
8. Ghosh D, Krokhin O, Antonovici M, Ens W, Standing KG, Beavis RC et al (2004) Lectin affinity as an approach to the proteomic analysis of membrane glycoproteins. J Proteome Res 3:841–850
9. Hart DA (1980) Lectins in biological systems: applications to microbiology. Am J Clin Nutr 33:2416–2425
10. Dai Z, Zhou J, Qiu SJ, Liu YK, Fan J (2009) Lectin-based glycoproteomics to explore and analyze hepatocellular carcinoma-related

glycoprotein markers. Electrophoresis 30: 2957–2966
11. Wang Y, Wu SL, Hancock WS (2006) Approaches to the study of N-linked glycoproteins in human plasma using lectin affinity chromatography and nano-HPLC coupled to electrospray linear ion trap–Fourier transform mass spectrometry. Glycobiology 16:514–523
12. Xiong L, Andrews D, Regnier F (2003) Comparative proteomics of glycoproteins based on lectin selection and isotope coding. J Proteome Res 2:618–625
13. North SJ, Hitchen PG, Haslam SM, Dell A (2009) Mass spectrometry in the analysis of N-linked and O-linked glycans. Curr Opin Struct Biol 19:498–506
14. He J, Liu Y, Xie X, Zhu T, Soules M, DiMeco F et al (2010) Identification of cell surface glycoprotein markers for glioblastoma-derived stem-like cells using a lectin microarray and LC-MS/MS approach. J Proteome Res 9:2565–2572
15. Goldstein IJ, Hayes CE (1978) The lectins: carbohydrate-binding proteins of plants and animals. Adv Carbohydr Chem Biochem 35: 127–340
16. Bradford MM (1976) A rapid and sensitive method for the quantitation of microgram quantities of protein utilizing the principle of protein-dye binding. Anal Biochem 72: 248–254
17. He J, Liu Y, Zhu TS, Xie X, Costello MA, Talsma CE et al (2011) Glycoproteomic analysis of glioblastoma stem cell differentiation. J Proteome Res 10:330–338
18. Liu H, Sadygov RG, Yates JR 3rd (2004) A model for random sampling and estimation of relative protein abundance in shotgun proteomics. Anal Chem 76:4193–4201
19. Griffin NM, Yu J, Long F, Oh P, Shore S, Li Y et al (2010) Label-free, normalized quantification of complex mass spectrometry data for proteomic analysis. Nat Biotechnol 28:83–89

Chapter 11

SILAC in Biomarker Discovery

Benjamin C. Orsburn

Abstract

Stable isotope labeling with amino acids in cell culture (SILAC) has become an extremely valuable tool in quantitative proteomics and in biomarker discovery. Incorporation of SILAC labels occurs when cells are passaged multiple times in media where the endogenous amino acids are replaced with the heavy isotope ones. During a typical experiment, cells from heavy and light strains are combined in equal ratios and all steps of protein extraction and digestion occur on these cells together, minimizing the number of external variables introduced during sample processing. Potential biomarkers are revealed during liquid chromatography-tandem mass spectrometry analysis by peptides that differ considerably in intensity between the two strains as revealed by the mass shift from the incorporated SILAC label. The protocol presented here describes how to perform a typical experiment using the SILAC technology for the search for biomarkers as revealed by differences in protein expression levels, as well as by phosphorylation, a common posttranslational modification.

Key words SILAC, Biomarkers, Phosphoproteomics, Phosphopeptide, FASP, FACE, TiO_2, IMAC

1 Introduction

Since its introduction to the scientific community at the beginning of this century, the use of stable isotopic labeled amino acids in cell culture (SILAC) has been applied to nearly model biological system (1). In this regard, the SILAC technology has been invaluable in the detection of biomarkers, particularly in cancer cell lines (2, 3). SILAC integration occurs when cells are repeatedly passaged in media containing isotopically labeled amino acids rather than naturally occurring ones. After multiple passages, the proteins in the cells become virtually devoid of the endogenous amino acids, as they have all been replaced by translational insertion of the labeled ones.

In drug mechanism studies, the same cell line will be passaged in the labeled media simultaneously with cells grown in normal media, producing a control and experimental cell line. Theoretically, the two cell lines produce identical proteins, with the exception of the SILAC labels. By treating one of the cell lines with a chemical agent, biomarkers indicating the prospective drug mechanism are

Ming Zhou and Timothy Veenstra (eds.), *Proteomics for Biomarker Discovery: Methods and Protocols*, Methods in Molecular Biology, vol. 1002, DOI 10.1007/978-1-62703-360-2_11,

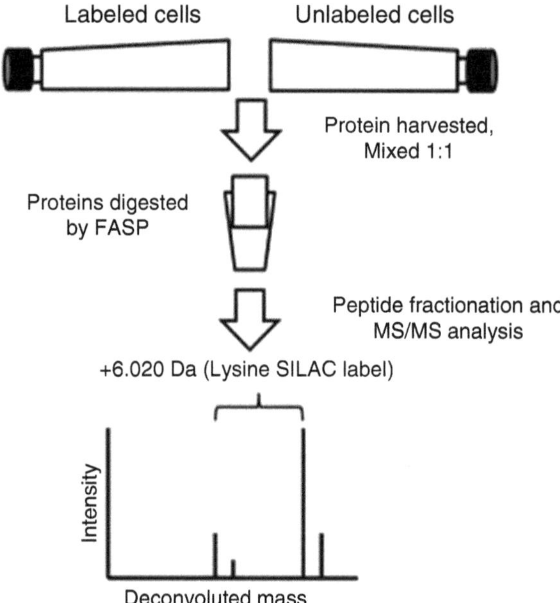

Fig. 1 Example processing and analysis scheme for biomarker regulated by translation/transcription

revealed during liquid chromatography-tandem mass spectrometry (LC-MS/MS) analysis by the differences in the peptide profile of one line compared to that of the other. These changes can be quantified by comparing the intensity counts of the peptides obtained from the treated cell line compared to that of the untreated one (4).

The real advantage of the SILAC technique over other labeling technologies is the fact that all downstream processing events, including protein digestion and protein/peptide separation and enrichment, occur simultaneously for both the control and treated samples. This is due to the fact that the samples are mixed either before, or shortly after, protein harvest. The proteins may be mixed by lysing identical numbers of cells, or by performing protein determination via BCA or similar biochemical determination following protein harvest. By combining these samples as early as possible in the sample processing method, the number of variables in the experiment may be drastically reduced.

SILAC has been applied successfully in a large number of biomarker studies such as the quantitative comparison of normal pancreatic cell lines to the Panc1 immortalized cancer cell line (5). In these studies, SILAC labels were used to identify biomarkers that were regulated via protein transcription or translation. Recently, this technology has also been successfully applied toward the identification of biomarkers that are regulated posttranslationally by methods such as phosphorylation (6). The following method will describe our methodology for identifying biomarkers that are regulated via transcription/translation and by phosphorylation; a scheme for these two methods is shown in Figs. 1 and 2, respectively.

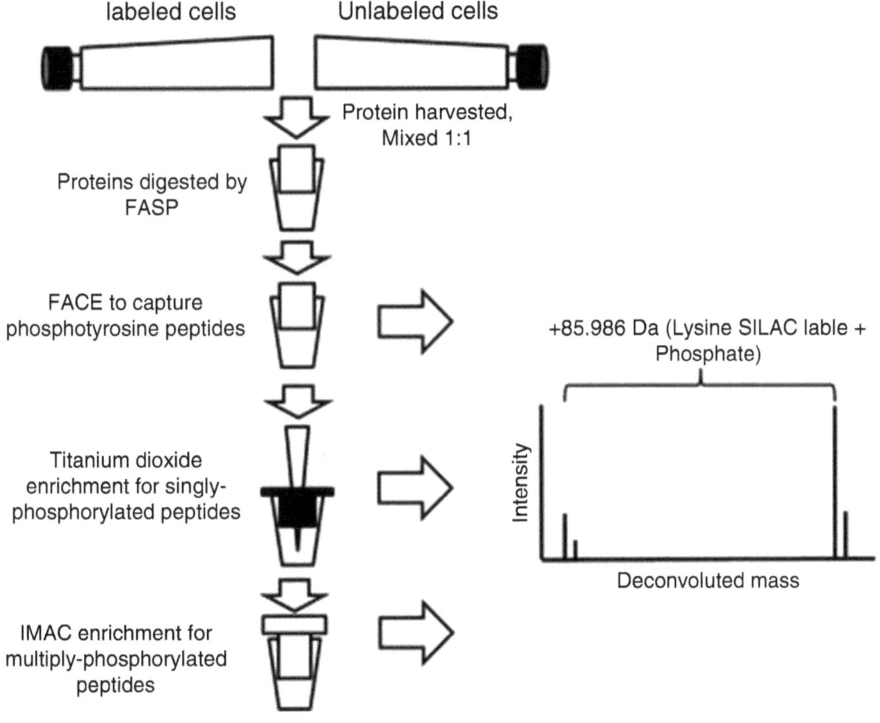

Fig. 2 Example processing and analysis scheme for phosphorylated biomarkers

2 Materials

1. RPMI media or DMEM, deficient for Lysine and Arginine (Invitrogen).
2. $^{13}C_6$ Lysine (Invitrogen).
3. $^{13}C_6{}^{15}N_4$ Arginine (Invitrogen).
4. Protein lysis buffer: 2 % SDS, 0.5 M Urea, 50 mM DTT in 20 mM Tris–HCl, pH 7.5.
5. 75 cm^3 culture plates (Falcon).
6. Trypsin, sequencing grade (Promega).
7. FASP chambers with 30 kDa M.W. cutoff filter (Protein Discovery).
8. FASP wash buffer, 8 M urea in 20 mM Tris–HCl, pH 7.5 (Protein Discovery).

Peptide Separation by ZipTip

1. C-18 or SCX packed tip ZipTips (Millipore).

Peptide Separation by OFFGEL

1. Low-resolution separation kit, pH 4–7 (Agilent).

Phosphopeptide Enrichment (Optional)

1. Antiphosphotyrosine 4 G10 antibody (Millipore).
2. Titanium dioxide phosphopeptide enrichment kits (Pierce).
3. IMAC phosphopeptide enrichment kits (Gallium or Fe-NTA, Pierce).
4. Graphite desalting columns (Pierce).

3 Methods

3.1 Passaging Cells in SILAC Media

1. Determine the appropriate number of cells for your experiment using unlabeled cells (see Note 1).
2. Seed early passage cells of interest in SILAC-labeled and unlabeled media at appropriate density for that cell type. We typically start at 10–20 % confluence for adherent cells.
3. Passage a minimum of 5 times to obtain complete labeling of your cells.
4. (Optional) Harvest, digest, and perform LC-MS/MS analysis on a small fraction of cells to verify complete integration of the SILAC labels (see Note 2).

3.2 Protein Harvest

1. If dosing cells with a chemical agent, add the agent to the cell media of the heavy cells and place the culture back into the incubator for the appropriate amount of time.
2. To terminate protein interactions, rapidly pour out the culture media and replace with approximately 20 mL of ice-cold PBS (see Note 3).
3. Dump off the PBS and replace it with fresh cold PBS; repeat this step twice.
4. Following third wash forcefully knock as much PBS as possible out of the culture flask and replace with lysis buffer.
5. Remove cells from the plate by scraping the plate with a tissue scraper. It is best to tilt the plate and work in one direction. After all of the cell suspension/lysis buffer has been pushed to one side, tilt the plate to the opposite side and repeat. The solution will be extremely viscous.
6. Remove the cell/lysis buffer mixture with a wide bore disposable pipette.

3.3 Protein Digestion

1. Place the tubes containing your protein/lysis solution in a 1 L glass beaker with water at approximately 80 °C for 10 min.
2. Sonicate the protein/lysis solution until homogenous. Typically 2–3 short pulses at 30 % intensity will be sufficient for mammalian cells (see Note 4).

3. Place crude protein extract into the top of assembled FASP chambers (see Note 5).
4. Concentrate by centrifugation at 13,000 × *g* for 15 min.
5. Wash proteins by adding 200 μL of FASP wash buffer and concentrating again (13,000 × *g* for 15 min).
6. Repeat step 5 twice.
7. Alkylate the cysteine residues by adding 100 μL freshly prepared 50 mM iodoacetamide solution to each chamber.
8. Incubate in the dark for 20 min.
9. Remove iodoacetamide by repeating steps 5 and 6.
10. Equilibrate the reaction chamber for digestion by adding 200 μL of 50 mM NH_4HCO_3 to the chamber and concentrating as performed previously. Perform this step twice (see Note 6).
11. Add freshly reconstituted trypsin in 50 mM NH_4HCO_3. Use a ratio of 1:20 trypsin to protein.
12. Allow to digest at 37 °C overnight.
13. The following day, centrifuge at 13,000 × *g* for 10 min to release digested peptides. Add 50 μL of 50 mM NH_4HCO_3 and centrifuge through to increase peptide yield.

3.4 Phosphopeptide Enrichment (Optional Steps)

1. Pool the eluted peptides.
2. Add antiphosphotyrosine antibody 4 G10 to the pooled peptides, in a ratio of 1:20 antibody to starting protein concentration. That is, if starting with 2 mg of total protein incubate this with 100 μg of 4 G10 antibody (see Note 7).
3. Incubate for 24 h at 4 °C with end-over-end rotation.
4. Separate the mixture into 200 μL aliquots and place each aliquot into a new FASP reaction chamber.
5. Release unbound peptides by centrifuging at 13,000 × *g* for 15 min. Save the unbound peptides for further enrichment steps via TiO_2 and/or IMAC enrichment (see Note 8).
6. Wash the bound peptides by adding 200 μL of 50 mM NH_4HCO_3 and centrifuging though the filter. Repeat this step once.
7. To elute the bound peptides, add 50 μL of 0.3 % TFA to each chamber and vortex for 30 s. Elute the peptides by centrifugation. Repeat this step twice, changing the TFA concentration on the second and third washes to 0.1 %.
8. Concentrate the phosphotyrosine peptides with vacuum centrifugation to <100 μL and desalt for LC-MS analysis with graphite spin column or C-18 ZipTip (see Note 9).
9. Perform TiO_2 enrichment of flowthrough from FACE capture by manufacturer's protocol.

10. Perform IMAC enrichment of flowthrough from TiO_2 enrichment by manufacturer's instructions.
11. Due to the complexity and relatively high yields of phosphopeptides enriched via TiO_2 or IMAC enrichment, these samples may require fractionation to obtain maximum sequence coverage. In contrast, due to the relatively low numbers of peptides obtained by FACE enrichment, fractionation prior to LC-MS/MS analysis is not recommended.

3.4.1 Peptide Fractionation by ZipTip (See Note 10)

1. Perform fractionation of peptides using ZipTip's per manufacturer's recommendations. For C-18 tips, we have found the best coverage results from sequentially eluting 7 fractions, with 5 %, 10 %, 15 %, 20 %, 30 %, 50 %, and 100 % acetonitrile, respectively.

3.4.2 Peptide Fractionation by OFFGEL (See Note 11)

1. Perform OFFGEL fractionation via manufacturer's instruction.
2. For proteins regulated by transcription/translation, use strips and electrolytes for pH range 3–10.
3. For enriched phosphopeptides, use strips and electrolytes for pH range 4–7.

3.5 Analysis

1. Perform LC-MS/MS analysis of your samples using preferred settings.
2. Process your raw data with the following dynamic modifications in addition to your normal processing settings: lysine +6.020 and arginine +10.008.
3. Both the MAXQUANT open source software program and Proteome Discoverer versions 1.2. and up (Thermo) are compatible with quantization of SILAC-labeled proteins (see Note 12).

4 Notes

1. Due to the relatively high cost of SILAC media, it may be prudent to determine the minimal appropriate number of cells for your experiment prior to this experiment, particularly if you will be performing phosphopeptide enrichment and analysis. This can be best performed by growing a population of labeled cells and performing all digestion and separation steps. As biomarkers are typically of relatively low copy number, an appropriate number of cells will be the smallest number of cells that still provides you with a large number of protein identifications. The relatively low number of phosphopeptides compared to unmodified peptides will require a greater starting material (often >1 mg, or one 75 cm^3 culture flask).
2. To determine the efficiency of the SILAC labeling of your cells, harvest, digest, and perform LC-MS/MS analysis on a small

Table 1
Appropriate buffer for your enzyme

Enzyme	Buffer	Buffer pH
LysC (*A. lyticus*)	100 mM Tris–HCl	9.2
GluC (*S. aureus*)	50 mM Tris–HCl	8.0
LysN (*G. frondosa*)	100 mM glycine	9.0

number of your SILAC-labeled cells. Process the samples in two separate ways: (a) without the SILAC labels (no modifications to lysine or arginine) and (b) with the SILAC mass labels (lysine, +6.020 Da; arginine, 10.008 Da) as static modifications. If you successfully integrated the SILAC label into your cells you will obtain no high-confidence sequenced fragments lacking this label, aside from ubiquitous contaminants such as keratin. To determine the percent of proteins that were successfully labeled, divide the number of labeled vs. unlabeled peptides identified by your processing scheme. If the percentage incorporation is <95 %, continue to passage the cells.

3. As time may be of the essence, particularly in drug mechanism studies or when looking for posttranslational modifications, preprepared 20 mL aliquots of PBS on ice will simplify and speed up your harvesting procedure.
4. Heat a beaker containing 400 mL of water in a microwave to 80 °C (~4 min). Place the tubes into the cooling liquid. Following sonication, place the sample back into the solution. No more than 10 min will be necessary to thoroughly reduce your proteins.
5. Best digestion will occur when each filter contains <200 μg of protein. Use multiple FASP chambers if necessary. If the protein is dilute, concentrate it in the chamber by adding up to 500 μL of protein at a time and centrifuging at 13,000 × *g* for 5-min intervals prior to beginning the digestion steps.
6. If using an enzyme other than trypsin replace the 50 mM NH_4HCO_3 solution in Subheading 3.3, steps 10–13, with the appropriate buffer for your enzyme as shown in Table 1 (7).
7. If you used an enzyme other than trypsin, dry your peptides thoroughly with vacuum centrifugation and resuspend in 50 mM NH_4HCO_3 and pH to 7.4 for optimal binding to the antibody.
8. As the 4 G10 antibody almost exclusively captures peptides with phosphotyrosine residues, further enrichment will be necessary to capture other phosphopeptides. Both titanium dioxide (TiO_2) and immobilized metal affinity chromatography

(IMAC) are useful technologies for the enrichment of phosphopeptides. We perform both enrichments due to the fact that TiO_2 preferentially enriches peptides with single phosphorylation sites, while IMAC is superior for enrichment of peptides with multiple phosphate residues (8). If performing both TiO_2 and IMAC enrichment, it will be best to perform TiO_2 first followed by IMAC enrichment of the flowthrough.

9. Since graphite spin columns often capture peptides that are too hydrophilic for C-18 ZipTips, we often perform ZipTip desalting on phosphopeptide samples followed by graphite desalting of the remaining solution containing the unbound peptides.
10. ZipTip fractionation is appropriate when the complexity of your samples is high, but the concentration of your proteins is relatively low. Since the maximum binding capacity of a C-18 ZipTip is 5 μg, larger amounts of peptides can be fractionated by this method by repeating the procedure and pooling similar elutions.
11. OFFGEL fractionation is most appropriate when you are separating a high concentration of complex sample. While the maximum limit of the machine is reported to be >3 mg, the best separations occur when peptides are loaded in the range of 50–1,000 μg.
12. When processing SILAC samples that have been enriched for phosphopeptides, a large number of your quantization values may be out of your dynamic range due to the fact that many phosphorylation events are essentially "ON/OFF" switches. This fact is not commonly integrated into SILAC processing software. In order to identify these highly differentially regulated biomarkers, allow missing channels to be set at the minimum signal intensity and also allow ratios of above maximum cutoffs. In Proteome Discoverer or Bioworks (Thermo), these phosphorylation events will be given ratios of 1,000-fold. Failure to change these settings will result in loss of these important qualitative phosphorylation events. Alternatively, you may wish to process each set of samples two ways: with and without the SILAC labels set as static modifications. Comparison of the lists of identified peptides will help flush out these extreme phosphorylation events.

References

1. Mann M (2006) Functional and quantitative proteomics using SILAC. Nat Rev Mol Cell Biol 7:952–958
2. Cuomo A, Moretti S, Minucci S, Bonaldi T (2011) SILAC-based proteomic analysis to dissect the "histone modification signature" of human breast cancer cells. Amino Acids 41:387–399
3. Everley PA, Krijgsveld J, Zetter BR, Gygi SP (2004) Quantitative cancer proteomics: stable isotope labeling with amino acids in cell culture (SILAC) as a tool for prostate cancer research. Mol Cell Proteomics 3:729–735
4. Matta A, Ralhan R, DeSouza LV, Siu KW (2010) Mass spectrometry-based clinical proteomics:

head-and-neck cancer biomarkers and drug-targets discovery. Mass Spectrom Rev 29:945–961

5. Gronborg M, Kristiansen TZ, Iwahori A et al (2006) Biomarker discovery from pancreatic cancer secretome using a differential proteomic approach. Mol Cell Proteomics 5:157–171
6. Lee HJ, Na K, Kwon MS, Kim H, Kim KS, Paik YK (2009) Quantitative analysis of phosphopeptides in search of the disease biomarker from the hepatocellular carcinoma specimen. Proteomics 9:3395–3408
7. Orsburn BC (2011) The handy little book of proteomics protocols. BTB, Baltimore, MD
8. Orsburn BC, Stockwin LH, Newton DL (2011) Challenges in plasma membrane phosphoproteomics. Expert Rev Proteomics 8:483–494

Chapter 12

Trypsin-Mediated $^{18}O/^{16}O$ Labeling for Biomarker Discovery

Xiaoying Ye, King C. Chan, DaRue A. Prieto, Brian T. Luke, Donald J. Johann Jr., Luke H. Stockwin, Dianne L. Newton, and Josip Blonder

Abstract

Differential $^{18}O/^{16}O$ stable isotopic labeling that relies on post-digestion ^{18}O exchange is a simple and efficient method for the relative quantitation of proteins in complex mixtures. This method incorporates two ^{18}O atoms onto the C-termini of proteolytic peptides resulting in a 4 Da mass-tag difference between ^{18}O- and ^{16}O-labeled peptides. This allows for wide-range relative quantitation of proteins in complex mixtures using shotgun proteomics. Because of minimal sample consumption and unrestricted peptide tagging, the post-digestion ^{18}O exchange is suitable for labeling of low-abundance membrane proteins enriched from cancer cell lines or clinical specimens, including tissues and body fluids. This chapter describes a protocol that applies post-digestion ^{18}O labeling to elucidate putative endogenous tumor hypoxia markers in the plasma membrane fraction enriched from a hypoxia-adapted malignant melanoma cell line. Plasma membrane proteins from hypoxic and normoxic cells were differentially tagged using $^{18}O/^{16}O$ stable isotopic labeling. The initial tryptic digestion and solubilization of membrane proteins were carried out in a buffer containing 60 % methanol followed by post-digestion ^{18}O exchange/labeling in buffered 20 % methanol. The differentially labeled peptides were mixed in a 1:1 ratio and fractionated using off-line strong cation exchange (SCX) liquid chromatography followed by on-line reversed-phase nano-flow RPLC-MS identification and quantitation of peptides/proteins in respective SCX fractions. The present protocol illustrates the utility of $^{18}O/^{16}O$ stable isotope labeling in the context of quantitative shotgun proteomics that provides a basis for the discovery of hypoxia-induced membrane protein markers in malignant melanoma cell lines.

Key words $^{18}O/^{16}O$ stable isotope labeling, Hypoxia biomarker, Melanoma, Quantitative shotgun proteomics, Mass spectrometry

1 Introduction

The Bence Jones protein was the first cancer diagnostic protein biomarker for multiple myeloma described by the English physician Henry Bence Jones in 1847. Since then, fewer than ten proteins have been approved by the FDA for cancer diagnosis indicating

Ming Zhou and Timothy Veenstra (eds.), *Proteomics for Biomarker Discovery: Methods and Protocols*, Methods in Molecular Biology, vol. 1002, DOI 10.1007/978-1-62703-360-2_12, © Springer Science+Business Media, LLC 2013

that we are still at the beginning of a long journey in the search for effective protein-based diagnostic cancer biomarkers. Advances in mass spectrometry (MS)-based proteomics provide promises to quicken the pace in the quest for biomarker development by facilitating method development for discovering protein biomarkers in a high-throughput fashion. Among different MS techniques, global quantitative shotgun proteomics has become a crucial component in biomarker research (1, 2).

Global quantitative profiling of bio-specimens by MS can be achieved using stable isotopic labeling or label-free methods (3). Stable isotopic labeling approaches provide a more accurate measurement of the differences in protein abundance since the control and treated sample/proteins are pooled, separated, and analyzed in a single run, thus preventing any interference that may result from the chromatographic separation or differences arising from ionization efficiency. Additionally, label-free methods are inexpensive and provide increased sensitivity of protein identification. Widely used isotopic labeling methods include isotope-code affinity tags (ICAT), isobaric peptide tags for relative and absolute quantification (iTRAQ), stable isotope labeling with amino acids in cell culture (SILAC), and $^{18}O/^{16}O$ labeling. Each of these methods has its advantages and limitations. $^{18}O/^{16}O$ labeling is unique because the introduction of isotopic tag is catalyzed by serine proteases such as trypsin, commonly used in bottom-up proteomic workflow. The major benefits are the absence of a chemical reaction that may result in the introduction of additional by-products into the clinical samples during the labeling reaction (4, 5). Other advantages, such as minimal sample consumption (low μg amounts), simple sample preparation, and applicability to limiting human clinical samples such as tissue biopsies and laser capture microdissected specimens, make $^{18}O/^{16}O$ labeling appealing in quantitative proteomics (6). Consequently, this technique has been used in the comparative profiling of complex protein mixtures in a wide range of biomarker studies (7–10).

We have applied the $^{18}O/^{16}O$ labeling approach to analyze plasma membrane proteins enriched from hypoxia-adapted malignant melanoma cells (11). Tumor hypoxia occurs when tumor cells are deprived of oxygen. It is well known that tumor hypoxia facilitates development of resistance to cancer chemotherapy and promotes an aggressive metastatic cancer phenotype (12). Typically, the assessment of tumor oxygenation status is measured using polarographic O_2 needle electrodes. However, technical limitations of this approach have required the development of "endogenous protein-based hypoxia markers" (13). Herein we describe a quantitative proteomic approach for the elucidation of hypoxia markers using differential $^{18}O/^{16}O$ stable isotopic labeling coupled with shotgun proteomics. The described workflow depicts a protocol that relies on post-digestion ^{18}O exchange/labeling of plasma membrane proteins isolated from a cancer cell line. This procedure can readily be applied to any complex protein mixture including tissue lysates, body fluids, or whole cell lysates.

2 Materials

Prepare all solutions using ultrapure water unless otherwise specified.

2.1 Cell Culture

1. Melanoma cell line B16F10 (Frederick National Laboratory for Cancer Research DCTD Tumor Repository, Frederick, MD) (see Note 1).
2. DMEM medium supplemented with 10 % heat-inactivated fetal bovine serum (FBS) and 100 U/mL of penicillin–streptomycin antibiotics.
3. Modular incubator chambers (Billups-Rothenberg, Del Mar, CA).

2.2. Plasma Membrane Isolation (see Note 2)

1. Hypotonic buffer: 10 mM HEPES, pH 7.5, 1.5 mM $MgCl_2$, 10 mM KCl, 1× complete protease inhibitor tablet, 1 mM sodium orthovanadate.
2. Prepared 37.2 % sucrose solution.
3. Dounce homogenizer (Kontes Glass Co., Vineland, NJ).
4. Centrifuge (Beckman Coulter, San Diego, CA).
5. SW28 rotor (Beckman Coulter, San Diego, CA).
6. 0.2 M sodium carbonate (Na_2CO_3), pH 11.

2.3 Solubilization and Tryptic Digestion of Membrane Proteins

1. Sequencing grade trypsin purchased from Promega (Madison, WI).
2. Tris 2-carboxyethylphosphine, Bond Breaker™, 0.5 M TCEP Solution, Neutral pH, from Pierce (Rockford, IL).
3. Bicinchoninic acid (BCA) protein assay reagent kit from Pierce (Rockford, IL).
4. Bovine serum albumin (BSA).
5. NanoDrop 1000 spectrophotometer (ThermoFisher Scientific, Wilmington, DE).
6. Lyophilization freeze dryer (ATR, Laurel, MD).

2.4 Post-digestion ^{18}O Exchange/Labeling

1. Sequencing grade trypsin (see Note 3).
2. $H_2{}^{18}O$ (95 % pure), from Cambridge Isotope Laboratories, Inc. (Andover, MA) (see Note 4).

2.5 Off-Line LC-SCX Fractionation

1. SCX-LC buffer A: 45 % (v/v) CH_3CN in H_2O.
2. SCX-LC buffer B: 45 % (v/v) CH_3CN containing 0.5 M ammonium formate (HCO_2NH_4), pH 3.

2.6 LC-MS/MS Analysis

1. Reversed-phase LC buffer A: 0.1 % formic acid (FA) in H_2O.
2. Reversed-phase LC buffer B: 0.1 % FA in ACN.

3 Methods

3.1 General Practice

The general experimental workflow is shown in Fig. 1. Tryptic digestion is widely used in the context of the bottom-up proteomics for peptide/protein identification. Hence, the ^{18}O post-digestion exchange/labeling takes advantage of trypsin's capability to incorporate a stable isotope mass-tag at the peptide's C-terminus. The 4 Da mass-tag provides a basis of relative quantitation using high-resolution and high-mass-accuracy MS.

3.2 Cell Growth and Hypoxia Treatment

1. Culture forty 75 cm^2 flasks of B16F10 cells to 75 % confluence in DMEM and 10 % FCS.
2. Divide the flasks into two groups.

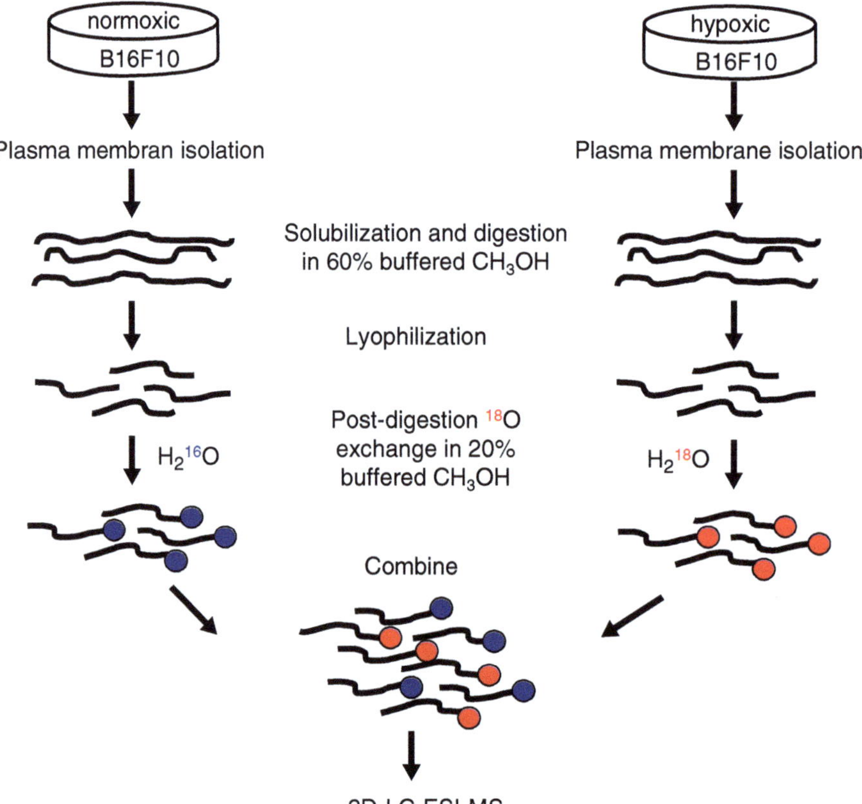

Fig. 1 Flow chart showing experimental design and sample preparation workflow for methanol-facilitated post-digestion ^{18}O exchange/labeling for shotgun quantitative proteomic profiling of the plasma membrane proteome enriched from normoxia- and hypoxia-adapted melanoma B16F10 cells

3. Subject one group to hypoxic conditions for 24 h. Induce hypoxia by exposing the cells to 1 % O_2 for 24 h using modular incubator chambers (see Note 5).
4. Culture the normoxic group under standard incubator conditions (5 % CO_2, 95 % air).

3.3 Plasma Membrane Isolation

1. After exposure, wash the cells three times with ice-cold PBS.
2. Scrape the cells into 20 mL of ice-cold hypotonic buffer.
3. Incubate the cell suspension on ice for 30 min to rupture cells.
4. Homogenize the cells with 50 passes in a dounce homogenizer.
5. Centrifuge at 1,000 × *g* for 10 min to remove intact cells or nuclei.
6. Layer the remaining supernatant onto 10 mL of 37.2 % sucrose solution.
7. Centrifuge at 96,467 × *g* in an SW28 rotor for 3 h at 4 °C.
8. Remove the white band of plasma membrane at the interface and resuspend it in ice-cold 0.2 M sodium carbonate, pH 11 (see Note 6).
9. After 30 min, centrifuge the suspension at 96,467 × *g* in an SW28 rotor for 1.5 h at 4 °C.
10. Store the membrane pellet at −80 °C until required.

3.4 Solubilization and Tryptic Digestion of Membrane Proteins

1. Disperse membrane pellet/sample in 1 mL of 50 mM NH_4HCO_3, pH 7.9.
2. Determine protein concentration using the BCA assay with BSA in 50 mM NH_4HCO_3 as a reference (see Note 7).
3. Aliquot equal amount (375 μg each) of normoxic and hypoxic samples.
4. Reduce proteins by adding TCEP Bond Breaker (1 mM final concentration) and incubate for 30 min at 37 °C with shaking.
5. Alkylate proteins by adding iodoacetamide (final concentration 15 mM) and incubate for 30 min at 37 °C with shaking.
6. Centrifuge at 100,000 × *g* for 1 h and wash the membranes pellets with ice-cold ultrapure water twice.
7. Re-solubilize the membrane pellet separately in 500 μL of 60 % (v/v) methanol in 40 % (v/v) 25 mM NH_4HCO_3, pH 7.9, facilitated by intermittent sonication. Five to ten cycles are usually sufficient to achieve solubilization (see Note 8).
8. Digest separately for 6 h using 1:20 trypsin-to-protein ratio at 37 °C with shaking.
9. Lyophilize both samples separately to dryness.

Table 1
Gradient setup for SCX-LC fractionation

Time interval (min)	Gradient (%B)
0–5	2
5–55	2–20
55–80	20–49
80–86	40–100
86–96	100

3.5 Differential $^{18}O/^{16}O$ Stable Isotope Labeling Using Post-digestion ^{18}O Exchange (See Note 9)

1. Make 1 mL of 2 M NH_4HCO_3 stock solution in $H_2{}^{18}O$ by solubilizing 158.12 mg of NH_4HCO_3 in 1 mL of $H_2{}^{18}O$ (see Note 10).
2. Make 500 μL of 50 mM NH_4HCO_3 in $H_2{}^{18}O$ by adding 11.25 μL of 2 M stock solution to 488.75 μL of $H_2{}^{18}O$.
3. Dissolve the digest from hypoxia-adapted cells for ^{18}O labeling peptides in 500 μL of 20 % methanol/50 mM NH_4HCO_3 in $H_2{}^{18}O$ (v/v) employing intermittent sonication in a water bath (see Note 11).
4. Briefly spin the sample using a tabletop microcentrifuge.
5. Prepare trypsin by adding 40 μL of 50 mM NH_4HCO_3 in $H_2{}^{18}O$ to the vial containing 20 μg of lyophilized trypsin.
6. Add trypsin using a 1:20 trypsin:protein ratio. Briefly vortex, and then spin the sample using a microcentrifuge.
7. Incubate for 6 h at 37 °C with shaking.
8. Quench trypsin activity by boiling the sample for 3 min.
9. Cool the sample to room temperature and add 0.4 % TFA (final concentration) to ensure quenching of trypsin activity (see Note 12).
10. In parallel, apply the same procedure (steps 1–9) to the digest from normoxic cells, but substituting $H_2{}^{16}O$ for $H_2{}^{18}O$ as a solvent.
11. Combine the samples.
12. Desalt the combined samples using C18 solid phase extraction column in accordance to a manufacturer instructions.
13. Lyophilize the sample after desalination.

3.6 SCX-LC Fractionation

1. Solubilize the lyophilized sample in 200 μL 45 % (v/v) ACN containing 0.1 % (v/v) FA.
2. Remove an aliquot of ~5–10 μg for screening SCX pre-run and adjust the gradient in accordance to pre-run (see Note 13).

Table 2
Gradient setup for nano-flow reversed-phase liquid chromatography (nanoRPLC)

Time interval (min)	Gradient (%B)
0–100	2–60
100–130	60–98
130–150	98

3. Load the sample onto the SCX column, set the flow rate to 50 μL/min, and use the adjusted SCX gradient as shown in Table 1.
4. Collect 96 fractions, one fraction every minute.
5. Pool into fractions in accordance to SCX chromatogram peak density and lyophilize to dryness (see Note 14).

3.7 Nano-Flow Reversed-Phase Liquid Chromatography (nanoRPLC)–Tandem Mass Spectrometry (MS/MS) Analysis

1. Dissolve each fraction in 25 μL 0.1 % formic acid in water and load 5 μL of the resulting peptide solution on a nanoRPLC column coupled online with an ion-trap MS/MS via electro spray ionization.
2. After injection wash the column for 30 min with 2 % B at 0.5 μL/min.
3. Elute peptides using optimized reversed-phase gradient as shown in Table 2.
4. Re-equilibrate the column with 2 % B for 30 min prior to subsequent sample loading at 0.5 μL/min.
5. Operate the MS instrument in a data-dependent mode where the five most intense ions detected in each FTICR-MS scan (*m/z* 200–2,000) (see Note 15) were selected for MS/MS in the ion trap (precursor selection from *m/z* 400 to *m/z* 2,000). A normalized collision energy of 36 % was employed for collision-induced dissociation (CID) along with dynamic exclusion of 90 s to reduce redundant selection of peptides for CID. The ESI voltage and the heated capillary temperature were set at 1.6 kV and 160 °C, respectively.

3.8 Data Analysis

1. Download a nonredundant mouse proteome database (http://www.ebi.ac.uk/integr8/EBI-Integr8-HomePage.do).
2. Search acquired mass spectra employing SEQUEST on a Beowulf 20-node parallel virtual machine cluster computer (ThermoElectron, San Jose, CA) against the nonredundant mouse database. Use 25 ppm mass tolerance for precursor ion and 0.5 Da for fragment ions. Include in the search the vari-

able modifications for methionine oxidation, cysteine carbamidomethylation, and +4.008 Da for ^{18}O-labeled C-termini.

3. Only consider fully tryptic peptides with up to two miscleavages as positively identified. Filter the identified peptides by delta correlation scores ((ΔCn) ≥ 0.08) and charge-state-dependent cross-correlation (Xcorr) criteria (≥1.9 for $[M+H]^{+1}$ peptides, ≥2.2 for $[M+2H]^{+2}$ peptides, and ≥3.1 for $[M+3H]^{+3}$ peptides).
4. Assess the false-positive rate of peptide identification by running the search against reversed mouse database using the same SEQUEST filtering criteria (Xcorr and ΔCn) (see Note 16).
5. Calculate heavy-to-light (i.e., $^{18}O/^{16}O$) peptide ratios using XPRESS algorithm implemented in the BioWorks™ package (Version 3.2) from ThermoElectron. To be consistent with precursor ion tolerance used in peptide identification step, set the parameters using the mass tolerance to 25 ppm, minimum threshold to 50,000, and the number of smoothing points to 5.
6. Calculate the average ratios and standard deviations of all positively identified proteins using peptides unique to each specific protein (see Note 17).
7. Manually evaluate peptide identification list, sequence coverage, MS/MS spectra, observed $^{18}O/^{16}O$ peptide ratios, and extracted ion chromatograms for peptides/proteins of interest (see Note 18). For peptides showing variable rate of ^{18}O incorporation consider employing an algorithm capable to accurately calculate $^{18}O/^{16}O$ ratios and eliminate artifacts caused by variable ^{18}O exchange (see Note 19).
8. Extract proteins with ratios greater than 1.71 or less than 0.58 (see Note 20).
9. Evaluate biological relevance using a variety of public available tools such as Ingenuity pathway, DAVID, and GeneGo (see Note 21).
10. Validate abundance change of proteins of interest using a complementary technique such as real-time PCR or western blot analysis (see Note 22).

4 Notes

1. Previous work indicated that the murine melanoma cell line B16F10 undergoes a profound in vitro response to hypoxia characterized by increased activity of the VEGF-A/KDR autocrine loop and enhanced cord formation on matrigel. Also, the B16F10 cells display vasculogenic mimicry when exposed to hypoxia, an in vivo hypoxia-driven phenomenon where blood

vessels are formed without endothelial cell involvement. Importantly, the aggressive behavior of B16F10 xenograft models is optimal for in vivo targeted validation. These combined properties were central to selecting B16F10 for proteome analysis (14).

2. The primary interface between the cell and the hypoxic environment is the plasma membrane. Thus an investigation of the plasma membrane sub-proteome is expected to identify potential tumor hypoxia markers. For example, the ubiquitous hypoxia marker, GLUT-1, is located at the plasma membrane. Additionally, hypoxia-inducible factor (HIF) activates apoptosis by inducing insulin-like growth factor-1 (IGF-1) signaling through the IGF-1 receptor (IGF-1R) also found at the plasma membrane (12). Therefore, this study targets the plasma membrane sub-proteome in the quest of potential hypoxia biomarkers.
3. In addition to free/unbound trypsin solution, immobilized trypsin has been proposed for ^{18}O labeling to prevent ^{18}O back-exchange and increase enzyme/substrate ratio (15, 16). Although effective for ordinary sized samples, immobilized trypsin is not suitable for digestion of amount-limited membrane protein specimens.
4. More expensive $H_2{}^{18}O$ of 99 % purity is also available from Cambridge Isotope Laboratories. We observed that the use of $H_2{}^{18}O$ of 99 % purity improves the homogeneity of ^{18}O exchange, although it cannot completely eliminate incomplete ^{18}O incorporation caused by differential enzyme affinity/dynamics observed for different amino acid residues.
5. It is important in biomarker research to establish/confirm the model and cross-validate significant MS results, which is crucial for correct data interpretation. In this study, the optimum conditions for proteome analysis were defined using several assays capable of assessing hypoxic conditions that include semiquantitative RT-PCR, supernatant ELISA, and pimonidazole hydrochloride reaction that form adducts visualized by immunocytochemistry using the 1Mab1 monoclonal antibody.
6. The extent of plasma membrane enrichment was assessed by comparing the crude B16F10 lysate with the sucrose gradient-purified material using western blotting. Equal amounts (20 μg) of protein were separated by SDS-PAGE, blotted onto PVDF membrane, and then probed with antibodies specific for antigens restricted to subcellular compartments including β-actin (cytosolic), cytochrome c (mitochondrial/cytosolic), GAPDH (cytosolic), PCNA (nucleus), G-58k (Golgi), and transferrin receptor TFR-1 (plasma membrane). Significant increases in TFR-1 reactivity after purification with a corresponding decrease in G-58K, GAPDH, cytochrome c, and β-actin confirmed plasma membrane enrichment.

7. In this experimental context, to accomplish the most accurate quantization, it is critical to precisely assay the protein content of normoxic and hypoxic sample before digestion and adjust their concentration accordingly.
8. The pivotal step in successful analysis of membrane proteins is their effective solubilization. The use of methanol-based buffers in the context of the post-digestion ^{18}O exchange/labeling eliminates the need for detergents or chaotropes that interfere with LC separations and peptide ionization. We have shown that the buffer containing 60 % MeOH is efficient in solubilizing membrane proteins for MS-based proteomics (17, 18). To enhance solubilization of amount-limited clinical specimens a MS-friendly acid-cleavable detergent (i.e., 0.1 % PPS) can be added to 60 % methanol (19).
9. During the post-digestion ^{18}O exchange, the ^{18}O labeling is decoupled from protein digestion step. This allows for targeted optimization of the ^{18}O exchange and minimizes $H_2{}^{18}O$ consumption (20).
10. It is important to keep all stock solutions/buffers containing the $H_2{}^{18}O$ hermetically closed at 4 °C in a desiccator.
11. We perform post-digestion ^{18}O exchange/labeling in 20 % (v/v) methanol buffer to enhance the homogeneity of ^{18}O incorporation (21). We have shown increased trypsin activity in the buffer containing 20 % (*v*/*v*) methanol compared to pure aqueous buffer using α-*N*-benzoyl-l-arginine ethyl ester (BAEE) assay (18). This finding is in agreement with observations reported by others, demonstrating the capability of mixed organic–aqueous solvents to increase the activity of trypsin/enzyme when compared to pure aqueous buffers (22).
12. It has been previously shown that heating followed by acidification is sufficient to completely quench residual trypsin activity and prevent oxygen back-exchange in ^{18}O-labeled samples (23). To eliminate methanol from digestion buffer it is important to make a hole on the tube cap prior to boiling step and facilitate transfer of methanol from the liquid phase to vapor phase (i.e., evaporate) that would take place during heating process.
13. This step is important for optimization/correction of the fractionation gradient since the optimal gradient may vary from sample to sample. Also, SCX columns prepared from different batches may behave differently. The effective screening analysis of 5–10 μg of peptide sample is attainable using an UV laser-induced fluorescence detector (24).
14. The total number of pooled SCX fractions for reverse phase LC is dependent on sample complexity and experimental goal. Fractions were pooled after base-peak chromatogram inspection based on peak density.

15. The use of high-resolution and high-mass-measurement-accuracy MS instrumentation (i.e., Fourier transform ion cyclotron resonance (FTICR) or Orbitrap) for MS scanning is desirable as it allows a detailed inspection of MS spectra, including respective isotopic manifolds depicting light and heavy peptide isotopomer, showing the difference of 2 Da for doubly charged peptide ion pair (see Figs. 2a and 3a).
16. False-positive analysis of identified peptides derived a false-positive rate of 3.25 % in this study. A false-positive rate of ≤5 % is desirable in global quantitative proteomic studies. If a high false-positive rate is detected, consider use of more stringent criteria to filter out low-quality mass spectra for peptide identification.
17. In proteomics a significant number of detected peptides do not uniquely identify only one protein within large mammalian proteome databases due to the numerous protein isoforms and relatively small size of identified tryptic peptides, which occasionally do not possess sufficiently divergent sequences to specifically identify a single protein. To accomplish accurate comparative quantitation at the protein level, it is important to exclude these peptides from quantitative analyses.
18. Figure 2b depicts unambiguous identification of the K.LAADEEENADNNMK.A ^{16}O-labeled peptide from a plasma membrane marker, transferrin receptor TFR-1, while Fig. 2c depicts unambiguous identification of the same ^{18}O-labeled isotopomer exemplified in 4 Da difference between y-fragment ions of light and heavy isotopomer, confirming exchange of both oxygens at the C-terminus.
19. The XPRESS software simply divides the peak area exclusively from $^{18}O_2$ dual-tagged peptide isotopomers over that of $^{16}O_2$ naturally tagged peptide species; thus the contribution of the mono-labeled peptides is overlooked. When the presence of these $^{18}O_1$ mono-labeled isotopomers is not taken into consideration for the calculations of $^{18}O/^{16}O$ ratios, a variable degree of underestimation of the peptide/protein ratio is to be expected. Hence, the average $^{18}O/^{16}O$ ratios obtained by the XPRESS are smaller than the expected real values in this scenario. This may cause significant errors especially when peptides/proteins have high heavy-to-light ratios. To address this issue we have developed an algorithm to correct for variable $^{18}O_2$ incorporation, which determines the efficiency of ^{18}O exchange reaction and recalculates $^{18}O/^{16}O$ ratios for peptides exhibiting variable ^{18}O exchange (25).
20. It has been determined that abundance ratios greater than 1.71 and less than 0.58 are the statistically significant limits using a nonlinear least-squares regression applied to a manually curated dataset. According to these criteria, in the dataset of 2,433 proteins quantified by at least two tryptic peptides in this study,

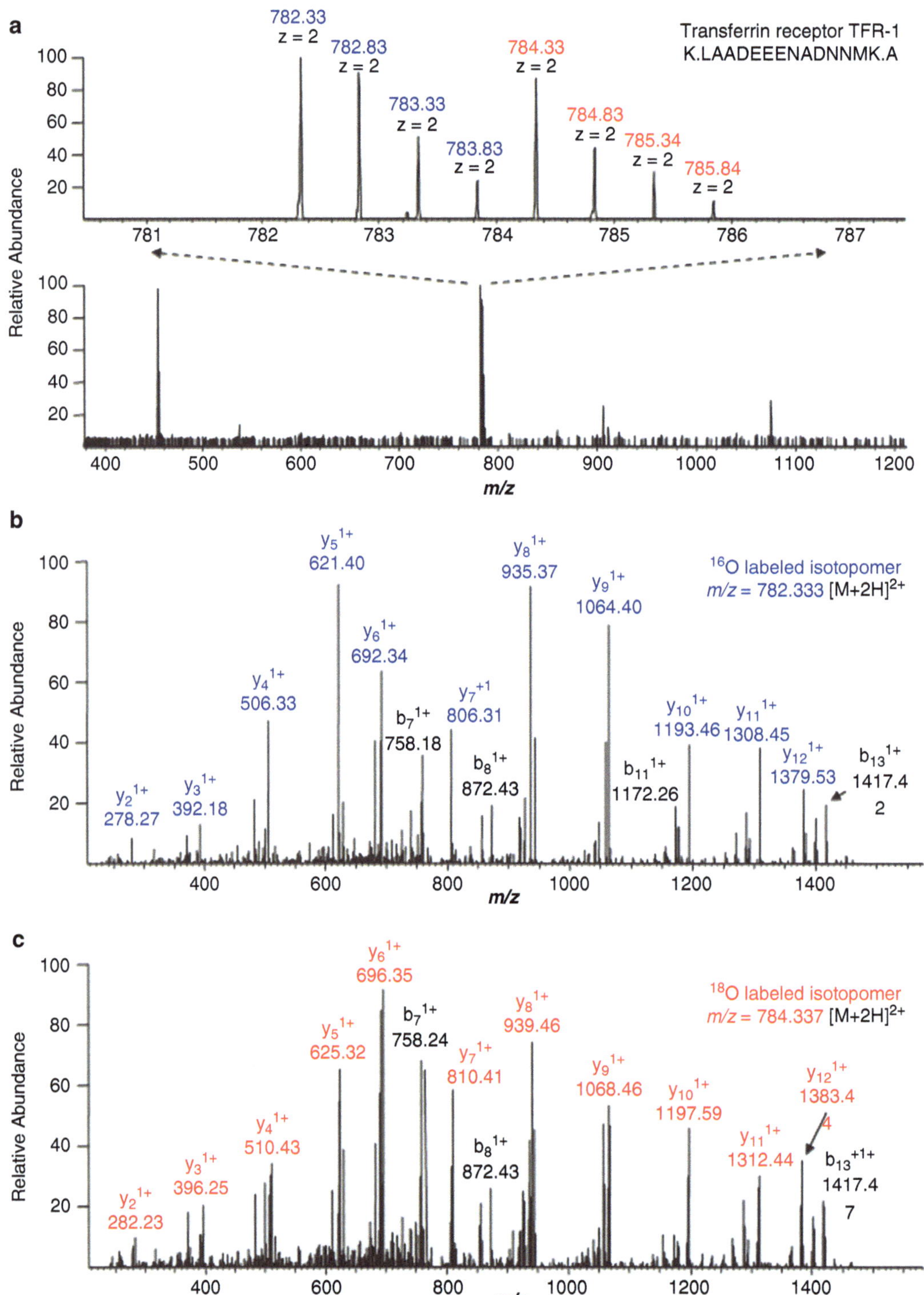

Fig. 2 Selected mass spectra of differentially labeled peptide isotopomers from a plasma membrane marker, transferrin receptor TFR-1: (**a**) zoomed portion of the FTICR mass spectrum of the $^{16}O/^{18}O$ differentially labeled doubly charged $[M+2H]^{2+}$ molecular ion pair (light, *m/z* 782.333; heavy, *m/z* 784.337) showing a typical difference of 2 *m/z*; (**b**, **c**) MS/MS fragmentation ion series of (**b**) light (*blue font*) ^{16}O-labeled and (**c**) heavy (*red font*) ^{18}O-labeled isotopomeric peptide showing typical 4 Da mass shifts between singly charged y-ion fragments, confirming ^{18}O-tagg at the C-terminus

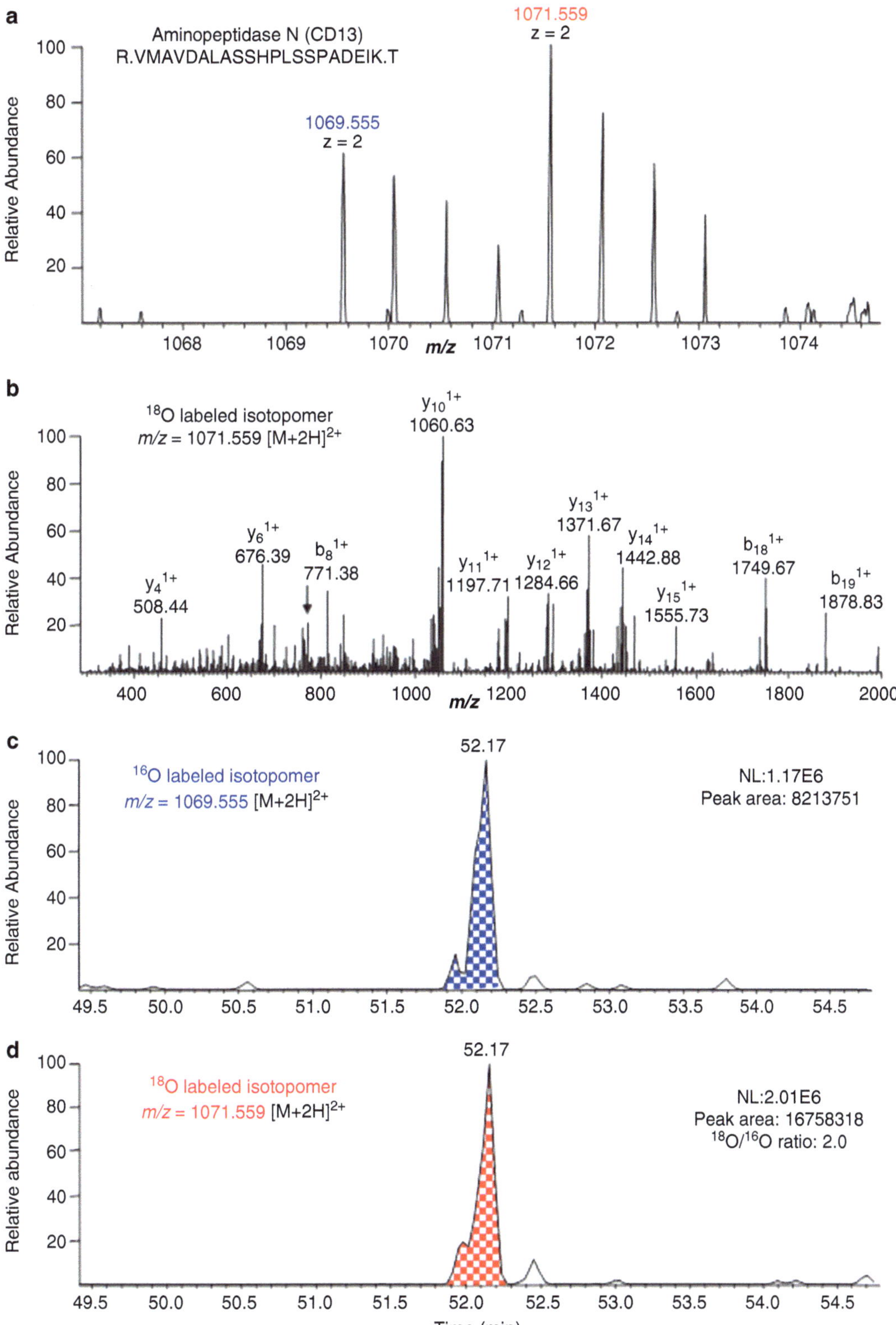

Fig. 3 Mass-spectrometry-based identification and quantitation of a selected peptide from aminopeptidase N (CD13): (**a**) zoomed portion of the FTICR mass spectrum of the $^{16}O/^{18}O$-labeled doubly charged $[M+2H]^{2+}$ molecular ion pair (light, *m/z* 1,069.555; heavy, *m/z* 1,071.559); (**b**) MS/MS fragmentation spectrum of heavy ^{18}O-labeled isotopomeric peptide R.VMAVDALASSHPLSSPADEIK.T; (**c**, **d**) extracted ion chromatogram of the light isotopomeric peptide (**c**) and heavy ^{18}O-labeled isotopomeric peptide (**d**) of the molecular ion pair shown in (**a**), indicating a higher relative abundance of the ^{18}O-labeled peptide ($^{18}O/^{16}O$ ratio 2.0)

Table 3
Increased relative concentration of the aminopeptidase N (CD13) observed in hypoxia-adapted B16F10 cells

Gene name	Protein	Peptide count	Mean	Standard deviation	Accession
Anpep	Aminopeptidase N	34	2.05	0.77	P97449

292 proteins showed increased expression, whereas 580 had decreased levels after hypoxia.

21. Quantitative proteomics is a powerful tool in biomarker research. However, the benefit to generate a large amount of information also makes it a liability in the search for cancer biomarkers. This step is thus critical to sift through proteomic data for important information. For instance, Fig. 3 and Tables 3 and 4 illustrate identification and quantitation of aminopeptidase N (CD13), showing a higher concentration of this gene product within the hypoxic plasma membrane proteome and its possible up-regulation in response to hypoxia. CD13 is involved in digestion, chemokine regulation, antigen presentation, and angiogenesis. Expression of APN/CD13 has been previously documented in melanoma and the aminoprotease inhibitor bestatin has been found to prevent melanoma angiogenesis in an APN/CD13-dependent fashion (26, 27). In light of this evidence, we speculate that the upregulation of CD13 is of considerable interest and subsequently put this protein on the list for further validation.
22. In multidimensional quantitative proteomics, logistical challenges in terms of sample quantity and instrument access often limit the number of experimental replicates that can be performed (21). This fact makes validation using orthogonal biological techniques essential. This is exemplified in the follow up study (28), validating previously observed changes in CD44 relative concentration using global 18O/16O labeling (21). The follow up study confirmed CD44 as principal mediator of the intoxication by clostridial iota-family toxins (28).

Acknowledgments

This project has been funded in whole or in part with federal funds from the National Cancer Institute, National Institutes of Health, under Contracts HHSN261200800001E and NO1-CO-12400. The content of this publication does not necessarily reflect the views or policies of the Department of Health and Human Services, nor does mention of trade names, commercial products, or organization imply endorsement by the United States Government.

Table 4
List of all heavy and light labeled peptides contributing to its quantification

ID	Peptide	Ratio	M + H	*z*	Xcorr	ΔCn
1	K.AVNQQTAVQPPATVR.T	2.96	1,579.8	2	3.63	0.48
2	K.AVNQQTAVQPPATVR[.T	2.51	1,583.8	2	4.32	0.45
3	K.DLMVLNDVYR.V	1.33	1,237.6	2	2.45	0.39
4	K.DLMVLNDVYR.V	1.33	1,237.6	2	3.34	0.45
5	K.ESKPYPEDPSC*TMTEFHSTPK.M	2.58	2,468	3	3.98	0.55
6	K.HFLLDSK[.S	2.97	863.47	1	2.10	0.16
7	K.LFENYGGGSFSFANLIQGVTR.R	2.44	2,277.1	2	5.67	0.66
8	K.LFENYGGGSFSFANLIQGVTR.R	1.99	2,277.1	2	5.28	0.63
9	K.LQNQLQTDLSVIPVINR.A	3.71	1,951	2	5.07	0.51
10	K.LQNQLQTDLSVIPVINR[.A	3.99	1,955.1	2	4.58	0.46
11	K.LQNQLQTDLSVIPVINR[.A	2.87	1,955.1	2	4.75	0.43
12	K.MIPITLALDNTLFLVK.E	2.71	1,802	3	3.77	0.47
13	K.MIPITLALDNTLFLVK.E	2.22	1,802	2	3.46	0.49
14	K.MIPITLALDNTLFLVK[.E	2.73	1,806	2	2.79	0.36
15	K.PYPEDPSC*TMTEFHSTPK[.M	2.21	2,127.9	2	4.41	0.49
16	K.TPDQIMELFDSITYSK.G	0.99	1,887.9	2	5.63	0.57
17	K.TPDQIMELFDSITYSK.G	1.68	1,887.9	2	5.12	0.57
18	R.ALEQALEK[.T	2.51	905.5	2	2.51	0.34
19	R.AQIIHDSFNLASAK.M	1.53	1,514.7	2	3.66	0.62
20	R.AQIIHDSFNLASAK[.M	1.52	1,518.8	2	3.22	0.42
21	R.ESSLVFDSQSSSISNK.E	1.99	1,714.8	2	4.35	0.62
22	R.ESSLVFDSQSSSISNK[.E	2.15	1,718.8	2	3.82	0.44
23	R.FSSEFELQQLEQFK.A	2.07	1,759.8	2	3.57	0.38
24	R.FSSEFELQQLEQFK.A	2.77	1,759.8	2	4.95	0.55
25	R.FSSEFELQQLEQFK.A	2.03	1,759.8	2	4.52	0.55
26	R.FSSEFELQQLEQFK[.A	1.02	1,763.8	2	4.96	0.43
27	R.FSSEFELQQLEQFK[.A	2.15	1,763.8	2	4.24	0.39
28	R.MLSSFLTEDLFK.K	1.61	1,430.7	2	2.95	0.48
29	R.TLDGTPAPNIDK.T	1.43	1,241.6	2	2.69	0.52
30	R.TLDGTPAPNIDKTELVER.T	1.23	1,969	2	3.40	0.51
31	R.VMAVDALASSHPLSSPADEIK.T	1.34	2,138	2	2.97	0.56
32	R.VMAVDALASSHPLSSPADEIK.T	0.87	2,138	2	4.84	0.51
33	R.VMAVDALASSHPLSSPADEIK[.T	1.24	2,142	2	4.17	0.54
34	R.VMAVDALASSHPLSSPADEIK[.T	1.15	2,142	2	4.37	0.50

Brackets at the C-terminus denote MS identified heavy ^{18}O tag

References

1. Issaq HJ, Waybright TJ, Veenstra TD (2011) Cancer biomarker discovery: opportunities and pitfalls in analytical methods. Electrophoresis 32:967–975
2. Cox J, Mann M (2011) Quantitative, high-resolution proteomics for data-driven systems biology. Annu Rev Biochem 80:273–299
3. Veenstra TD (2007) Global and targeted quantitative proteomics for biomarker discovery. J Chromatogr B Analyt Technol Biomed Life Sci 847:3–11
4. Desiderio DM, Kai M (1983) Preparation of stable isotope-incorporated peptide internal standards for field desorption mass spectrometry quantification of peptides in biologic tissue. Biomed Mass Spectrom 10:471–479
5. Yao X, Freas A, Ramirez J, Demirev PA, Fenselau C (2001) Proteolytic ^{18}O labeling for comparative proteomics: model studies with two serotypes of adenovirus. Anal Chem 73:2836–2842
6. Ye X, Luke B, Andresson T, Blonder J (2009) ^{18}O stable isotope labeling in MS-based proteomics. Brief Funct Genomic Proteomic 8:136–144
7. Gao Y, Gopee NV, Howard PC, Yu LR (2011) Proteomic analysis of early response lymph node proteins in mice treated with titanium dioxide nanoparticles. J Proteomics 74:2745–2759
8. Loftheim H, Asberg A, Reubsaet L (2010) Accelerated ^{18}O-labeling in urinary proteomics. J Chromatogr A 1217:8241–8248
9. Qian WJ, Petritis BO, Kaushal A et al (2010) Plasma proteome response to severe burn injury revealed by ^{18}O-labeled "universal" reference-based quantitative proteomics. J Proteome Res 9:4779–4789
10. Qian WJ, Monroe ME, Liu T et al (2005) Quantitative proteome analysis of human plasma following in vivo lipopolysaccharide administration using ^{16}O/^{18}O labeling and the accurate mass and time tag approach. Mol Cell Proteomics 4:700–709
11. Stockwin LH, Blonder J, Bumke MA et al (2006) Proteomic analysis of plasma membrane from hypoxia-adapted malignant melanoma. J Proteome Res 5:2996–3007
12. Harris AL (2002) Hypoxia–a key regulatory factor in tumour growth. Nat Rev Cancer 2:38–47
13. Vaupel P, Mayer A (2007) Hypoxia in cancer: significance and impact on clinical outcome. Cancer Metastasis Rev 26:225–239
14. Rybak SM, Sanovich E, Hollingshead MG et al (2003) "Vasocrine" formation of tumor cell-lined vascular spaces: implications for rational design of antiangiogenic therapies. Cancer Res 63:2812–2819
15. Lopez-Ferrer D, Hixson KK, Smallwood H, Squier TC, Petritis K, Smith RD (2009) Evaluation of a high-intensity focused ultrasound-immobilized trypsin digestion and ^{18}O-labeling method for quantitative proteomics. Anal Chem 81:6272–6277
16. Sevinsky JR, Brown KJ, Cargile BJ, Bundy JL, Stephenson JL Jr (2007) Minimizing back exchange in ^{18}O/^{16}O quantitative proteomics experiments by incorporation of immobilized trypsin into the initial digestion step. Anal Chem 79:2158–2162
17. Blonder J, Hale ML, Lucas DA et al (2004) Proteomic analysis of detergent-resistant membrane rafts. Electrophoresis 25:1307–1318
18. Blonder J, Conrads TP, Yu LR et al (2004) A detergent- and cyanogen bromide-free method for integral membrane proteomics: application to Halobacterium purple membranes and the human epidermal membrane proteome. Proteomics 4:31–45
19. Ye X, Johann DJ Jr, Hakami RM et al (2009) Optimization of protein solubilization for the analysis of the CD14 human monocyte membrane proteome using LC-MS/MS. J Proteomics 73:112–122
20. Yao X, Afonso C, Fenselau C (2003) Dissection of proteolytic ^{18}O labeling: endoprotease-catalyzed ^{16}O-to-^{18}O exchange of truncated peptide substrates. J Proteome Res 2:147–152
21. Blonder J, Hale ML, Chan KC et al (2005) Quantitative profiling of the detergent-resistant membrane proteome of iota-b toxin induced vero cells. J Proteome Res 4:523–531
22. Klibanov AM (2001) Improving enzymes by using them in organic solvents. Nature 409:241–246
23. Storms HF, van der Heijden R, Tjaden UR, van der Greef J (2006) Considerations for proteolytic labeling-optimization of ^{18}O incorporation and prohibition of back-exchange. Rapid Commun Mass Spectrom 20:3491–3497
24. Chan KC, Muschik GM, Issaq HJ (2000) Solid-state UV laser-induced fluorescence detection in capillary electrophoresis. Electrophoresis 21:2062–2066
25. Ye X, Luke BT, Johann DJ Jr et al (2010) Optimized method for computing (18)O/(16)O ratios of differentially stable-isotope labeled peptides in the context of postdigestion (18)O exchange/labeling. Anal Chem 82:5878–5886
26. Aozuka Y, Koizumi K, Saitoh Y, Ueda Y, Sakurai H, Saiki I (2004) Anti-tumor angiogenesis

effect of aminopeptidase inhibitor bestatin against B16-BL6 melanoma cells orthotopically implanted into syngeneic mice. Cancer Lett 216:35–42

27. Saiki I, Fujii H, Yoneda J et al (1993) Role of aminopeptidase N (CD13) in tumor-cell invasion and extracellular matrix degradation. Int J Cancer 54:137–143

28. Wigelsworth DJ, Ruthel G, Schnell L, et al (2012) CD44 Promotes intoxication by the clostridial iota-family toxins. PLoS One 7:e51356

Chapter 13

Two-Dimensional SDS-PAGE Fractionation of Biological Samples for Biomarker Discovery

Thierry Rabilloud and Sarah Triboulet

Abstract

Two-dimensional electrophoresis is still a very valuable tool in proteomics, due to its reproducibility and its ability to analyze complete proteins. However, due to its sensitivity to dynamic range issues, its most suitable use in the frame of biomarker discovery is not on very complex fluids such as plasma, but rather on more proximal, simpler fluids such as CSF, urine, or secretome samples. Here, we describe the complete workflow for the analysis of such dilute samples by two-dimensional electrophoresis, starting from sample concentration, then the two-dimensional electrophoresis step per se, ending with the protein detection by fluorescence.

Key words 2D-PAGE, Fluorescence dyes, Image analysis, Sample preparation, Secretome, Biological fluids

1 Introduction

Proteomics in biomarker discovery suffers, maybe more acutely than other areas of proteomics, from two very important caveats. The first one is the necessity of analyzing rather large series of samples in order to take into account the interindividual variability, and this is a very important requisite for the robustness of the biomarkers. The second caveat lies in the definition of the correct sample. Plasma and serum seem at a first glance as natural choices, as many clinical assays are usually performed on these body fluids. However, the effect of their tremendous complexity on the efficiency of the biomarker discovery process has been very often underestimated, and applying standard proteomic techniques to such complex fluids is generally not an efficient process (1). Thus, more recent work has focused either on proximal fluids (2) such as CSF (3–7) or urine (8–10) or even samples of secreted proteins obtained from a relevant tissue or cell type model (11–15). In this case, the implicit rationale is that it will be much easier to find tissue leakage specific markers in the

Ming Zhou and Timothy Veenstra (eds.), *Proteomics for Biomarker Discovery: Methods and Protocols*, Methods in Molecular Biology, vol. 1002, DOI 10.1007/978-1-62703-360-2_13,

secreted proteins, where they will be much more enriched than in plasma. Then, the presence and specificity of these putative biomarkers can be assessed in the clinically relevant sample (generally a body fluid like serum or plasma) by specific techniques, either antibody-based (16) or mass spectrometry-based (17).

However, from a proteomic point of view, all these samples differ from serum and plasma in the fact that they are much more protein-poor and interfering substances-rich, and there will be a major issue to solve in sample preparation for proteomic studies. Then, once the sample has been correctly prepared, there is a second crucial choice, i.e., the proteomic platform that will be used for biomarker discovery. Put rather bluntly, there is no discovery proteomic platform today that combines sensitivity and reproducibility. 2D gel-based proteomics is by far the most reproducible platform in discovery proteomics, as exemplified by the much higher demands for gel-based publications in terms of sample numbers as compared to the demands for publishing shotgun proteomics experiments (18). However, 2D gel-based proteomics lacks sensitivity (19) and ignores some classes of proteins such as membrane proteins (20). Conversely, shotgun proteomics techniques are much more sensitive and are not biased against different protein classes. However, they cannot handle easily the large sample series required for biomarker discoveries.

As to the targeted proteomics techniques, such as SRM/MRM, they need first a target to search for, i.e., to have a predetermined analyte to search for. Moreover, when applied to complex body fluids such as plasma or serum, they are prone to artifacts resulting from spurious degradation of the peptides themselves (21–23).

In this context, 2D gel-based proteomics is still a choice to consider in a biomarker discovery setup, taking advantage of its high reproducibility and its ability to resolve intact proteins. However, because of its rather poor sensitivity, 2D gel-based proteomics must be applied on carefully selected samples (typically not serum or plasma), which must be then carefully handled and concentrated. It is therefore not surprising that one of the very few success stories in proteomic discovered biomarkers has been made by 2D gel electrophoresis on CSF (24), and has led to deep clinical investigation of the serendipitous biomarkers discovered in this way (the 14-3-3 proteins) in several brain pathologies (6, 25–27).

In addition, 2D gel-based proteomics offer unique capabilities when performing reverse probing, i.e., analyzing a patient immunological response to a pathogen to define pathogenomic profiles, as exemplified for candidiasis (28, 29).

Thus, this chapter will deal with two different aspects. The first one will be sample preparation, with a focus on sample cleaning and protein concentration, and the second will be the 2D gel technology itself, with a focus on detection methods.

2 Materials

All solutions are made in ultrapure water (>15 MΩ/cm).

2.1 Protein Precipitation/Resolubilization

1. NLS solution: this solution consists of 10 % (w/v) *N*-lauroyl sarcosinate in water. Dissolution is best obtained by using hot water. The solution must be kept at room temperature, but must be changed every 2 weeks.
2. TCA solution: this solution consists of 100 % (w/v) trichloroacetic acid in water. This can be either purchased ready-made (as a 6.1 M solution of trichloroacetic acid) or made by adding 450 ml of water to a 1 kg bottle of trichloroacetic acid (see Note 1).
3. Tetrahydrofuran, kept in the refrigerator.
4. Protein solubilization solution: this solution contains 7 M urea, 2 M thiourea, 4 % CHAPS (w/v), 20 mM spermine base, and 5 mM Tris(carboxyethyl)phosphine (see Note 2). This solution is best prepared by sonication in a sonicating bath. It can be kept for months frozen at −20 °C and thawed before use.
5. Carrier ampholytes, 3–10 range, sold as a 40 % (w/v) solution. Stable for months at +4 °C.

2.2 Isoelectric Focusing

1. IPG strips. They are available from several suppliers and in various pH ranges. Kept frozen at −20 °C. For resolution reasons, it is recommended to use long (>15 cm) strips.
2. IPG rehydration solution: this solution contains 7 M urea, 2 M thiourea, 4 % CHAPS (w/v), 0.4 % (w/v) carrier ampholytes, and 200 mM dithiodiethanol (see Note 3). This solution is best prepared by sonication in a sonicating bath. It can be kept for months frozen at −20 °C and thawed before use.
3. IPG rehydration chamber. This is usually a grooved methacrylate block where the widths and lengths of the grooves can accommodate the IPG strips.
4. IPG running device: various chambers exist from different suppliers to run IPG strips. Some combine the temperature control device, the power supply in a single device where the rehydration step and the running steps can be carried out sequentially without handling the strips in between. Older devices where the rehydration chamber, running chamber, cooling device, and power supply are separated can however be used with equal success, but they usually require some strip handling between rehydration and running.

2.3 Equilibration and SDS-PAGE

1. 20 % SDS solution: prepared by adding 200 g of pure SDS to 800 ml very hot water (>90 °C) under fast magnetic stirring. Is stable for months at room temperature, but SDS may crystallize

out if the temperature drops. Should crystallization occur, rewarm at 37 °C and shake until redissolved.

2. 60 % (v/v) glycerol solution. Stable for months at room temperature.
3. Stacking buffer: 120 g/l of Tris and 0.8 M HCl.
4. Standard gel buffer: 130 g/l of Tris and 0.6 M HCl (see Note 4).
5. Alternate gel buffer 1 for low molecular weight proteins: 150 g/l of Tris and 0.6 M HCl.
6. Alternate gel buffer 2 for high molecular weight proteins: 110 g/l of Tris and 0.6 M HCl.
7. Acrylamide solution: 30 % w/v acrylamide and 0.8 % w/v methylene bisacrylamide. Best to purchase ready-made and is stable at +4 °C for months.
8. Ammonium persulfate solution: 10 % w/v ammonium persulfate in water. Made fresh every week and kept at room temperature.
9. TEMED (tetramethylethylenediamine): use as pure liquid. Keep at room temperature and stable for months.
10. Water-saturated butanol. Mix equal volumes of water and 2-butanol. Shake well and let the phases separate. Store at room temperature.
11. Multi-gel casting chamber. This accessory usually comes as a bundle with multi-gel running cells (see below).
12. Multiplate gel running cell. Especially for biomarker discovery, it is recommended to run large series of gels in parallel. Several types of such multi-gel cells are commercially available. They must be coupled to very powerful cooling device and powerful power supplies (not necessary to develop more than 300–500 V, but 10–15 W/gel are necessary) (see Note 5).
13. Tank buffer: Tris 6 g/l, taurine 25 g/l, SDS 1 g/l.
14. Equilibration buffer: for 100 ml, mix 12.5 ml stacking buffer, 12.5 ml 20 % SDS, 50 ml 60 % glycerol, and 36 g urea. Make the day of use or at the earliest the day before. Requires about 30 min to dissolve at room temperature but must not be warmed for faster dissolution.
15. Sealing agarose: for 100 ml, weigh 1 g of low-melting agarose and add 12.5 ml stacking buffer, 2 ml 20 % SDS, and water up to 100 ml. Dissolve by heating to >90 °C (e.g., by short pulses in a microwave oven), and then add bromophenol blue (0.04 % w/v final concentration). Aliquot in 10 ml portions and keep at 4 °C.

2.4 Protein Detection and Image Processing

1. Either commercially available or homemade solutions can be used.
2. Recommended commercial stains are Flamingo from Bio-Rad and Krypton from Thermo-Fisher (30) (see Note 6).
3. For preparing the homemade fluorescent probe concentrate (31), prepare a solution containing 20 mM ruthenium chloride, 60 mM bathophenanthroline disulfonate, disodium salt, and 400 mM ammonium formate (preferably from a concentrated, titrated solution). Either reflux for 3 days or cook for 3 days in an oven set at 95 °C next to a beaker containing water to saturate the oven with water vapor and thus limit the losses in volume by evaporation. The solution should turn very deep orange red. Reconstitute at the initial volume with water, and keep in a bottle at 4 °C. Stable for months at this temperature.
4. For image acquisition, a laser scanner is ideal. Flamingo is best excited at 515 nm (emission at 535 nm), Krypton at 520 nm (emission at 580), and the ruthenium complex at 488 nm (emission at 600 nm). If laser scanners are not available, a chamber containing a 302 nm UV table, coupled with either a CCD camera or a simple digital camera and suitable filters (UV + yellow or orange to cut the purple emission of the UV tubes) can also be used. In any case, a spatial resolution of at least 10 pixels/mm is required.
5. For image analysis, dedicated software is indispensable. Second-generation software using image warping as the initial step are highly recommended.
6. Fixing solution 1 (Flamingo and Krypton stains): 10 % (v/v) acetic acid and 30 % (v/v) ethanol. Prepare just before use.
7. Fixing solution 2 (ruthenium complex stain): 1 % (v/v) of 85 % phosphoric acid and 30 % (v/v) ethanol. Prepare just before use.
8. Staining solution (ruthenium complex stain): 1 μM ruthenium complex in 1 % (v/v) of 85 % phosphoric acid and 30 % (v/v) ethanol. Prepare just before use.

3 Methods

3.1 Sample Preparation

1. The sample preparation described here is convenient for fluid samples (not tissue or cellular samples) that are relatively poor in their protein concentration (down to the microgram/ml range) and rather rich in interfering substances such as salts. It is derived from work carried out on secreted proteins (32). However, it must be noted that the samples must contain no detergent (see Note 7).

2. If the sample contains more than 0.1 M salt, dilute it with water to bring the salt concentration below this threshold. Cool the sample on ice, and add lauroyl sarcosinate to 0.1 % (w/v) final concentration. Mix well, and then add TCA to 7.5 % (w/v) final concentration. A voluminous white precipitate forms immediately.
3. Let the proteins precipitate for 2 h on ice, and then centrifuge at 10,000 × *g* at 4 °C for 10 min. During this time, cool some tetrahydrofuran on ice. Carefully remove the supernatant, and then add 1–2 ml of cold tetrahydrofuran on the voluminous white pellet. The pellet should literally dissolve in the solvent (or leave very fine particles if the starting amount of proteins is high).
4. Centrifuge once again at 10,000 × *g* at 4 °C for 10 min. Remove the supernatant. As the pellet is in many occasions completely invisible, it is highly advisable to mark the outside of the tubes in the centrifuge, i.e., where the pellet is. This avoids aspiration of the pellet when removing the supernatant.
5. Repeat the cold tetrahydrofuran wash once. Remove the solvent, and then let the tubes dry open at room temperature for 10–15 min (see Note 8).
6. Then add a minimal volume of protein solubilization solution (100–300 μl) and let the protein dissolve by sonication for 30 min in a sonication bath (see Note 9). Recover the solution, and measure the protein content by a Bradford-type assay. Then add carrier ampholytes to 0.4 % (w/v) final concentration and store the sample at −20 °C.

3.2 Isoelectric Focusing

1. Take the amount of sample required to have 200 μg of proteins (as read by a Bradford assay against BSA as a standard). Dilute the sample with IPG rehydration solution to reach the adequate strip rehydration volume (see Note 10). If the volume of added IPG rehydration solution represents less than 50 % of the final volume, add pure dithiodiethanol to reach a final concentration higher than 100 mM. Add some tracking dye (see Note 11) so that the solution is lightly colored.
2. Add each sample to a groove of the rehydration chamber, and then add the IPG strip (see Note 12). If necessary, cover with mineral oil to prevent evaporation of the water and urea crystallization.
3. Let rehydrate overnight (see Note 13) at room temperature.
4. Then if necessary, transfer the strips in the running chamber. Check the contacts with the electrodes. Run the IPG strips under constant voltage with low-voltage initial steps (see Note 14). A recommended running program is the following: 100 V for 1 h, then increase from 100 to 300 V over 15 min; then 300 V

for 3 h, then increase from 300 to 1,000 V over 1 h; then 1 h at 1,000 V, increase to 3,500 V over 1 h; and 3,500 V constant for at least 15 h (longer times are not detrimental). The movement of the tracking dye becomes usually visible during the 300 V plateau, and is more and more visible at the higher voltage phases. No movement of the dye indicates an absence of contact between the IPG gel and the electrodes. The tracking dye should collect at the acidic (anodal) part of the gradient. If a dyed zone extending beyond the ends of the gradients remains, this is indicative of a high-conductivity zone (poor salt removal), and it announces poor-resolution 2D gels at the end.

3.3 SDS Gel Casting

1. Ideally, the SDS gels must be cast the day before their use to ensure complete and reproducible polymerization. Gel sizes between 150 × 200 mm and 200 × 250 mm are recommended. Below the 150 × 200 size, the 2D gels lack resolution, as the total resolution is proportional to the gel surface in 2D electrophoresis. Above the 200 × 250 size, the gels become too fragile and too many gels are broken during the final stages of the experiment. The minimal thickness to accommodate an IPG strip is 1 mm, but 1.5 mm-thick gels are recommended.
2. Mount the gels' assemblies in the multi-gel casting chamber. In order to be able to differentiate each and every gel of the setup, it is recommended to add a distinctive mark cut from thin filter paper at the bottom of each gel assembly, between the glass plates, so that this mark will be embedded in the polyacrylamide gel during polymerization (see Note 15). Separate each gel assembly from its neighbors by inserting a thin polycarbonate plastic foil (usually supplied with the casting chamber) between the gel assemblies. When the gels assemblies are mounted, close the casting chamber.
3. Many casting chambers allow infusion of the gel solution from the bottom, and this is the recommended procedure to use. This will however require a funnel and silicone tubing to allow casting by gravity. In this case, attach a short piece of silicone tubing (ca 10 cm) to the inlet of the casting chamber, and provide with a stopper. Attach a long piece of silicone tubing (ca 50 cm) to the outlet of the funnel and provide with a stopper. Stoppers that allow easy control of the flow by progressive squeezing of the silicone tube are highly recommended. Then unite the two pieces of tubing with a plastic linker.
4. The amount of gel mix needed will of course depend on the size of the gels and on the number of gels cast. From the theoretical data (gel dimensions x number of gels), calculate the theoretical volume and add 20 % more to take into account the volume that will be lost between and around the gels in the casting chamber. This gives the practical volume needed.

5. When using the solutions described in Subheading 2, 10 % acrylamide gels are easily prepared by mixing 1/6 of the practical volume of concentrated gel buffer, 1/3 of the practical volume of 30 % acrylamide solution, and 1/2 of the practical volume of distilled water (see Note 16). Using the standard gel buffer will lead to a mass window spanning the 15–200 kDa window, with optimal resolution in the 15–40 kDa range and some crowding above. If a higher resolution of the medium and high molecular weight range is desired, use alternate buffer 2. This will lead however to the loss of proteins below 20 kDa. Conversely, if low molecular weight proteins (down to 5 kDa) are of interest, use alternate buffer 1. The use of alternate buffer 1 will however increase the crowding in the >30 kDa region of the gel.
6. When both the gel assemblies and the gel mix are ready, cast the gels. Under slow magnetic stirring (no big whirls), add 0.5 μl of TEMED and 5 μl of 10 % ammonium persulfate solution (in this order) per ml of gel mix. Mix for 20 s and pour in the casting chamber, avoiding trapping any air bubbles. Leave a 5 mm space at the top of the gels, and then overlay each gel with 1 ml of water-saturated butanol (upper phase in the bottle). Then close completely the two stoppers, disconnect the linker between the two pieces of tubing, and collect the excess gel mix present in the funnel into a beaker. Clean the funnel and the tubing by running distilled water through them and let dry.
7. Let the gel polymerize, and check for polymerization in the extra gel mix present in the beaker. When the gels are polymerized, open the casting chamber and recover each gel assembly (plates + gel). Clean from adhering gel particles under running tap water, rinse with distilled water, and pile in a closed box, separating once again the gel assemblies each one from the other by a clean thin polycarbonate foil. These separating foils are cleaned with water and ethanol exactly as the glass plates (see Note 15) and can be reused for months.
8. Put the closed box containing the gels in the cold room until use. The gels should be used 1–3 days after polymerization.

3.4 IPG Strip Equilibration, Transfer to SDS Gel, SDS Gel Running

1. At the end of the IPG run, the power is switched off, the paper strips at the ends of the gels (when present) are removed, and the mineral or silicon oil covering the strips is poured out. It is replaced by at least 10× the strips' volume of equilibration buffer. Equilibration is carried out for 20 min at room temperature, ideally under reciprocal ("ping-pong") shaking at ca. 45 strokes per minutes. During the equilibration period, the cooling of the second dimension gels is started and the agarose is melted, e.g., in a microwave oven. The agarose should be rather hot when used, to prevent premature setting of the agarose gel.

2. The second-dimension gels are also put out of the cold room, the groove at their tops is carefully dried with a lint-free tissue, and the gels are mounted on their supports to receive the strips.
3. For each gel, pick a strip with tweezers. X-shaped tweezers, which open only when the fingers exert a pressure on them, are ideal for this purpose. Still holding the strip by one of its ends, insert the other end of the strip in the groove on top of the second dimension gel, pipet 1 ml of hot agarose in the groove, and complete the transfer of the strip. All air bubbles should be carefully removed before the gel sets. If a side loading of, e.g., a molecular weight standard is desired, a Teflon "tooth" is inserted on the side of the second dimension gel at this stage.
4. Using the paper shapes present at the bottom of each gel, note which strip (and therefore which sample) is loaded on top of which second-dimension gel.
5. When the agarose gel has set, the Teflon tooth can be removed, leaving an empty space to load the control on the side. For example, a 1:1 mixture of sealing agarose and molecular weight standard can be loaded, and this is let to set again.
6. The gels are then mounted in the multi cell, and the tank buffer is poured in the cell.
7. Gels are run at 25 V constant voltage for 1 h, and then under constant power up to the end of the run, until the tracking dye reaches the bottom of the gels (see Note 17).

3.5 Protein Detection

1. At the end of the run, turn off the electric power and the cooling device. Remove the gels' cassettes from the multi-gel cell. Wearing gloves, open one gel cassette. Turn the glass plate supporting the gel upside down above the box containing the fixing solution. Fixing solution 1 is used for Flamingo and Krypton stains, and fixing solution 2 for the ruthenium complex stain. Fix the gels for at least 1 h, and then continue with the staining process. Follow the manufacturer's instructions for Flamingo and Krypton stains.
2. For staining with the ruthenium complex, stain overnight with the staining solution, and then destain for 4–6 h with fixing solution 2. Rinse with water for 5–10 min before image acquisition.
3. Scan with the laser scanner or acquire an image on a UV table with the suitable camera, and then process the images with the image analysis software.
4. Figure 1 shows an example of the results that can be obtained on a secretome sample using the methods described in this chapter.

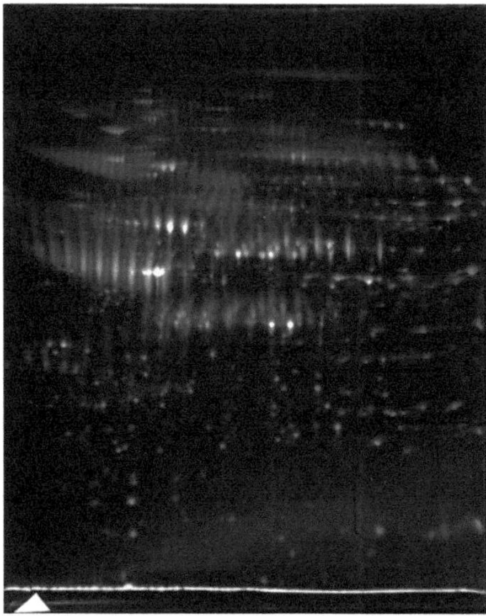

Fig. 1 J774 murine macrophage cells were seeded in a T175 flask (at 100,000 cells/ml) and grown up to confluence in Ultradoma medium supplemented with 1 % bovine fetal serum (3 days). The growth medium was then removed, and the cell layer was gently rinsed three times with PBS at 37 °C, and then three times with serum-free Ultradoma medium at 37 °C. 25 ml of serum-free Ultradoma medium were then added and the cells were incubated at 37 °C for 24 h. The conditioned culture medium was then collected, centrifuged for 5 min at 1,000 × *g* to remove floating cells, and then centrifuged for 20 min at 10,000 × *g* to remove smaller debris. The supernatant was then processed using the methods described in this chapter (TCA-sarkosyl precipitation, analysis by 2D gels, spot detection by fluorescence) to yield the image described in the figure

4 Notes

1. Trichloroacetic acid is highly hygroscopic in its solid state. Thus, it is highly advisable to process a complete and new bottle every time to prepare the 100 % TCA solution.
2. This solution provides a basic pH and a high denaturing power, ideal for resolubilizing proteins after precipitation. In addition, it is fully compatible with Bradford-type protein assays.
3. Dithiodiethanol provides an almost ideal blocking of thiol groups, without needing any alkylation step (33). It increases resolution of the basic proteins in the IEF dimensions and simplifies the equilibration process. If desired, thiol alkylation can be performed after protein detection and spot excision.
4. Most of the 2D gels run in the world use the classical glycine-based system (34). However, we believe that the taurine-based system (35) offers distinct advantages. First, as the gel buffer

operates quite close to the pKa of Tris, precise control of the pH is easier and offers thus better reproducibility over the long term. Second, it is much easier to tune the resolution just by changing the pH of the gel buffer, either in the high molecular weight range or in the low molecular weight range (35). Third, as the ionic strength of the buffer (0.1 M) is higher than the one of the Laemmli system, the resolution is slightly higher and the binding of SDS is also higher.

5. Temperature control during SDS-PAGE is of great importance to maximize resolution and reproducibility. First of all, the higher the migration speed, the higher the resolution, but only if the gels do not heat. Thus, it is very important to evacuate the heat generated during migration by the Joule effect. Due to the geometry of most multi-gel cells, this requires both a high cooling power and a powerful pump to overcome the pressure and flow drops in the multi-gel cells. As the final result, the optimal migration parameters should be determined empirically and depend both on the geometry of the electrophoresis cell and on the power of the cooling apparatus. Second, it should be kept in mind that the Tris buffers are among those that show the most important changes of pH with the temperature (0.3 pH units/10 °C). Thus, temperature stability from run to run is essential to ensure maximal reproducibility of the migration.
6. In addition to good compatibility with mass spectrometry, important constraints are imposed on the gel detection process. Ideally, the protein detection process should be sensitive, linear over several orders of magnitude to be able to detect changes both for low- and high-abundance proteins, and homogeneous from one protein to another to avoid biases. Recent work has shown that these specifications are best met by fluorescent detection using Flamingo and Krypton stains (30), which operate by environment-sensitive fluorescent probes (the probe is not fluorescent in water but fluoresces when bound to proteins).
7. The rationale of the protocol relies on the binding of the lauroyl sarcosinate to the proteins and to the co-precipitation of the proteins and of the lauroyl sarcosinate as its free acid under very acidic conditions. Other detergents (except bile salts) will be soluble under acidic conditions and keep both the lauroyl sarcosinate and the proteins soluble even under acidic conditions. The same solubilizing effect is obtained for samples containing high concentrations of chaotropes (urea, thiourea, guanidine salts). However in this case, efficient precipitation can be obtained if the sample is diluted with water to bring the chaotrope concentration below 1 M.

8. In this protein precipitation process, the lauroyl sarcosinate acts as a carrier and has a dual role. Its first role is to carry down the protein precipitate, as proteins do not precipitate well with TCA when they are too dilute (below 1 mg/ml) and without carrier. The second role of NLS is to decrease the protein–protein interactions within the pellet, which greatly helps protein resolubilization of the final stage. Tetrahydrofuran has been selected as a washing solvent on a multifactorial rationale. First, it is an excellent solvent of both TCA and NLS in its free acid form. Second, it is a very poor solvent for proteins, so that they do not redissolve prematurely at this stage. Third, tetrahydrofuran is slightly miscible with water, so that traces of the initial supernatant will not form a separate phase during the washing process. This ensures complete removal of the initial acidic aqueous supernatant. Fourth, tetrahydrofuran will be able to remove lipids and other hydrophobic substances that may have coprecipitated with the NLS.
9. A sample preparation protocol based on precipitation and resolubilization requires that both steps are efficient. Thus, as the proteins are very severely precipitated and denatured by TCA, they must be redissolved in highly solubilizing solutions, such as concentrated SDS (for SDS-PAGE or shotgun-based protocols), or concentrated chaotropic solutions for 2D PAGE-based protocols. Solubilization in intermediate urea concentrations (e.g., 6 M urea as in ref. 36) is inefficient and leads to severe protein losses that are not encountered when the sample is properly resolubilized (32).
10. The optimal final rehydration volume depends on the characteristics of the strips used. It has been experimentally determined (37) that optimal rehydration occurs for final acrylamide concentrations in the rehydrated strip slightly higher than 3 %. Most commercial strips are cast as 4 % gels with a 0.5 mm thickness. Then the optimal rehydration volume in microliters is given by the following formula: strip length (gel part only) in millimeters × strip width in millimeters × 0.65.
11. A tracking dye can be very useful to check for any migration problem. Any anionic dye with no affinity to proteins is suitable. Examples include bromophenol blue, bromocresol green, Orange G, and Chicago sky blue. The use of several different dyes within a single experiment decreases the probability to change inadvertently the strip order. However, it should be kept in mind that many of these tracking dyes have pH indicator properties. Thus, bromophenol blue, bromocresol green, and Orange G take all a similar yellow-orange hue at acidic pH.

12. The ideal pH range for the IPG strip cannot be determined theoretically. It is thus advisable to start with a wide pH gradient (e.g., 3–10 linear) and to adapt the pH range according to the protein density per pH unit. It should also be kept in mind that running purely basic gradients (e.g., 7–10) is more difficult than running gradients covering also an acidic part. Thus, it can be interesting to run each sample twice, on a 3–10 pH gradient to have a good resolution of the basic proteins, and a 4–7 gradient gel to have adequate resolution of the acidic proteins, which are in most cases more numerous than neutral or basic ones.
13. It has been described that application of a small voltage (50 V) during the rehydration step improved protein entry into the pH gradient. This is however possible only when the rehydration and running step take place in the same chamber and thus depends on the apparatus used.
14. Low-voltage initial steps are required in order to remove smoothly all low molecular weight, charged chemicals (e.g., salts, buffers, etc.) without generating too much Joule heat that would be detrimental to resolution. The program proposed here has been empirically determined. It should be kept in mind that isoelectric focusing with IPG, on a gradient that is at least 2 pH units wide, requires at least 100 Vh/cm^2, where the numerator is the integration of volts by time and the denominator the square of the length of the IPG gel.

    Some manufacturers recommend to apply, after the cleaning step, a defined power per strip (generally 50 μW/strip), as this ensures the highest possible volt.hours in a defined time frame. However, a purely voltage-limited program is safer. If, in a series of samples analyzed in parallel, some are more conductive than others, a collective watt-based program will lead to most of the power passing through the more conductive samples ($P = U^2/R$). Thus, the more conductive samples will dissipate too much power, and consequently too much heat, which will decrease resolution. Moreover, when using a collective watt-based program, the running profile will be different (in volt.hours) from one series of samples to another. This is not the case with a conservative volt-based program as the one described here.
15. The cleanliness of the glass plates is essential to obtain high-quality 2D gels. At the end of each run, remove every gel particle from the glass plates by brushing them under hot tap water. Then rinse each plate in distilled water and let dry. Do not use any detergent to clean the plates, as it is highly likely that some of the detergent will stay on the glass plates despite rinsing and will interfere with the SDS electrophoresis. Just before

use, clean again the glass plates with water and then 95 % alcohol, using a lint-free paper tissue. If the plates are really clean, the alcohol rinsing process should produce some wiping noise. With this cleaning scheme, however, the plates will become dirty over a time frame of several months, and the gel resolution will start to deteriorate. This is easily seen when "tails" begin to appear ahead of the most intense spots. When this happens, clean thoroughly the glass plates with a mildly abrasive dish cleaning powder and a sponge, and then rinse profusely under hot tap water, and finally with distilled water.

16. Unpolymerized acrylamide is toxic. Wear suitable protective clothes and gloves (preferably powder-free nitrile gloves) when handling gels or acrylamide-containing solutions.
17. The first, initial low-voltage step is intended to let the SDS elute completely the proteins from the strip. Once the SDS front has passed the strip, the elution power is minimal, so that vertical trailing is induced. The gels should be then run at maximum speed to limit diffusion, and this maximal speed depends from the gel size, the multi-gel cell geometry, and the performances of the cooling system. It must therefore be determined empirically for each system, but manufacturers of multi-gel cells generally make useful suggestions in their instructions for use.

References

1. Lescuyer P, Hochstrasser D, Rabilloud T (2007) How shall we use the proteomics toolbox for biomarker discovery? J Proteome Res 6:3371–3376
2. Teng PN, Bateman NW, Hood BL et al (2010) Advances in proximal fluid proteomics for disease biomarker discovery. J Proteome Res 9:6091–6100
3. Choi YS, Choe LH, Lee KH (2010) Recent cerebrospinal fluid biomarker studies of Alzheimer's disease. Expert Rev Proteomics 7:919–926
4. Kroksveen AC, Opsahl JA, Aye TT et al (2011) Proteomics of human cerebrospinal fluid: discovery and verification of biomarker candidates in neurodegenerative diseases using quantitative proteomics. J Proteomics 74:371–388
5. Maurer MH (2010) Proteomics of brain extracellular fluid (ECF) and cerebrospinal fluid (CSF). Mass Spectrom Rev 29:17–28
6. Zanusso G, Fiorini M, Ferrari S et al (2011) Cerebrospinal fluid markers in sporadic Creutzfeldt-Jakob disease. Int J Mol Sci 12:6281–6292
7. Zhang J (2007) Proteomics of human cerebrospinal fluid—the good, the bad, and the ugly. Proteomics Clin Appl 41:805–819
8. Casado-Vela J, del Pulgar TG, Cebrian A et al (2011) Human urine proteomics: building a list of human urine cancer biomarkers. Expert Rev Proteomics 8:347–360
9. Julian BA, Suzuki H, Suzuki Y et al (2009) Sources of urinary proteins and their analysis by urinary proteomics for the detection of biomarkers of disease. Proteomics Clin Appl 3:1029–1043
10. Mischak H, Kolch W, Aivaliotis M et al (2010) Comprehensive human urine standards for comparability and standardization in clinical proteome analysis. Proteomics Clin Appl 4:464–478
11. Schaaij-Visser TB, Proost N, Nagel R et al (2011) Secretome proteomics to identify indicators for lung cancer treatment response prediction and monitoring. J Thorac Oncol 6:S1011
12. Makridakis M, Roubelaids MG, Bitsika V et al (2010) Analysis of secreted proteins for the study of bladder cancer cell aggressiveness. J Proteome Res 9:3243–3259
13. Makridakis M, Vlahou A (2010) Secretome proteomics for discovery of cancer biomarkers. J Proteomics 73:2291–2305
14. Luo XY, Liu YS, Wang R et al (2011) A high-quality secretome of A549 cells aided the

discovery of C4b-binding protein as a novel serum biomarker for non-small cell lung cancer. J Proteomics 74:528–538
15. Caccia D, Domingues LZ, Micciche F et al (2011) Secretome compartment is a valuable source of biomarkers for cancer-relevant pathways. J Proteome Res 10:4196–4207
16. Sarkissian G, Fergelot P, Lamy PJ et al (2008) Identification of Pro-MMP-7 as a serum marker for renal cell carcinoma by use of proteomic analysis. Clin Chem 54:574–581
17. Roessler M, Rollinger W, Mantovani-Endl L et al (2006) Identification of PSME3 as a novel serum tumor marker for colorectal cancer by combining two-dimensional polyacrylamide gel electrophoresis with a strictly mass spectrometry-based approach for data analysis. Mol Cell Proteomics 5:2092–2101
18. Celis JE (2004) Gel-based proteomics: what does MCP expect? Mol Cell Proteomics 3:949
19. Rabilloud T, Chevallet M, Luche S et al (2011) Two-dimensional gel electrophoresis in proteomics: past, present and future. J Proteomics 73:2064–2077
20. Rabilloud T (2009) Membrane proteins and proteomics: love is possible, but so difficult. Electrophoresis 30(Suppl 1):S174–S180
21. Yi JZ, Liu ZX, Craft D et al (2008) Intrinsic peptidase activity causes a sequential multi-step reaction (SMSR) in digestion of human plasma peptides. J Proteome Res 7:5112–5118
22. Hoofnagle AN (2010) Peptide lost and found: internal standards and the mass spectrometric quantification of peptides. Clin Chem 56:1515–1517
23. Bystrom CE, Salameh W, Reitz R et al (2010) Plasma renin activity by LC-MS/MS: development of a prototypical clinical assay reveals a subpopulation of human plasma samples with substantial peptidase activity. Clin Chem 56:1561–1569
24. Hsich G, Kinney K, Gibbs CJ et al (1996) The 14-3-3 brain protein in cerebrospinal fluid as a marker for transmissible spongiform encephalopathies. N Engl J Med 335:924–930
25. Ladogana A, Sanchez-Juan P, Mitrova E et al (2009) Cerebrospinal fluid biomarkers in human genetic transmissible spongiform encephalopathies. J Neurol 256:1620–1628
26. Bartosik-Psujek H, Archelos JJ (2004) Tau protein and 14-3-3 are elevated in the cerebrospinal fluid of patients with multiple sclerosis and correlate with intrathecal synthesis of IgG. J Neurol 251:414–420
27. Colucci M, Roccatagliata L, Capello E et al (2004) The 14-3-3 protein in multiple sclerosis: a marker of disease severity. Mult Scler 10:477–481
28. Pitarch A, Nombela C, Gil C (2009) Proteomic profiling of serologic response to Candida albicans during host-commensal and host-pathogen interactions. Methods Mol Biol 470:369–411
29. Pitarch A, Nombela C, Gil C (2011) Prediction of the clinical outcome in invasive candidiasis patients based on molecular fingerprints of five anti-Candida antibodies in serum. Mol Cell Proteomics 10(M110):004010
30. Maass S, Sievers S, Zuhlke D et al (2011) Efficient, global-scale quantification of absolute protein amounts by integration of targeted mass spectrometry and two-dimensional gel-based proteomics. Anal Chem 83:2677–2684
31. Aude-Garcia C, Collin-Faure V, Luche S et al (2011) Improvements and simplifications in in-gel fluorescent detection of proteins using ruthenium II tris-(bathophenanthroline disulfonate): the poor man's fluorescent detection method. Proteomics 11:324–328
32. Chevallet M, Diemer H, Van Dorssealer A et al (2007) Toward a better analysis of secreted proteins: the example of the myeloid cells secretome. Proteomics 7:1757–1770
33. Luche S, Diemer H, Tastet C et al (2004) About thiol derivatization and resolution of basic proteins in two-dimensional electrophoresis. Proteomics 4:551–561
34. Laemmli UK (1970) Cleavage of structural proteins during the assembly of the head of bacteriophage T4. Nature 227:680–685
35. Tastet C, Lescuyer P, Diemer H et al (2003) A versatile electrophoresis system for the analysis of high- and low-molecular-weight proteins. Electrophoresis 24:1787–1794
36. Dowell JA, Johnson JA, Li LJ (2009) Identification of astrocyte secreted proteins with a combination of shotgun proteomics and bioinformatics. J Proteome Res 8:4135–4143
37. Sanchez JC, Hochstrasser D, Rabilloud T (1999) In-gel sample rehydration of immobilized pH gradient. Methods Mol Biol 112:221–225

Chapter 14

Informatics of Protein and Posttranslational Modification Detection via Shotgun Proteomics

Jerry D. Holman, Surendra Dasari, and David L. Tabb

Abstract

Frequently, proteomic LC-MS/MS data may contain sets of modifications that evade identification during standard database search. For many laboratories, the standard technique to seek posttranslational modifications (PTMs) adds a short list of specified mass shifts to database search configuration. This technique provides information for only the specified PTMs, takes substantial time to run, and drives false discoveries upward through an exponential expansion of search space. This protocol describes a more structured approach to blind PTM discovery through reducing protein lists, targeting attention to a data-driven list of mass shifts, and seeking the resulting short list of modifications through targeted search.

Key words Shotgun proteomics, Proteome informatics, Posttranslational modification, Protein identification, Sequence tagging, Database search, Software

1 Introduction

Algorithms to identify peptides from LC-MS/MS were introduced in 1994 (1), and the ability to identify PTMs by these algorithms was added only 1 year later (2). In essence, the database search strategy trawls a protein sequence database for candidate peptides of the approximate mass of the precursor ion for an MS/MS. These candidates are then compared to the observed MS/MS by predicting what fragment ions should be observed for each sequence and comparing to the observed set. Configuring these tools to find PTMs enables them to convolute these peptide sequences with a small, defined set of mass shifts associated with particular residues (such as potentially adding 16 Da to methionine residues), creating an exponential expansion of the search space. To curtail the time required to seek a wide variety of PTMs, Craig and Beavis introduced the "refined search" technique in 2003 (3). This approach first determines the set of unmodified peptides found in a mixture, produces a reduced protein set, and then seeks a wide

Ming Zhou and Timothy Veenstra (eds.), *Proteomics for Biomarker Discovery: Methods and Protocols*, Methods in Molecular Biology, vol. 1002, DOI 10.1007/978-1-62703-360-2_14, © Springer Science+Business Media, LLC 2013

variety of modifications for those proteins. In practice, this technique mixes two different kinds of search results in a single result set, complicating the determination of which spectra have been identified successfully.

Although initially published in the same year as database search, the sequence tagging strategy took longer to fully automate than database search (4). The approach infers a partial sequence from each tandem mass spectrum (5) and then uses these inferred tags as filters to reduce the set of peptide sequences compared to each spectrum. Where database search employs only the observed precursor mass to select compared peptides, tagging is able to use the set of inferred partial sequences as well as the masses flanking this tag to select comparisons. Mass shifts can then be introduced in the flanking mass regions to reflect PTMs. The InsPecT algorithm was the first to automate all elements of this process (6). Tsur et al. leveraged the ability to infer sequences from spectra in a "blind search" for modifications of unknown mass and specificity (7). This protocol emulates a blind search technique based in sequence tagging that was described by Dasari in the context of toxicological proteomics (8).

This implementation integrates the MyriMatch database search engine (9), the DirecTag sequence inference engine (10), the TagRecon tag matcher (11), and the IDPicker protein assembly environment (12, 13). The use of the first three is coordinated through the BumberDash user interface (14), and IDPicker features its own graphical user interface. This protocol deploys this strategy in the context of formalin-fixed, paraffin-embedded tissues rather than the toxicological data set described earlier, but the principal features are the same: (1) build a reduced set of proteins that can be confidently identified from unmodified peptides, (2) infer a reduced set of PTMs for which substantial MS/MS evidence can be found, and (3) conduct a limited search for PTMs employing sequence tag information.

2 Materials

This protocol describes an informatics method for detecting protein posttranslational modifications from shotgun proteomics data sets. For this, readers will need the BumberDash search software suite, MS/MS data files, a protein sequence database, and IDPicker (version 3.0) software for results filtering and protein assembly.

2.1 BumberDash Search Software Suite

The BumberDash suite incorporates the MyriMatch and DirecTag-TagRecon software. MyriMatch is a database search engine designed for shotgun protein identification. DirecTag-TagRecon is a sequence tagging-based search engine optimized for finding unexpected modifications in peptides using shotgun proteomics data.

Both these tools accept two types of inputs: a raw MS/MS data file and a protein sequence database. The software accepts MS/MS data in a variety of instrument native and derived formats (see Note 1 for a full list of acceptable formats). Protein sequences are read from a FASTA-formatted file.

MyriMatch derives peptide sequences from the protein database, predicts fragmentation spectra, and compares them to the experimental MS/MS. The quality of each peptide-spectrum match (PSM) is assessed using the intensity and mass error associated with matched fragments. Finally, the software records the top five matches for each MS/MS to an identification file in either pepXML or mzIdentML format.

DirecTag-TagRecon software starts by generating short amino acid tags for each MS/MS in the data file. The software matches the tags to candidate peptide sequences from the protein database. Mass mismatches between the tag and candidate sequence are interpreted as modifications. Theoretical tandem spectra are predicted for candidate peptides and compared to the experimental MS/MS. PSMs are scored, and identifications are recorded to standard formats following the same method employed by MyriMatch.

2.2 IDPicker Software

IDPicker (version 3.0) filters raw identifications from the search software to a confident set using the target-decoy strategy. The software reads the identifications from pepXML or mzIdentML files. IDPicker computes false discovery rates (FDR) of all identifications and filters the results to meet a user-defined target FDR. Finally, the software assembles a minimal protein list that encompasses all filtered peptides using the rules of parsimony. IDPicker writes the results to a SQL database and presents them to the user in a flexible graphical user interface. The software allows the user to arrange the results into an experimental hierarchy (biological replicates, technical replicates, controls, experiments, et cetera). The software can also create protein, peptide, and spectral identification reports in text format. These reports are useful for downstream analysis like spectral counting-based protein differentiation.

BumberDash and IDPicker (version 3.0) software are available for download free of charge from the Internet Web site http://fenchurch.mc.vanderbilt.edu. To employ this protocol, readers should download and install the software on a local computer.

2.3 LC-MS/MS Data Sets

We demonstrate this protocol with a shotgun proteomic data set generated from formalin-fixed paraffin-embedded (FFPE) colon tissues (15). Tissues were stored in FFPE for 1, 3, 5, or 10 years. Independently, tissues were fixed in formalin for 0, 1, 2, or 4 days. Proteins in each fixation were reduced with dithiothreitol, alkylated with iodoacetamide, and digested with trypsin. Resulting peptide mixtures were analyzed via LC-MS/MS in replicates on a Thermo LTQ mass spectrometer. All the raw data files can be downloaded from the Proteome Commons Tranche Web site https://proteomecommons.org/dataset.jsp?i=77376. Readers can select subsets of the data for analysis to save computation time.

3 Methods

Figure 1 illustrates the three-stage workflow for detecting unexpected PTMs from shotgun proteomics data sets. First, we characterize the protein content of the sample with a simple MyriMatch database search (protein identification). Resulting PSMs are filtered with IDPicker software at 2 % FDR. Proteins with at least two unique peptide identifications are included in a subset FASTA database. Next, we query the proteins in the subset database for unanticipated PTMs using DirecTag-TagRecon software (blind PTM search). Peptide and PTM identifications are filtered with IDPicker software at a stringent 2 % FDR. Confident PTMs are identified using IDPicker software (see Subheading 3.3). Finally, we refine the PTM identification results with a directed PTM search. In this step, we re-query the subset FASTA with DirecTag-TagRecon software configured to look for only the confident PTMs. This (optional) refinement step is designed to improve the sensitivity and specificity of the PTM search. Identifications from the directed PTM search are filtered with IDPicker software at 2 % FDR.

3.1 Protein Identification

1. Launch the BumberDash software.
2. Click on the bottommost row or go to "File->New Job" to add a new job. This opens an Add Job dialogue (see Fig. 2).
3. In the Add Job dialogue, select "MyriMatch- Database Search" as the "Type of Search" (default choice in BumberDash).
4. Name the search job and check the box if results should be stored as a subfolder in the output directory. This step may be omitted if desired (see Note 2).
5. Select the "Input Files" using the "Browse" button. BumberDash accepts MS/MS data files in mzXML, mzML, mgf, mz5, or any vendor raw format. The user can hold down ctrl to select multiple data files.
6. Change the "Output Directory" to the desired location. By default, BumberDash writes the output files to the folder containing the "Input Files."
7. Select a "FASTA database" containing the protein sequences for the search. BumberDash accepts the protein databases in FASTA format.

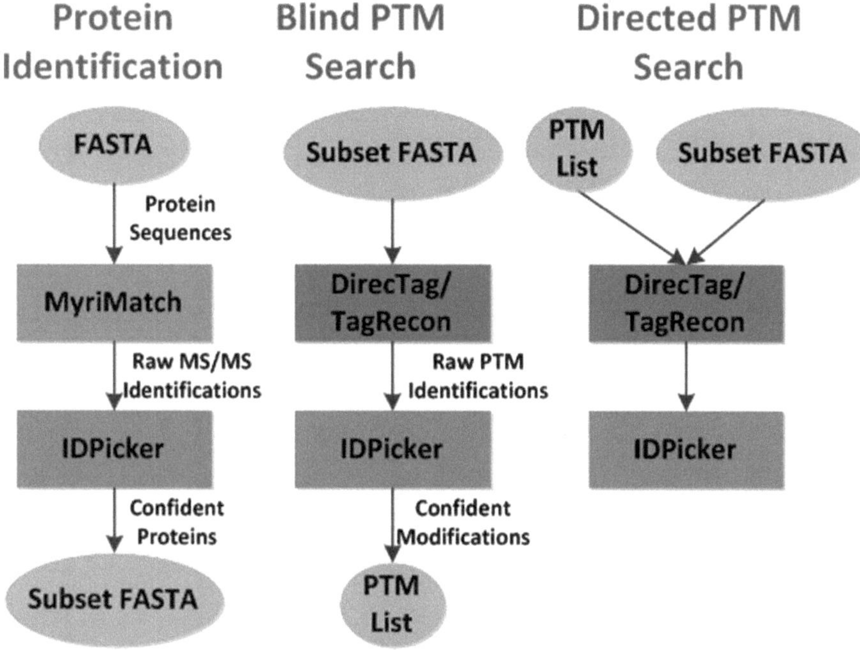

Fig. 1 PTM identification workflow

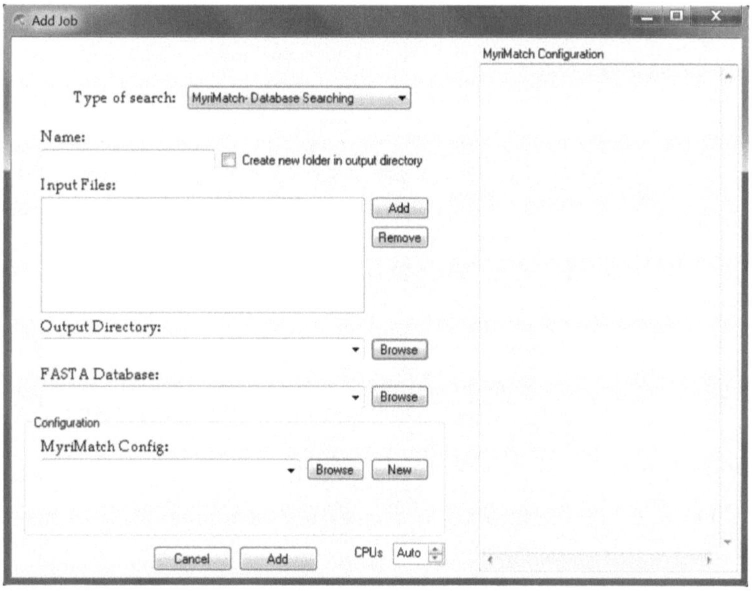

Fig. 2 Add Job dialogue of BumberDash

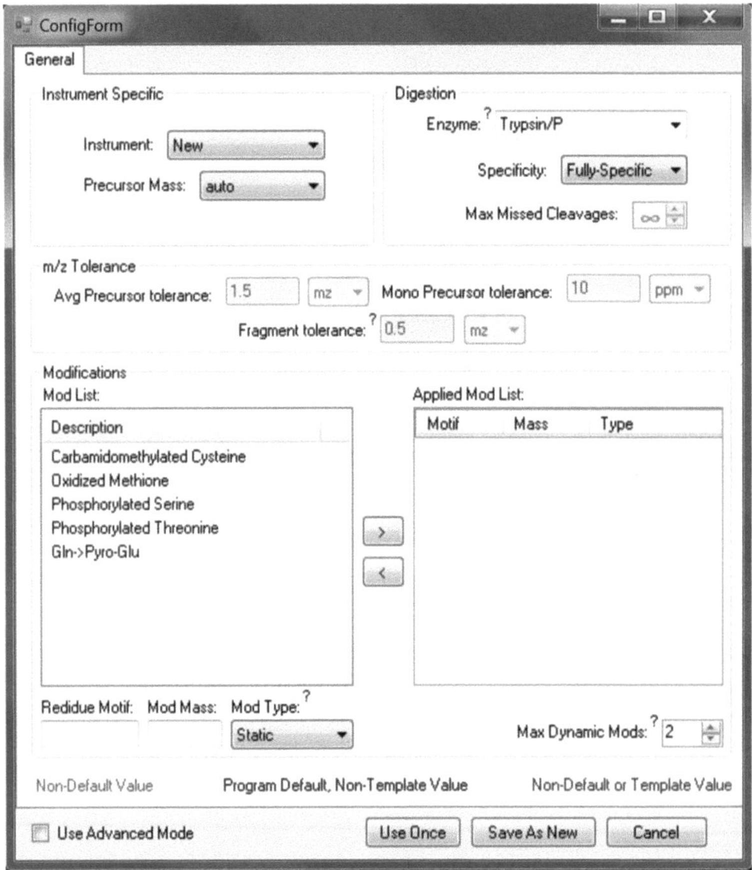

Fig. 3 BumberDash configuration editor

8. If you have a premade MyriMatch configuration file, select it with the "Browse" button. Otherwise, create a new configuration file using the "New" button next to the configuration drop-down (see Note 3).
9. Clicking the "New" button will start the MyriMatch configuration editor (Fig. 3). Make changes to the desired parameters. You can either save the changes either temporarily or permanently. Click the "Use Once" button if you want to use the configuration for the current job only. Click "Save As New" and supply a filename if you want to retain the configuration for future searches. Return to the Add Job dialogue.
10. Append the job to the search queue with the "Add" button. Figure 4 shows an example of BumberDash job queue. Users can track the status of jobs in the queue. Jobs can also be cancelled (click the "X" button) from the queue (see Note 4).

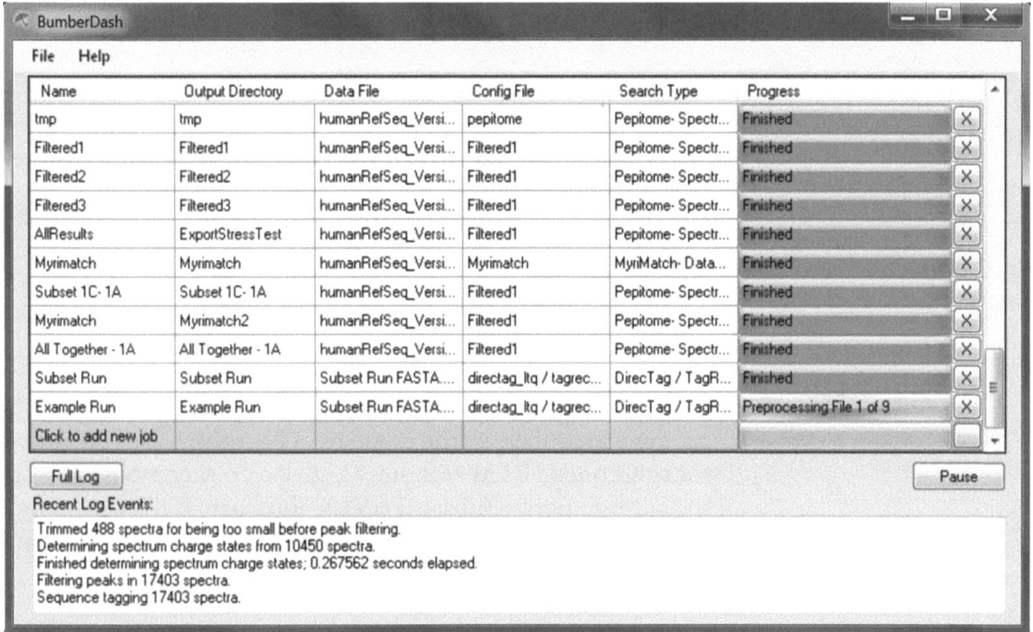

Fig. 4 BumberDash job queue

11. Wait for the MyriMatch job to finish. Launch IDPicker (version 3.0) software to process the database search results. If IDPicker 3 is already installed on the computer it can be launched from BumberDash's File->Run menu.
12. Import the resulting identification files (pepXML) into IDPicker. The pepXML files are found in the "Output Directory" of the corresponding BumberDash job. To import the files, launch the IDPicker results navigator with "File->Import Files" option. IDPicker results navigator has two panels. The left panel navigates the folder structure containing the results files. The right panel shows the pepXML files found in a folder selected in the left panel. To import files, browse to the output folder containing the pepXML files in the left panel and click the ">" button to see the files in the right panel. Make sure that the desired files are checked and click "Open."
13. After this step, IDPicker will display an "Import Settings" dialogue (see Fig. 5). These settings are used by the software to filter the PSMs present in the pepXML files. Set the "Database" to the FASTA file used for the database search. Change "Max FDR" setting to "0.02," "Max Rank" to "1," and "Qonverter Settings" to "MyriMatch Optimized." Click "OK" to start creating the IDPicker report and wait for it to finish.
14. The filtered report contains information about the proteins, peptides, and PSMs identified in the data set. We need to create

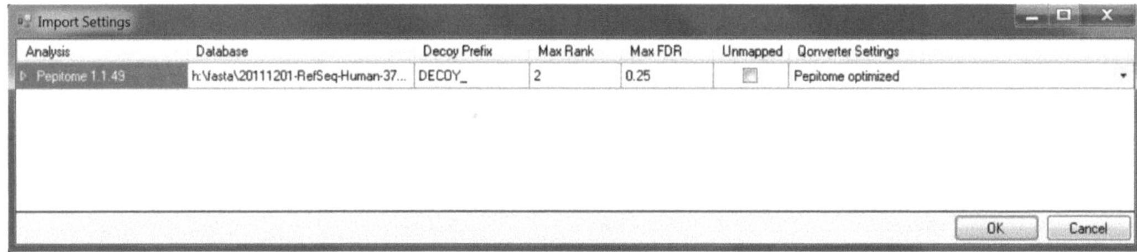

Fig. 5 IDPicker (version 3) "Import Settings" dialogue

a small subset FASTA database containing sequences of proteins identified in the data set. This subset FASTA is used for subsequent PTM searches (Fig. 1). To accomplish this, go to File->Export->Subset FASTA, instruct IDPicker to add decoy sequences to the generated database, and save the database to a location of your choice.

3.2 Blind PTM Search for Finding Unexpected Sequence Modifications

1. Load BumberDash and bring up the Add Job dialogue (see Subheading 3.1, step 2).
2. Select "DirecTag/TagRecon- Sequence Tagging" as the "Type of Search."
3. Name the job, select the input MS/MS data files, and specify an output directory (see Subheading 3.1, steps 4–6). The input files selected for this search should be identical to those selected for the protein identification step (see Subheading 3.1).
4. Select the subset FASTA produced at the end of Subheading 3.1 as input "FASTA database."
5. Next, we need to configure DirecTag and TagRecon separately. Both these configurations are shown as separate drop-down boxes in the Add Job dialogue (Fig. 2).
6. If you want to use a premade DirecTag configuration file click on the "Browse" button next to the DirecTag configuration drop-down box. Otherwise, click the "New" button to start the configuration editor (Fig. 3). Make changes to the desired parameters. You can save the changes either temporarily or permanently. Click the "Use Once" button if you want to use the configuration for the current job only. Click "Save As New" and supply a filename if you want to retain the configuration for future searches. Return to the Add Job dialogue (see Note 5).
7. Repeat the previous step to configure TagRecon. Ensure that the "Explain Unknown Mass Shifts As" option is set to "blind-ptms." Return to the Add Job dialogue.
8. Add the job to the search queue (see Subheading 3.1, step 10) and wait for completion (see Note 6).

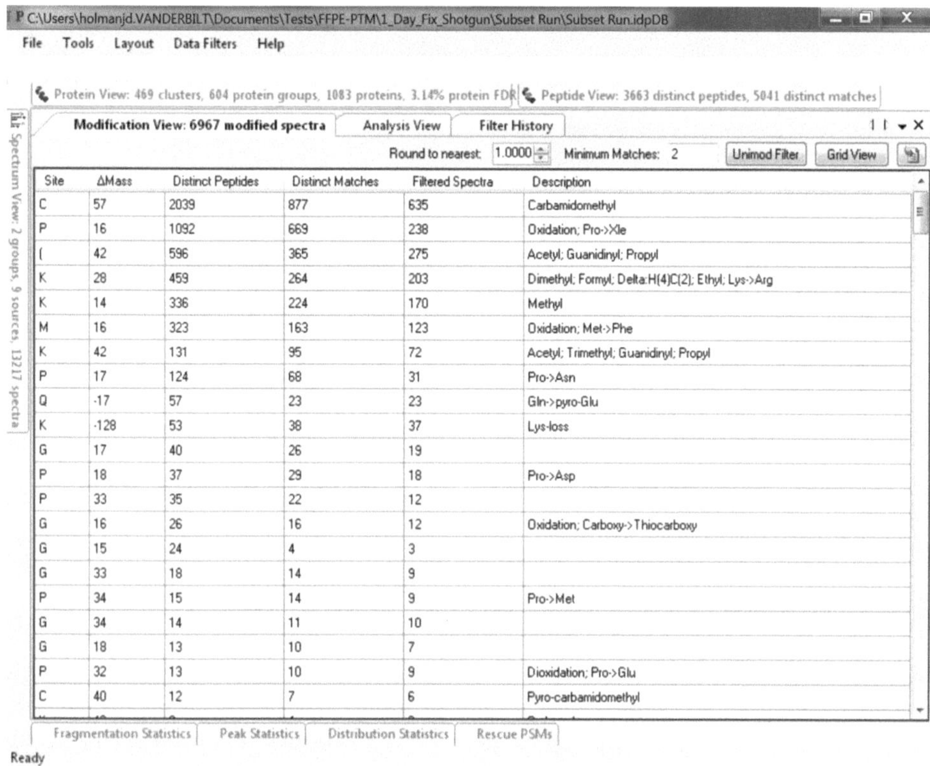

Fig. 6 IDPicker 3 modification detail view

9. Launch IDPicker (version 3.0) software to process "Blind PTM" search results (see Subheading 3.1, step 11). Add the search pepXML files to IDPicker (see Subheading 3.1, step 12).
10. Configure the "Import Settings" of IDPicker as described in Subheading 3.1, step 12 with two important changes. Set the "Database" to the subset FASTA file used in the search and change the "Qonverter Settings" to "TagRecon Optimized." Start the IDPicker results filtering using the "OK" button.
11. The "Blind PTM" IDPicker report will contain a summary of proteins, peptides, PSMs, and PTMs present in the data set. The PTM summary (Fig. 6) is located in the modification form (by default located in the lower right quarter of the IDPicker GUI). We need to use this form to identify a list of confident PTMs present in the sample (see Subheading 3.3). These confident PTMs are used in the final "Directed PTM" search (Fig. 1) (see Note 7).

3.3 Detecting Confident Mass Shifts Using IDPicker

1. From the modification grid, switch to detail view by clicking the button in the upper-right corner. This should show the list of possible modifications identified along with data associated with the modification.

Fig. 7 Unimod filter popup

2. Click the "Unimod Filter" button to bring up a list of all possible Unimod explanations for the data. If fewer modifications appear than expected it may be necessary to increase the "Round to Nearest" number (see Note 8).
3. Select the posttranslational checkbox and click outside the pop-up to show only modifications identified as possibly being posttranslational (Fig. 7). Generally modifications which are associated with many spectra and which contribute to at least two peptides are less likely to be erroneous results.
4. Choose three to eight modifications which are of greatest interest based on the data. Record the mass shifts and residue characters to use in the next step.

3.4 Run-Directed Posttranslational Modification Search

1. Once more load BumberDash and start a "DirecTag/TagRecon- Sequence Tagging" job.
2. The input files, subset FASTA, and DirecTag configuration should remain the same. Select a new job name and output folder if desired.
3. In the TagRecon configuration, "ExplainUnknownMassShiftsAs" should be set to "preferredptms." Modifications obtained from the previous step should be added to the configuration by entering the residue motif (or character), the modification delta mass, and the mod type set to "PreferredPTM."
4. Run the job and load the results in IDPicker 3 (see Note 9).

4 Notes

1. BumberDash supports mzML, mzXML, MGF, Agilent, Bruker FID/YEP/BAF, Thermo RAW, Waters RAW, MGF, MS2/CMS2/BMS2, and mzIdentML file formats.
2. Job naming is optional, if no name is given when files are selected a name will be generated automatically based on the top-level directory of the input files when they are selected. If the new folder box is checked then a new folder will be generated in the output directory in which to store the result files. On the main screen jobs which have not yet been run but are set to produce a subfolder have an output directory ending with a "+." Jobs which have been or are in the process of being run will have a "*" at the end of the output directory indicating that it points to the subfolder that has been produced for the job. The full output directory can be viewed in the tooltip of the abbreviated output directory cell.
3. BumberDash comes preloaded with recommended settings for a few different instrument types, which can be loaded by selecting an option in the Instrument drop-down menu. Setting the specificity to semi-specific (or placing "MinTerminiCleavages = 1" in the configuration file) will yield a much better data set; however it will also greatly increase the required time compared to a fully specific ("MinTerminiCleavages = 2") search. For the data from the Sprung article (15), search times on a Core 2 Quad CPU were approximately 5 h per file for fully specific searches and 35 h per file for semi-specific searches.
4. BumberDash runs at a "Below Normal" priority level, so it should not throttle the CPU usage of other programs running on the computer. To further allow BumberDash to run in the background it minimizes to the system tray, not taking up space in the task bar. To restore a minimized instance of BumberDash double-click the icon in the tray.
5. We recommend adding carbamidomethylation of cysteine as a static modification to DirecTag runs. This change causes Cys residues to account for 160 Da rather than 103 Da, reflecting their modification by iodoacetamide.
6. For the Sprung data set (15) on the Core 2 Quad CPU test machine, this process took about two hours per file when a subset database produced by a fully specific search was used. The more comprehensive subset database produced by a semi-specific search brought the required time up to 8 h per file.
7. Other forms in the IDPicker window can be minimized by clicking the pin on the top-right section of the panel title bar.

Modifications may be highlighted based on how many times they were observed in the given data set: green for more than 10, blue for more than 50, and red if the modification was seen over 100 times. To show only common modifications click "Unimod Filter" in the top right and check all category boxes, leaving "Show Hidden" unchecked. If a modification number is bold it contains a list of possible Unimod explanations in the tooltip.

8. Initially only the Unimod annotations labeled as "unhidden" will appear in the pop-up box. To show all identified modifications in the list click the checkbox in the upper-left corner. Rounded modification masses must exactly match the Unimod entries to be recognized. Increasing the "Round to Nearest" number will increase the likelihood a modification is recognized by Unimod, but decrease the overall specificity.
9. For best results run job again with "Database Search" and selected modifications as dynamic mods, using the subset FASTA. Both the database search and tag search results can be loaded into IDPicker 3 at the same time to create a combined report.

Acknowledgments

The algorithms described in this protocol were developed through support to all three authors by R01 CA126218. In addition, J.D.H. was supported by U01 CA08402.

References

1. Eng JK, McCormack AL, Yates JR (1994) An approach to correlate tandem mass spectral data of peptides with amino acid sequences in a protein database. J Am Soc Mass Spectrom 5:976–989
2. Yates JR, Eng JK, McCormack AL, Schieltz D (1995) Method to correlate tandem mass spectra of modified peptides to amino acid sequences in the protein database. Anal Chem 67:1426–1436
3. Craig R, Beavis RC (2003) A method for reducing the time required to match protein sequences with tandem mass spectra. Rapid Commun Mass Spectrom 17:2310–2316
4. Mann M, Wilm M (1994) Error-tolerant identification of peptides in sequence databases by peptide sequence tags. Anal Chem 66:4390–4399
5. Tabb DL, Saraf A, Yates JR (2003) GutenTag: high-throughput sequence tagging via an empirically derived fragmentation model. Anal Chem 75:6415–6421
6. Tanner S, Shu H, Frank A et al (2005) InsPecT: identification of posttranslationally modified peptides from tandem mass spectra. Anal Chem 77:4626–4639
7. Tsur D, Tanner S, Zandi E, Bafna V, Pevzner PA (2005) Identification of post-translational modifications by blind search of mass spectra. Nat Biotechnol 23:1562–1567
8. Dasari S, Chambers MC, Codreanu SG et al (2011) Sequence tagging reveals unexpected modifications in toxicoproteomics. Chem Res Toxicol 24:204–216
9. Tabb DL, Fernando CG, Chambers MC (2007) MyriMatch: highly accurate tandem mass spectral peptide identification by multivariate hypergeometric analysis. J Proteome Res 6:654–661
10. Tabb DL, Ma Z-Q, Martin DB, Ham A-JL, Chambers MC (2008) DirecTag: accurate sequence tags from peptide MS/MS through statistical scoring. J Proteome Res 7:3838–3846

11. Dasari S, Chambers MC, Slebos RJ, Zimmerman LJ, Ham A-JL, Tabb DL (2010) TagRecon: high-throughput mutation identification through sequence tagging. J Proteome Res 9:1716–1726
12. Zhang B, Chambers MC, Tabb DL (2007) Proteomic parsimony through bipartite graph analysis improves accuracy and transparency. J Proteome Res 6:3549–3557
13. Ma Z-Q, Dasari S, Chambers MC et al (2009) IDPicker 2.0: improved protein assembly with high discrimination peptide identification filtering. J Proteome Res 8:3872–3881
14. Holman JD, Ma Z-Q, Tabb DL (2012) Identifying proteomic LC-MS/MS data sets with Bumbershoot and IDPicker, Current protocols in bioinformatics/editorial board, Andreas D. Baxevanis. Chapter 13, Unit 13.17
15. Sprung RW Jr, Brock JWC, Tanksley JP et al (2009) Equivalence of protein inventories obtained from formalin-fixed paraffin-embedded and frozen tissue in multidimensional liquid chromatography-tandem mass spectrometry shotgun proteomic analysis. Mol Cell Proteomics 8:1988–1998

Chapter 15

Quantitation of Met Tyrosine Phosphorylation Using MRM-MS

Zhaojing Meng, Apurva K. Srivastava, Ming Zhou, and Timothy Veenstra

Abstract

Phosphorylation has long been accepted as a key cellular regulator of cell signaling pathways. The recent development of multiple-reaction monitoring mass spectrometry (MRM-MS) provides a useful tool for measuring the absolute quantity of phosphorylation occupancy at pivotal sites within signaling proteins, even when the phosphorylation sites are in close proximity. Here, we described a targeted quantitation approach to measure the absolute phosphorylation occupancy at Y1234 and Y1235 of Met. The approach is utilized to obtain absolute occupancy of the two phosphorylation sites in the full-length recombinant Met. It is further applied to quantitate the phosphorylation state of these two sites in SNU-5 cells treated with a Met inhibitor.

Key words Met, Phosphorylation, Multiple-reaction monitoring, Mass spectrometry, Quantitation

1 Introduction

The gene *MET* encodes hepatocyte growth factor (HGF) receptor, Met, a tyrosine kinase receptor with HGF as its only known ligand. MET signaling plays pivotal roles in various processes, including tissue remodeling and cancer metastasis (1). Met in its active mature form is a disulfide-linked heterodimer consists of an extracellular 50 kDa α-chain and a transmembrane 145 kDa β-chain with an extracellular-domain HGF binding site and a cytoplasmic tyrosine kinase domain (Fig. 1). Phosphorylation of Met is intimately involved in both Met function and signaling. Phosphorylation at Y1234 and/or Y1235 of Met is known to be crucial for its activation and regulation (2). Therefore, the phosphorylation status of Met kinase domain at the sites is routinely checked by Western blot; however, this method is unable to distinguish between the two sites due to their proximity.

Ming Zhou and Timothy Veenstra (eds.), *Proteomics for Biomarker Discovery: Methods and Protocols*, Methods in Molecular Biology, vol. 1002, DOI 10.1007/978-1-62703-360-2_15, © Springer Science+Business Media, LLC 2013

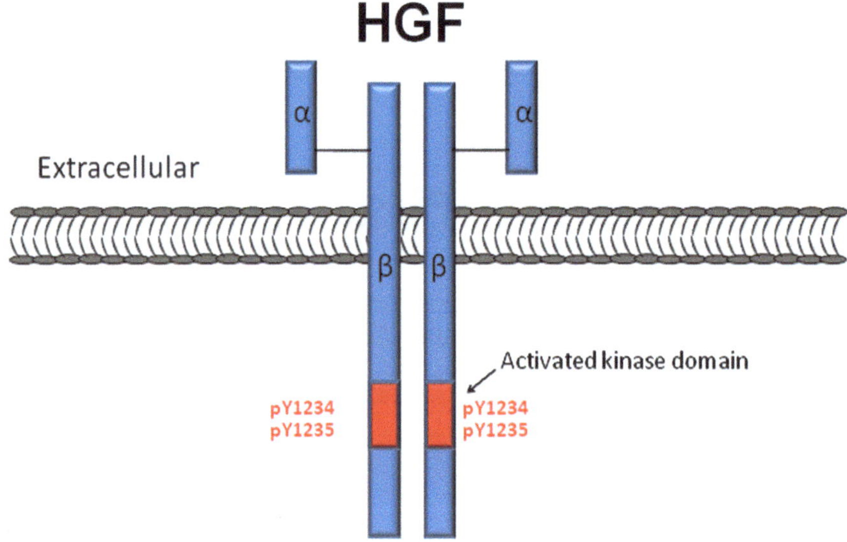

Fig. 1 MET tyrosine kinase receptor structure illustrating the activated kinase domain after binding hepatocyte growth factor. The phosphorylated residues targeted in this assay, Y1234 and Y1235, are located in the kinase domain of Met

Quantitation assays based on multiple-reaction monitoring mass spectrometry (MRM-MS) in combination with stable-isotope-labeled internal standards have recently been extensively investigated as an alternative to antibody-based protein quantitation (3). Here, we present a MRM-MS-based quantitation assay (in combination with stable-isotope-labeled phosphopeptides as internal standards) to determine the tyrosine phosphorylation occupancy at the two sites individually. The approach was initially developed to quantify phosphorylation occupancy at the two tyrosine sites of recombinantly expressed full-length Met protein. The LC-MRM assay was further utilized to directly analyze cell lysates expressing Met (gastric cancer SNU-5 cells) treated with the dual MET/ALK inhibitor PF-02341066 to demonstrate its applicability to quantify the inhibitor effect on phosphorylation occupancy of the two sites.

2 Materials

2.1 Drug-Treated SNU-5 Cell Lysate

1. Gastric cancer SNU-5 cells.
2. Met inhibitor PF-02341066.
3. Lysis buffer: 10 mM Tris–HCl, pH 7.4, 100 mM NaCl, 1 mM EDTA, 1 mM EGTA, 1 mM NaF, 20 mM $Na_4P_2O_7$, 2 mM Na_3VO_4, 1 % Triton X-100, 10 % glycerol, 0.1 % SDS, 0.5 % deoxycholate.

2.2 Protein Digestion and Sample Preparation

1. Trypsin, modified sequencing grade (Promega, Madison, WI).
2. PPS Silent surfactant (Protein Discovery, San Diego, CA).
3. FASP digestion kit (Protein Discovery, San Diego, CA).
4. Synthetic isotope-labeled and non-labeled peptides (New England Peptide, Gardner, MA).
5. *Escherichia coli* cell lysate digest (generated in house).

2.3 Liquid Chromatography-Mass Spectrometry

1. HPLC buffer A: 0.1 % formic acid.
2. HPLC buffer B: methanol (MeOH).
3. Xbridge BEH 300 C18 NanoEase LC column, particle size: 3.5 μm, column size: 300 μm×150 mm (Waters, Milford, MA).
4. Agilent 1200 series nano-pump system (Agilent Technologies, Inc., Palo Alto, CA).
5. TSQ Vantage mass spectrometer (Thermo Scientific, Fremont, CA).

2.4 Data Processing and Visualization

1. Xcalibur (Thermo Scientific).
2. Skyline (Dr. Michael MacCoss, University of Washington, Seattle, WA, USA).

3 Methods

3.1 Target Peptide Determination

MRM assay development for protein quantitation is initiated by selecting 3–5 target peptides per protein that possess predefined criteria (e.g., 8–25 amino acids long, not containing easily modified amino acids, do not contain residues known to be posttranslationally modified, have well-defined termini) and focusing on those peptides that give the best sensitivity and specificity for assay optimization. In this case since we were interested in specific phosphorylated sites, the peptides to be used in the assay were already defined. The Met amino acid sequence around the two tyrosine sites is LARDMYDKEYYSVHNKTG, with the two phosphorylated tyrosine residues in bold. After investigation using recombinant Met, the peptide DMYDKE**YY**SVHNK (containing one tryptic miscleavage site) was determined to be the target sequence for MRM assay development (see Note 1). Both isotope-labeled and non-labeled synthetic peptide sets of the sequences (peptides containing no phosphorylation, with Y1234 phosphorylated and with Y1235 phosphorylated) were ordered.

3.2 MRM Assay Generation and Optimization

The MS/MS spectra of triply charged peptides DMYDKE(pY)YSVHNK and DMYDKEY(pY)SVHNK acquired on a linear ion trap (FT-LTQ, Thermo Scientific) are shown in Fig. 2. Similar fragmentation patterns can also be obtained using a TSQ Vantage

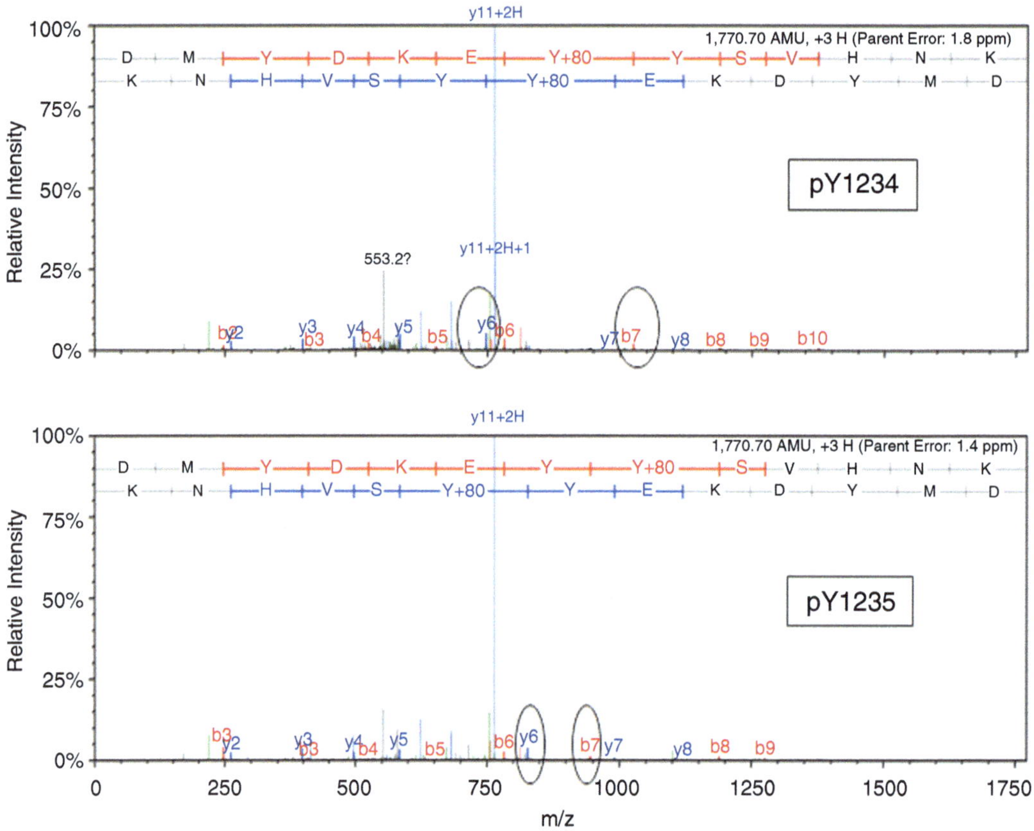

Fig. 2 Fragmentation spectra of triply charged peptides DMYDKE(pY[1234])YSVHNK and DMYDKEY(pY[1235])SVHNK visualized using Scaffold software. The two peptides show almost identical fragmentation patterns

mass spectrometer (data not shown). The two fragmentation patterns are almost identical due to the proximity of the two phosphorylation sites. The only fragment ions that differ between the two peptides are y6 and b7 ions as denoted within the spectra, which are too weak for sensitive MRM assay development. Therefore, instead of focusing on optimizing the transitions' intensity that could distinguish the two phosphopeptides, we chose to optimize the LC method to separate the two phosphopeptides prior to MRM-MS analysis. Baseline separation of the two phosphopeptides was achieved by optimizing both the gradient and flow rate using the synthetic peptides. Subsequently, MRM assay generation and optimization focused on transitions that provide the highest sensitivity for both phosphopeptides.

There are two main instrument parameters that must be optimized for MRM assay development using a TSQ Vantage once a stable ESI spray is established. These parameters are the S-lens voltage for obtaining the optimal precursor ion signal and the collision energy, which is ramped, while monitoring the targeted fragment ions to determine the fragment ions that provide the highest signal intensity. Both parameters can be optimized using direct infusion.

1. Peptide stocks are prepared by dissolving 1 mg of dry peptide in 5 % CAN. The concentrations of the peptide stock solutions are determined using amino acid analysis (see Note 2).
2. Prepare a peptide mixture containing both the light and heavy labeled peptides in 25 % MeOH/0.1 % formic acid using stock peptide solutions. Directly infuse this sample at 4 μl/min into the mass spectrometer using a syringe pump or LC system (see Note 3).
3. Operate the TSQ Vantage in profile mode with positive polarity: ESI voltage, 3,000 V; sheath gas, 5; capillary temperature, 270 °C; and Q2 collision gas pressure, 1.5 mbar. Both Q1 and Q3 are set for unit resolution.
4. Once a stable ESI signal is established, the S-lens voltage is tuned automatically while the precursor ion signal is monitored. As shown in Fig. 3a, the signals plateau between 130 and 150 V. In this assay, a S-lens voltage of 140 V was used for all the peptides monitored.
5. Figure 3b shows the final optimization of collision energies on the four most intense ions each determined in preliminary collision energy optimization of pY1235 light–heavy pair. The light and heavy peptides were mixed at different concentrations to enable better monitoring of the collision energy optimization curves. As shown in Fig. 3b, the optimum collision energies for different transitions were very similar for both heavy and light peptides. Finally, three transitions with the highest intensities were used in the final MRM assay (see Note 4).
6. The parameter values used in the optimized MRM assay to quantitate the pY1234 and pY1235 peptides are shown in Table 1.
7. Quantitate the peptides using the peak area ratio of the light version of the peptide over the known amount heavy version standard peptide spiked into the sample. Peak area is generated from total ion currents from all transitions monitored or selected transitions depending on situation (see Note 5).

3.3 Sample Preparation Procedure for In-Solution Digestion of Recombinant Protein

1. Dissolve 1 mg PPS detergent in 100 μl 50 mM ammonium bicarbonate to make a 1 % (w/v) PPS stock.
2. Take 10 μl of the simple recombinant protein sample solution and spike in 10 μl of the heavy peptide standards (see Note 6) diluted in 25 mM ammonium bicarbonate from heavy peptide stocks.
3. Add in 3 μl of the 1 % (w/v) PPS stock, vortex well, and spin down in a microcentrifuge.

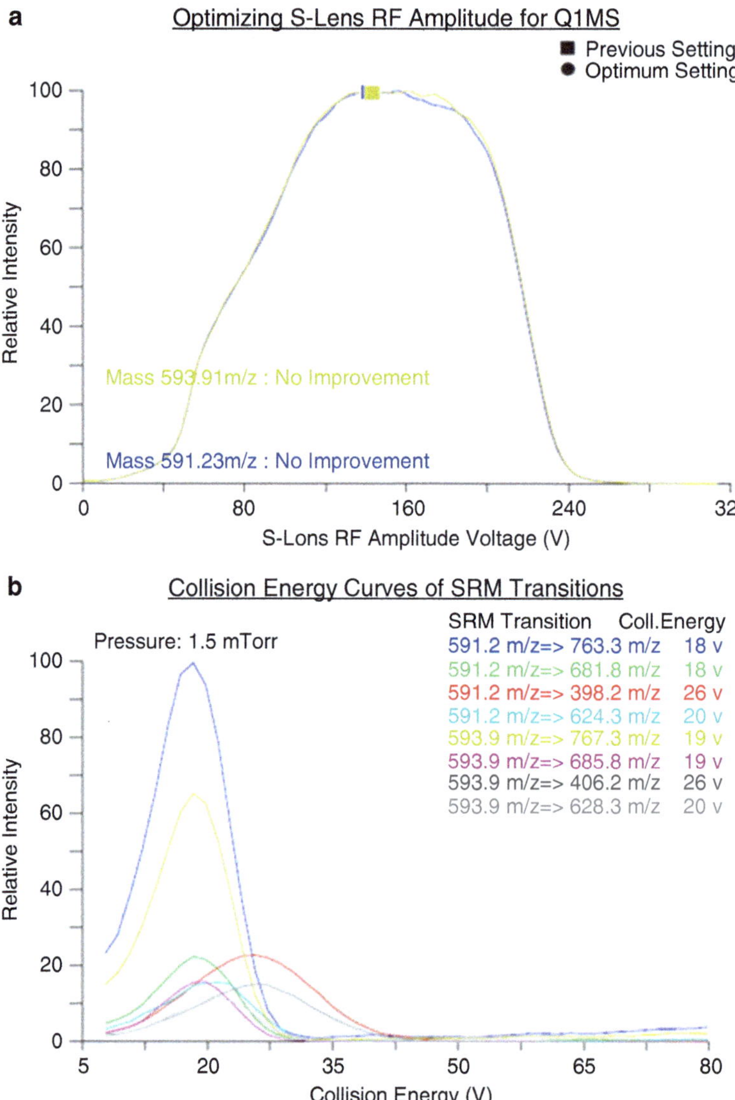

Fig. 3 Instrument parameters tuned for compound optimization of triply charged ions *m/z* 591.23 and 593.91 in MRM assay development. (**a**) S-lens RF amplitude tuning for precursor ion signal optimization. (**b**) Collision energy tuning for maximum fragment ion signal intensity

4. Add 1.5 μl of 100 mM DTT in 50 mM ammonium bicarbonate and incubate at 95 °C for 10 min to reduce the protein.
5. Cool the sample mixture to room temperature, add 3 μl of 100 mM IAA in 50 mM ammonium bicarbonate, and incubate at room temperature for 20 min in the dark to alkylate the protein (see Note 7).
6. Neutralize extra IAA by adding another 1.5 μl of 100 mM DTT in 50 mM ammonium bicarbonate to the solution and incubate at room temperature for another 10 min in the dark.

Table 1
Optimized mass spectrometry parameters used for the MRM assay to quantitate pY1234 and pY1235 in Met

Peptide sequence	Q1 (parent *m/z*)	Q3 (product *m/z*)	S-lens (V)	CE (V)	Dwell time (ms)
DMYDKE(pY)YSVHNK	591.23 (3+)	763.32 (Y_{11}, 2+)	140	19	50
DMYDKE(pY)YSVHNK	591.23 (3+)	681.79 (Y_{10}, 2+)	140	19	50
DMYDKE(pY)YSVHNK	591.23 (3+)	398.21 (Y_3, 1+)	140	26	50
DMYDKE(pY)YSVHNK	593.91 (3+)	767.33 (Y_{11}, 2+)	140	19	50
DMYDKE(pY)YSVHNK	593.91 (3+)	685.80 (Y_{10}, 2+)	140	19	50
DMYDKE(pY)YSVHNK	593.91 (3+)	406.23 (Y_3, 1+)	140	26	50
DMYDKEY(pY)SVHNK	591.23 (3+)	763.32 (Y_{11}, 2+)	140	19	50
DMYDKEY(pY)SVHNK	591.23 (3+)	681.79 (Y_{10}, 2+)	140	19	50
DMYDKEY(pY)SVHNK	591.23 (3+)	398.21 (Y_3, 1+)	140	26	50
DMYDKEY(pY)SVHNK	593.91 (3+)	767.33 (Y_{11}, 2+)	140	19	50
DMYDKEY(pY)SVHNK	593.91 (3+)	685.80 (Y_{10}, 2+)	140	19	50
DMYDKEY(pY)SVHNK	593.91 (3+)	406.23 (Y_3, 1+)	140	26	50

7. Add 1 μl trypsin solution reconstituted in 25 mM ammonium bicarbonate to the protein sample to create a trypsin to protein ratio of 1:50. Digest at 37 °C for 2 h while leaving the rest of the reconstituted trypsin solution on ice.
8. Add one more microliter trypsin solution (prepared in step 7) to the sample mixture and digest at 37 °C for an additional 2 h.
9. Following the vendor's protocol, PPS surfactant component is cleaved by adding 5 μl of 2 N HCl to the ~30 μl sample mixture. Vortex the sample and incubate at room temperature for 1 h.
10. Extract the peptides from the sample using a C18 ZipTip, reconstituted in 0.1 % TFA, and following the manufacturer's instructions.

3.4 Sample Preparation Procedure for In-Solution Digestion of Cell Lysates

1. 30 μg of cell lysates from SNU-5 cells treated with PF-02341066 (control (untreated), 10 and 100 nM PF-02341066 treated) are processed following filter-aided digestion procedure developed by the laboratory of Dr. Mattias Mann's group (4). A commercially available FASP digestion kit is used according to vendor protocol until the step just prior to trypsin digestion (see Note 8).
2. Add 100 fmol heavy labeled peptide standards mixture of pY1234 and pY1235 spiked into 65 μl 50 mM ammonium bicarbonate onto each of the three spin columns.

3. Add 10 μl of 0.06 μg/μl trypsin dissolved in 50 mM AMB to each spin column and mix the spin columns at 700 rpm for 1 min
4. Digest the samples overnight at 37 °C. Elute the tryptic peptides the following day according to the FASP digestion kit protocol.
5. Further clean up the peptides using Pepclean C18 spin columns, reconstituted in 0.1 % TFA. The peptides are now ready to be analyzed using the LC-MRM-MS assay.

3.5 Calibration Curve Generation and Quantitation Procedure

A standard curve is generated by spiking different levels of light peptide standards with a set level of heavy peptides into a matrix (~0.2–0.5 μg/μl) for accurate quantitation. The matrix used (e.g., serum, plasma, cell lysate, urine, etc.) to prepare the calibrant mixtures should be similar to the intended sample matrix both in property and in complexity (see Note 9).

1. Spike the light peptide standards into a 0.5 μg/μl sample matrix in 0.1 % TFA to final concentrations of 20, 5, 2, 0.2, and 0 fmol/μl (blank). Spike the heavy labeled peptide standards into the samples at a constant final level of 2 and 0.2 fmol/μl for calibration curve generation (see Note 10).
2. Inject a 5 μl sample onto LC column. Set up the LC to operate at a flow rate of 4 μl/min. After the initial 20 min wash step at 5 % B, initiate a shallow gradient of 0.275 %/min B buffer increasing from 16 % B to 27 % B around the elution window of the two phosphopeptides to achieve the baseline separation needed for their quantitation.
3. Operate the TSQ Vantage mass spectrometer using the same instrument parameters shown earlier in Subheading 3.2. Data acquisition is performed using MRM method with optimized S-lens voltage, collision energy, transition dwell time, and specific transitions shown in Table 1. Figure 4 shows a sample MRM trace of the two baseline-separated heavy labeled phosphopeptides spiked in *E. coli* tryptic digest peptide matrix (see Note 11).
4. Run each calibrant sample in triplicate using the optimized LC-MRM-MS assay.
5. Create a data processing method for quantitating the peptides using a representing raw data file and Xcalibur software. In the processing method, parameters involved in peak detection and peak integration have to be optimized according to the raw file. In addition, internal standards, retention time, and calibration levels are also specified in the processing method.
6. Batch process all raw files of the calibrant and sample analyses using the processing method generated in step 5. Calibration

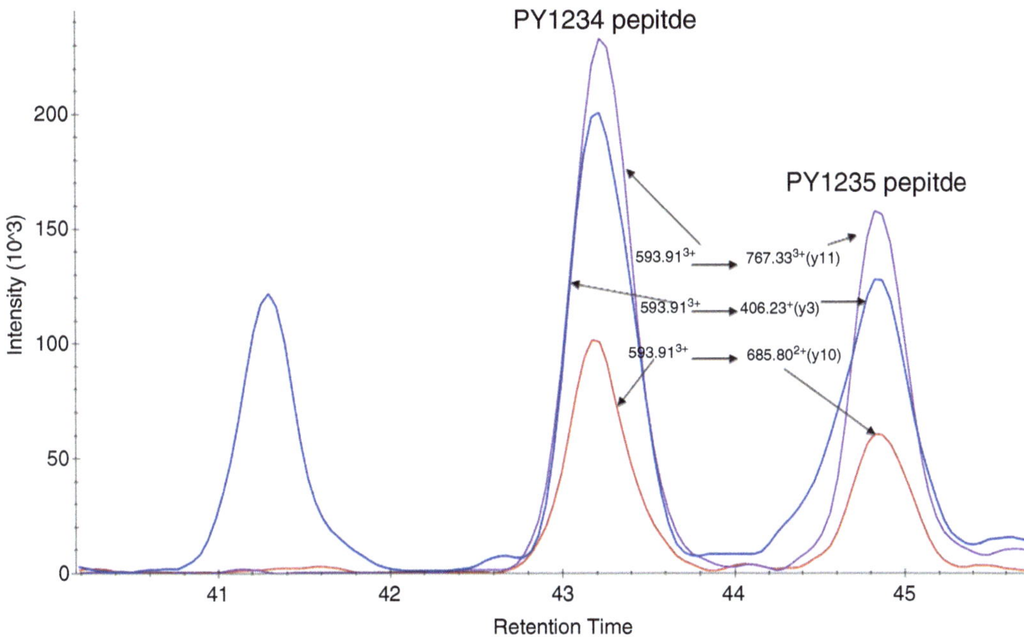

Fig. 4 MRM transition traces of baseline-separated heavy labeled synthetic phosphopeptides DMYDKE(pY^{1234}) YSVHNK and DMYDKEY(pY^{1235})SVHNK. Three identical transitions were monitored for both peptides

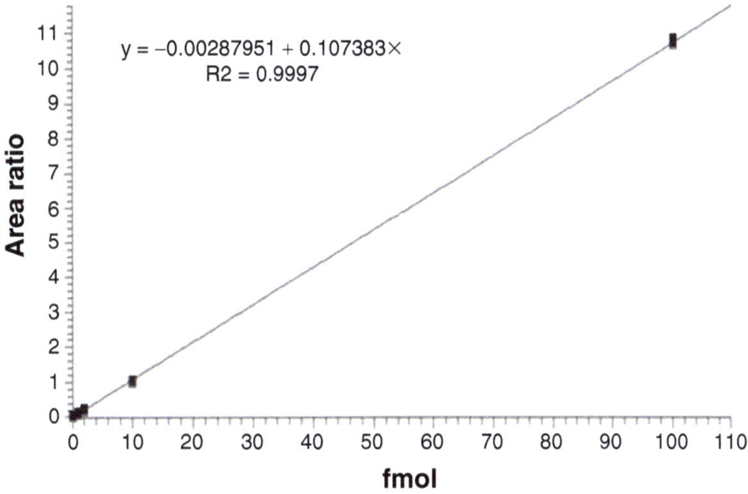

Fig. 5 Sample calibration curve generated for DMYDKEY(pY^{1235})SVHNK using a tryptically digested *E. coli* lysate as matrix

curve and quantitation result can then be visualized using Quan function of the Xcalibur software (see Note 12). Figure 5 shows a typical calibration curve generated for pY1235 peptide where $1/x$ weighing factor is used. Table 2 shows detailed information regarding the calibration curve including standard deviation (see Note 13).

Table 2
Sample pY1235 details of data used and specifics of calibration curve used to quantitate pY1235

Calibrant	Replicates	Specified (fmol)	Light/heavy area ratio	Calculated (fmol)	Percentage difference	Percentage RSD-AMT
A	A1	0.000	0.005	0.077	1.00	13.14
	A2	0.000	0.005	0.070	1.07	
	A3	0.000	0.007	0.090	1.00	
B	B1	1.000	0.113	1.078	7.83	8.98
	B2	1.000	0.096	0.919	-8.15	
	B3	1.000	0.098	0.935	-6.46	
C	C1	2.000	0.206	1.949	-2.54	2.59
	C2	2.000	0.213	2.008	0.38	
	C3	2.000	0.202	1.907	-4.66	
D	D1	10.000	1.041	9.717	-2.83	2.43
	D2	10.000	1.048	9.785	-2.15	
	D3	10.000	1.001	9.351	2.43	
E	E1	100.000	10.85	101.065	1.07	0.61
	E2	100.000	10.783	100.444	0.44	
	E3	100.000	10.719	99.843	-0.16	

7. pY1234 and pY1235 of in-house expressed recombinant Met proteins generally show about 23 % occupancy at pY1234 and 5 % occupancy at pY1235 (see Note 13).
8. pY1235 of the SNU-5 cell lysate is below detection limit in the whole cell lysate matrix and an enrichment step is being investigated to be included for the quantitation assay (see Note 14). pY1234 occupancy of the 10 nM drug-treated SNU-5 cell lysate does not show significant change compared to control. pY1234 occupancy of the 100 nM drug-treated SNU cell lysate decreased to about 7 % compared to 21 % in control cell lysate (see Note 13).

4 Notes

1. LARDMYDKEYYSVHNKTG is the sequence around the two tyrosine sites we are measuring. EYYSVHNK would have been the tryptic peptide target for tyrosine phosphorylation quantitation assay development if we only used information from an in silico digestion. However, after investigation using recombinant Met, the peptide sequence DMYDKE**YY**SVHNK with one miscleavage site was shown to be the major tryptic digestion product. Therefore, for targeted analysis assay development, especially when a specific location is involved, preliminary studies

using recombinant proteins determine the sequence targets for the specific sites are more reliable than in silico calculations.

2. Conduct amino acid analysis (AAA) of the synthetic peptide solutions to determine the absolute peptide concentration. This step is pivotal for accurate quantitation later. We have obtained drastically different results from AAA analysis from different labs for the same peptide samples. Therefore, care must be taken both to provide multiple aliquots for the same samples. Replicate measurements should be recorded and the results checked carefully to make sure quantitation is accurate.
3. To optimize by direct infusion, the peptides should be in a solution close to their LC elution composition (~25 % MeOH in this case) and delivered using flow rate that will be used in the final assay.
4. For the third transition, fragment ions *m/z* 398.2 and 406.2 are used as they provide slightly better signals. However, fragment ions *m/z* 624.3 and 628.3 could be used instead if transition specificity becomes a concern due to background contamination. These transitions have higher masses compared to the isolated precursor ions; therefore they tend to have better specificity compared to transitions with masses that are lower than the precursor ions.
5. If one of the transitions has interference from background ions preventing it from providing accurate quantitation information, the XIC of the corresponding transition pair could be discarded from both heavy and light peptide peak areas generated.
6. Heavy peptide standards are always spiked in before tryptic digestion to compensate for sample loss during the digestion and sample cleanup process. Spiking in heavy peptide standards at this step also compensates for the presence of trypsin miscleavage site. A heavy isotope-labeled full-length protein would be the ideal correction factor; however, a full-length protein was not possible in our case as we are interested in phosphorylation site quantitation.
7. The alkylation step is important even though the peptides targeted do not contain cysteine residues. Control experiments were conducted in which both samples were reduced by DTT but only one sample was alkylated. The non-alkylated sample showed a lot more incomplete digestion compared to the alkylated sample.
8. Filter-aided sample preparation is used for cell lysate sample preparation not only to reduce the amount of detergent present but also due to eliminate other lysis buffer components (e.g., protein denaturant, etc.) prior to tryptic digestion.
9. Using the same matrix for the calibrants simulates the measurement of the target in the real biological sample enabling more accurate quantitation. Using the matrix also minimizes

sample loss at low calibrant levels providing better linear calibration curves compared to without any matrix. Therefore, a simple tryptic peptide solution from a single protein digest (such as BSA digest) or a simple tryptic peptide solution can be used as matrix when generating calibration curve for recombinant Met tyrosine phosphorylation quantitation and an *E. coli* lysate tryptic digest mixture can be used to generate calibration curve for endogenous Met in cell lysates.

10. With 5 μl injections, the calibration curve is going to have about a two orders of magnitude dynamic range up to 100 fmol on column. This range is sufficient to measure phosphorylated peptides that normally exist in low stoichiometry. The calibration curve, however, can be extended using higher concentration of calibrants.
11. As seen in Fig. 4, baseline separation of the two phosphopeptides was achieved using the optimized LC method. Monitoring multiple transitions for the same precursor ion not only provided higher quantitation sensitivity but also assisted in distinguishing the correct peaks from background.
12. Other software could also be used for calibration curve generation as long as peak area of the various MRM result can be generated. Xcalibur is routinely used in our laboratory as it generates calibration curve and standard deviation information automatically. We also use Skyline software to visualize the MRM data (5). This software has been beneficial in designing MRM assays, but it does not generate calibration curve automatically within the software.
13. Phosphorylation occupancy results in this study was based on monitoring quantitation of unmodifed, pY1234 and pY1235 version of the same peptide sequence. The assay is being further developed with monitoring different peptides of Met for more accurate phosphorylation occupancies at the two sites.
14. pY1235 typically has a much lower stoichiometry than pY1234. Including a simple phosphopeptide enrichment step before LC-MRM-MS analysis not only enriches phosphopeptides but also simplifies the complexity of the sample matrix. The combination of both should provide the sensitivity needed for pY1235 quantitation from cell lysate.

Acknowledgments

This project has been funded in whole or in part with federal funds from the National Cancer Institute, National Institutes of Health, under Contract HHSN261200800001E. The content of this publication does not necessarily reflect the views or policies of the

Department of Health and Human Services, nor does mention of trade names, commercial products, or organizations imply endorsement by the United States Government.

References

1. Trusolino L, Bertotti A, Comoglio PM (2010) MET signaling: principles and functions in development, organ regeneration and cancer. Nat Rev Mol Cell Biol 11:834–848
2. Gentile A, Trusolino L, Comoglio PM (2008) The Met tyrosine kinase receptor in development and cancer. Cancer Metastasis Rev 27:85–94
3. Meng Z, Veenstra TD (2011) Targeted mass spectrometry approaches for protein biomarker verification. J Proteomics 74:2650–2659
4. Winiewski JR, Zougman A, Nagaraj N, Mann M (2009) Universal sample preparation method for proteome analysis. Nat Methods 6:359–362
5. Maclean B, Tomazela DM, Shulman N et al (2010) Skyline: an open source document editor for creating and analyzing targeted proteomics experiments. Bioinformatics 26:966–968

Chapter 16

Preparation of Human Serum for Prolactin Measurement by Multiple Reaction Monitoring Mass Spectrometry

Timothy J. Waybright, Xia Xu, Jessica M. Faupel-Badger, and Zhen Xiao

Abstract

The measurement of the protein hormone prolactin (PRL) in biological samples has developed over the years into a routine clinical assay aiding the diagnosis of multiple medical conditions. PRL is known to exist in multiple isoforms circulating throughout the body. Current methodologies for measuring the PRL levels typically involve a variety of immunoassays. However, most of these tests are not capable of distinguishing between the different isoforms. To address this need, we have developed a highly specialized method employing multiple reaction monitoring mass spectrometry (MRM-MS) capable of monitoring seven distinct peptides from two of the most common prolactin isoforms (the 23 kDa PRL and its 16 kDa N-terminal cleavage product). Since serum is the main source of clinical specimen for the measurement of prolactin isoforms, the method described in this chapter is focused on the approach to processing whole serum samples for prolactin analysis via reversed-phase liquid chromatography (RPLC) and MRM-MS.

Key words Prolactin, Mass spectrometry, Multiple reaction monitoring, Liquid chromatography

1 Introduction

The protein prolactin is a hormone that is primarily secreted from the anterior pituitary gland via specialized lactotropic cells. Although it is largely associated with multiple roles in lactation after pregnancy, its function can be traced to over 300 different biological activities (1). The secreted prolactin has been detected in various parts of the body, including the brain, placenta, uterus, mammary glands, and also in milk (2–4). Besides its physiological role in lactation, prolactin has been indicated in various pathological conditions such as epileptic seizures (5), breast cancer (6–13), prostate cancer (14), anorexia nervosa and obesity (15), and erectile dysfunction (16).

Prolactin exists in the body as different sized isoforms, consisting of a "big big" form, or macroprolactin (>100 kDa) (17), a "big" species (approximately 40–60 kDa), the predominant 23 kDa

Ming Zhou and Timothy Veenstra (eds.), *Proteomics for Biomarker Discovery: Methods and Protocols*, Methods in Molecular Biology, vol. 1002, DOI 10.1007/978-1-62703-360-2_16, © Springer Science+Business Media, LLC 2013

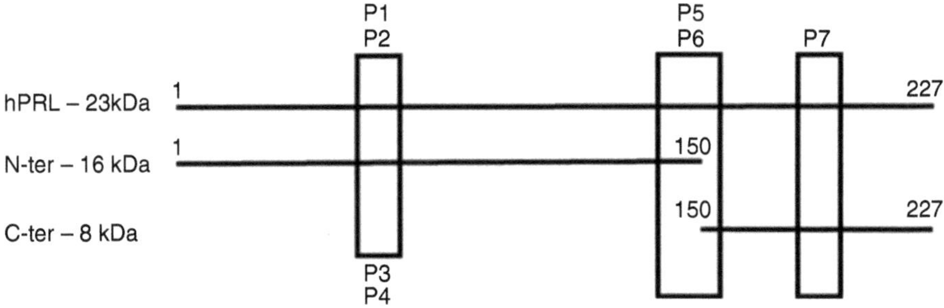

Fig. 1 Seven distinct peptides that showed ample LC-MS signals were selected from various regions of recombinant human prolactin (hPRL) to represent the most common isoforms

monomer (18), and a truncated 16 kDa N-terminal cleavage product (19–23). The two larger proteins are thought to have minimal roles, if any, in important biological functions. The majority of studies have therefore focused on the 23 kDa species and, in recent years, on the 16 kDa form.

In clinical and laboratory studies where prolactin levels are required, the challenges are compounded due to the limitation of the present techniques to precisely and quantitatively differentiate the various isoforms in biological specimens. Instead, the total prolactin level is usually reported based on immunoassay method employed. Immunoblotting, another method that can distinguish between the different isoforms, is not adaptable to processing samples in a high-throughput manner. Also, there is some uncertainty whether all isoforms of prolactin are being detected to satisfy the investigators' interests.

Our laboratory has developed a novel method whereby the peptide fragments of the 23- and 16-kDa human prolactin species can be evaluated and measured simultaneously using high-performance liquid chromatography (HPLC) coupled with multiple reaction monitoring-tandem mass spectrometry (MRM-MS/MS) (Fig. 1). This method affords a number of advantages: (1) the ability to measure the desired peptide(s) of interest via reproducible retention times, parent ions, and transition ions; (2) the use of internal and external calibrants to calculate and monitor coefficient of variations (CVs); (3) the adaptability of the procedure to monitor not only prolactin peptides but at the same time other compounds of interest during the same analysis; and (4) an automated analytical platform with the ability to process samples virtually 24 h a day.

2 Materials

2.1 Serum Sample Collection and Procurement

2.1.1 Blood Collection

The blood draw and serum sample collection procedures were approved by the NCI/NIH Institutional Review Board.

2.1.2 Serum Procurement

1. 1.5 mL Safe-Lock Polypropylene tubes (Eppendorf, Hauppauge, NY).
2. 1 mL pipette with tips (Rainin Instrument, Oakland, CA).
3. Falcon Blue Max 15 mL Polypropylene conical tubes (Becton Dickinson, Franklin Lakes, NJ).
4. Centrifuge.

2.2 Enzymatic Digestion and Peptide Extraction

2.2.1 Enzymatic Digestion

1. Recombinant human Prolactin protein (National Hormone & Peptide Program, Harbor-UCLA Medical Center, Torrance, CA) (aliquoted into 20 μL per vial and frozen at −80 °C).
2. 25 mM Ammonium bicarbonate (NH_4HCO_3) at pH 8.4 (Sigma, St. Louis, MO).
3. Ultrapure water (double distilled, deionized >18 Ω, NANOPure Diamond water system, Barnstead International, Dubuque, IA).
4. Porcine sequencing grade modified trypsin (Promega, Madison, WI).
5. 100 mM Bond Breaker™ TCEP (Thermo Scientific, Rockford, IL).
6. Pooled human male serum (Sigma, St. Louis, MO).
7. Third-trimester human female serum (Bioreclamation, Hicksville, NY).
8. 37 °C Incubator.

2.2.2 Peptide Extraction

1. Seven stable isotope-labeled (K* = universally labeled Lysine, $^{13}C_6$, $^{15}N_2$-K; R* = universally labeled arginine, $^{13}C_6$, $^{15}N_4$-R) human prolactin peptides (SI-PRL) were purchased from New England Peptide (Gardner, MA), resuspended at 2 ng/μL in 50 % methanol with 0.1 % formic acid, aliquoted into 110 μL per vial, and frozen at −80 °C (Fig. 1).
2. Trifluoroacetic acid (TFA) (Sigma, St. Louis, MO).
3. Vortex.
4. Methanol (MeOH) (EMD Chemicals, Gibbstown, NJ).
5. Acetonitrile (Fisher Scientific, Waltham, MA).
6. 3M Empore™ High Performance Extraction Disk Plate (3 M, St. Paul, MN).
7. 1 mL 96-well polypropylene plate (VWR, Randor, PA).
8. Plate vacuum manifold system (Orochem Technologies, Lombard, IL).
9. 2 mL wide-opening snap-clear vials with caps and rack (Fisher Scientific, Waltham, MA).
10. 1 mL pipette with tips (Rainin Instrument, Oakland CA).
11. 200 μL multichannel pipette with tips (Rainin Instrument, Oakland CA).

12. 20 μL pipette with tips (Rainin Instrument, Oakland CA).
13. Speed Vac.
14. Centrifuge.

2.3 Final Sample Preparation

1. Methanol (MeOH) (EMD Chemicals, Gibbstown, NJ).
2. Ultrapure water (double distilled, deionized >18 Ω, NANOPure Diamond water system, Barnstead International, Dubuque, IA).
3. Autosampler vials (Wheaton, Millville, NJ).
4. Formic acid (FA) (Sigma, St. Louis, MO).
5. 20 μL pipette with tips (Rainin Instrument, Oakland CA).
6. Sonicator.
7. Centrifuge.
8. Vortex.
9. Speed Vac.

3 Methods

3.1 Serum Sample Collection and Procurement

Serum samples are obtained from various sources (see Note 1).

3.1.1 Blood Collection

3.1.2 Serum Procurement

1. Blood samples in the collection tubes are placed at room temperature to clot for 30 min, and then centrifuged for 1 h at 5 °C to separate serum from blood cells (see Note 2).
2. Clarified sera are transferred as 1 mL aliquots and placed into 1.5 mL polypropylene tubes. Ten to fifteen aliquots can be generated.
3. To make aliquots in larger volumes, the remaining clarified sera are carefully transferred and put into 15 mL conical tubes.
4. All samples are frozen at −80 °C until analysis.

3.2 Enzymatic Digestion and Peptide Extraction

3.2.1 Enzymatic Digestion

1. Remove the appropriate samples from the freezer and allow to thaw on ice (see Note 3).
2. To 20 μL of male control serum, add 1 μL of a 1 ng/μL solution of recombinant prolactin.
3. To 25 μL of all other samples, add 30 μL of 25 mM ammonium bicarbonate (see Note 4).
4. To all of the samples, add 2 μL of the 100 mM TCEP solution.

5. Prepare the trypsin solution by adding 200 μL of 25 mM ammonium bicarbonate to each vial.
6. Vortex the trypsin solution gently and sonicate.
7. Based on the estimated amount of proteins in serum, add 25 μL of the trypsin solution to the male serum vial.
8. Similarly, add 44 μL of the trypsin solution to all of the other sample vials.
9. Vortex all of the vials gently, and then centrifuge briefly to ensure all of the liquid is settled to the bottom of the vial.
10. Place the vials in a covered box in the incubator overnight, shaking gently overnight at 37 °C.

3.2.2 Peptide Extraction

1. After the overnight tryptic digestion, remove samples from the incubator.
2. Remove the appropriate number of vials of stable-isotope-labeled prolactin peptides from freezer and thaw at room temperature.
3. To each vial of the digested serum samples, add 25 μL of the stable isotope solution.
4. To the male serum vial, add 330 μL of 0.1 % trifluoroacetic acid.
5. To all of the other vials, add 375 μL of trifluoroacetic acid.
6. Vortex the digested samples gently and centrifuge to ensure all of the liquid is settled at the bottom of the vial.
7. Place a 3M Empore™ High Performance Extraction Disk Plate in the vacuum manifold, and secure a waste container in the collection part of the apparatus (see Note 5).
8. To pre-equilibrate the disk plate, add 1 mL of methanol the appropriate number of wells.
9. Gently apply vacuum to the plate, making sure not to pull a vacuum too quickly, or to let the vacuum pull through empty wells for prolonged period of time (see Note 6).
10. After the methanol has gone through, stop vacuum and add 1 mL of 0.1 % trifluoroacetic acid (TFA) to the wells.
11. Gently reapply vacuum to the plate, making sure not to pull a vacuum too quickly, or to let the vacuum pull through empty wells for prolonged period of time.
12. Stop the vacuum after the TFA solution passes through the wells. Remove the waste container from the vacuum manifold and replace it with a 96-well polypropylene plate.
13. Transfer the samples from the tubes to individual wells in the disk plate, taking care not to cross-contaminate the samples in different wells.

14. Gently apply vacuum to the plate, making sure not to pull a vacuum too quickly, or to let the vacuum pull through empty wells for prolonged period of time.
15. Repeat steps 13 and 14 so that the samples are allowed to pass through the cartridge five times (see Note 7).
16. After the fifth passing, stop the vacuum. Remove the 96-well plate and replace with the waste container.
17. To wash the disks, add 1 mL of 0.1 % trifluoroacetic acid to each well.
18. Gently apply vacuum to the plate, making sure not to pull a vacuum too quickly, or to let the vacuum pull through empty wells for too long.
19. Repeat steps 17 and 18 twice so that the wells have been washed three times (see Note 8).
20. After the third wash, remove the waste container and replace with a new 96-well receiving plate to collect eluants.
21. To elute peptides, add 1 mL of methanol to each well.
22. Gently apply vacuum to the plate, making sure not to pull a vacuum too quickly, or to let the vacuum pull through empty wells for prolonged period of time. Stop the vacuum after the methanol solution passes through the disks.
23. Depending upon the number of receiving wells used for the methanol elution, a new 96-well plate may be required for the next step of elution. Or rotate the plate that is already in the vacuum manifold to make empty receiving wells available for the next step.
24. To further elute peptides, add 800 μL of 80 % acetonitrile/0.1 % trifluoroacetic acid to each well of the disk plate.
25. Gently apply vacuum to the disk plate. During this step, you may increase the vacuum flow and run the wells dry to make sure all of the liquid has been pulled off of the cartridge.
26. Combine eluants from steps 22 and 25. Transfer the samples to the wide-opening snap-clear vials (see Note 9).
27. Place the vials in the Speed Vac system and lyophilize to complete dryness. If samples are not processed in the next step immediately after drying, tightly cap and store at −80 °C.

3.3 Final Sample Preparation

1. Remove the samples from the Speed Vac or from the freezer and bring to room temperature.
2. To each sample, add 100 μL of methanol and vortex gently.
3. Sonicate each sample for approximately 1 min and gently vortex again.
4. Transfer each sample to an autosampler vial (see Note 10).

5. Place the vials in the Speed Vac system and lyophilize to complete dryness.
6. After the samples are dry, remove them from the Speed Vac.
7. To each sample add 20 μL of 5 % MeOH/95 % 0.1 % formic acid (starting mobile phase for sample analysis).
8. Gently mix with repeated aspirations by the pipette (see Note 11).
9. Cap the vials and place in a centrifuge.
10. Spin the samples for 30 min at 1,000 × *g*.
11. Remove the samples from centrifuge after spinning.
12. Transfer 16 μL of each sample to a separate autosampler vial and cap. The sample is now ready for MRM-MS/MS analysis (see Notes 12 and 13).

4 Notes

1. Generally each laboratory has its own source and standard operating procedure of obtaining blood draws directly, or from an intermediary clinic. For this study, samples are obtained from informed donors in the NCI Research Donors Program and sent to our laboratory within 30 min of collection.
2. The blood samples may be centrifuged to clarify the serum in various ways. We chose to centrifuge the samples in the collection tubes due to the convenience to remove the precipitated blood cells that we have no further use.
3. This part of our study involves two extra samples that we run with every batch. The first of these is a pooled male human serum sample purchased from Sigma. We use 20 μL of this sample as a control, to which we add 1 ng of the recombinant human prolactin (hPRL) as a reference. The second sample is serum from a healthy female donor in the third trimester of pregnancy. Since most female serum samples from this period of pregnancy contain very high level of prolactin, this second sample is treated as the positive control for unknown samples and we expect to obtain a significantly high measurement of prolactin in this sample than in most unknown samples.
4. The ammonium bicarbonate is added to ensure that the pH of the samples is sufficiently basic at 8.4 for trypsin to be effective.
5. Initial method development for this step was performed with single cartridges made of the same extraction disk. The only difference between the methods development disks and the plate disks is that the plate disks are slightly smaller than that of individual cartridges. The comparison using the same sample

showed no appreciable difference between the extraction efficiencies of the cartridge and the plate approaches.

6. The technical notes for the plate disks recommend a flow rate of about 2 mL/min. Based on our observations, this recommendation is a conservative rate, and a slightly higher flow rate does not adversely affect the extraction results.
7. Multiple passing of the samples through the same cartridge is highly recommended. Preliminary studies (data not shown) showed that the sample needed to be passed through the cartridge a minimum of three times to obtain reliable, reproducible results.
8. After the last wash, we recommend using a clean Kim-Wipe tissue cloth to dab the bottom of the cartridge plate gently. This would help remove all of the excess wash liquid from the samples that might otherwise be transferred to the sample vials and carry with it unwanted salts.
9. Since the concentration of prolactin is potentially very low in most serum samples, care must be taken to minimize sample loss throughout the sample preparation steps.
10. An additional 100 μL of methanol may be used to wash the vial.
11. When mixing with the pipette, you would notice that there may be particles at the bottom of the vial. While this does not affect the prolactin peptide solubility, you must ensure that mixing is thorough.
12. The volume transferred (16 μL) ensures that there will be no particulates transferred from one vial to the other. These may clog the LC lines and cause unwanted problems during analysis. The injection of one half of the sample (8 μL) per analysis is recommended.
13. Due to the low abundance of prolactin in serum, our laboratory has begun to explore methods by which sample concentrations can be further enhanced, such as immunodepletion of the high-abundant proteins with multiple affinity removal system (MARS) columns, or partial enrichment by filtration through molecular weight cutoff filters. These steps could potentially increase the serum protein digestion efficiency, extend the HPLC column lifetime, and improve the prolactin signal strength during MRM detection.

Acknowledgments

This project has been funded in whole or in part with federal funds from the National Cancer Institute, National Institutes of Health, under Contract HHSN261200800001E. The content of this publication does not necessarily reflect the views or policies of the Department of Health and Human Services, nor does mention of trade names, commercial products, or organizations imply endorsement by the United States Government.

References

1. Bole-Feysot C, Goffin V, Edery M, Binart N, Kelly PA (1998) Prolactin (PRL) and its receptor: actions, signal transduction pathways and phenotypes observed in PRL receptor knockout mice. Endocr Rev 19:225–268
2. Freeman ME, Kanyicska B, Lerant A, Nagy G (2000) Prolactin: structure, function, and regulation of secretion. Physiol Rev 80:1523–1631
3. Corbacho AM, Macotela Y, Nava G et al (2000) Human umbilical vein endothelial cells express multiple prolactin isoforms. J Endocrinol 166:53–62
4. Wennbo H, Törnell J (2000) The role of prolactin and growth hormone in breast cancer. Oncogene 19:1072–1076
5. Chen DK, So YT, Fisher RS (2005) Use of serum prolactin in diagnosing epileptic seizures. Neurology 65:668–675
6. Clevenger CV, Furth PA, Hankinson SE, Schuler LA (2003) The role of prolactin in mammary carcinoma. Endocr Rev 24:1–27
7. Tworoger SS, Eliassen AH, Sluss P, Hankinson SE (2007) A prospective study of plasma prolactin concentrations and risk of premenopausal and postmenopausal breast cancer. J Clin Oncol 25:1482–1488
8. Tworoger SS, Hankinson SE (2006) Prolactin and breast cancer risk. Cancer Lett 243: 160–169
9. Tworoger SS, Hankinson SE (2008) Prolactin and breast cancer etiology: an epidemiologic perspective. J Mammary Gland Biol Neoplasia 13:41–53
10. Vonderhaar BK (1998) Prolactin: the forgotten hormone of human breast cancer. Pharmacol Ther 79:169–178
11. Vonderhaar BK (1999) Prolactin involvement in breast cancer. Endocr Relat Cancer 6: 389–404
12. Faupel-Badger JM, Sherman ME, Garcia-Closas M et al (2010) Prolactin serum levels and breast cancer: relationships with risk factors and tumor characteristics among pre- and postmenopausal women in a population-based case–control study from Poland. Br J Cancer 103:1097–1102
13. Linher-Melville K, Zantinge S, Sanli T et al (2011) Establishing a relationship between prolactin and altered fatty acid β-Oxidation via carnitine palmitoyl transferase 1 in breast cancer cells. BMC Cancer 11:56–70
14. Rouet V, Bogorad RL, Kayser C et al (2010) Local prolactin is a target to prevent expansion of basal/stem cells in prostate tumors. Proc Natl Acad Sci USA 107:15199–15204
15. Baranowska B, Radzikowska M, Wasilewska-Dziubinska E, Roguski K, Morowiec M (2006) The role of VIP and somatostatin in the control of GH and prolactin release in anorexia nervosa and in obesity. Ann NY Acad Sci 921:443–455
16. Paick JS, Yan JH, Kim SW, Ku JH (2006) The role of prolactin levels in the sexual activity of married men with erectile dysfunction. BJU Int 98:1269–1273
17. Hattori N (1996) The frequency of macroprolactinemia in pregnant women and the heterogeneity of its etiologies. J Clin Endocrinol Metab 81:586–590
18. Smith CR, Norman MR (1990) Prolactin and growth hormone: molecular heterogeneity and measurement in serum. Ann Clin Biochem 27:542–550
19. Clapp C, González C, Macotela Y (2006) Vasoinhibins: a family of N-terminal prolactin fragments that inhibit angiogenesis and vascular function. Front Horm Res 35:64–73
20. Clapp C, Martial JA, Guzman RC, Rentier-Delure F, Weiner RI (1993) The 16-kilodalton N-terminal fragment of human prolactin is a potent inhibitor of angiogenesis. Endocrinology 133:1292–1299
21. Piwnica D, Touraine P, Struman I et al (2004) Cathepsin D processes human prolactin into multiple 16 K-like N-terminal fragments: study of their antiangiogenic properties and physiological relevance. Mol Endocrinol 18: 2522–2542
22. Faupel-Badger JM, Ginsburg E, Fleming JM et al (2010) 16-kDa prolactin reduces angiogenesis, but not growth of human breast cancer tumors *in vivo*. Horm Cancer 1:71–79
23. Hilfiker-Kleiner D, Kaminski K, Podewski E et al (2007) A cathepsin D-cleaved 16-kDa form of prolactin mediates postpartum cardiomyopathy. Cell 128:589–600

Chapter 17

Label-Free Quantitative Shotgun Proteomics Using Normalized Spectral Abundance Factors

Karlie A. Neilson, Tim Keighley, Dana Pascovici, Brett Cooke, and Paul A. Haynes

Abstract

In this chapter we describe the workflow used in our laboratory for label-free quantitative shotgun proteomics based on spectral counting. The main tools used are a series of R modules known collectively as the Scrappy program. We describe how to go from peptide to spectrum matching in a shotgun proteomics experiment using the XTandem algorithm, to simultaneous quantification of up to thousands of proteins, using normalized spectral abundance factors. The outputs of the software are described in detail, with illustrative examples provided for some of the graphical images generated. While it is not strictly within the scope of this chapter, some consideration is given to how best to extract meaningful biological information from quantitative shotgun proteomics data outputs.

Key words Shotgun proteomics, Label-free, Quantitative proteomics, Spectral counting, Normalized spectral abundance factors

1 Introduction

The field of shotgun proteomics has changed considerably in recent years as it has become less descriptive and more quantitative. Nowadays it has become increasingly common to see published studies which contain an abundance and richness of data which was once thought unattainable. There are many papers in the literature which include thousands of detailed individual protein measurements within a given cellular system, each of which includes the identity and relative amount of the protein in question. This can be performed for multiple samples, such as numerous points across a developmental or stress-imposition time course. The output of such experiments can be overwhelmingly large, but successful analysis of such data sets can reveal trends at the "big picture" level which are not discernible by other means.

In our laboratory we employ label-free quantitation using normalized spectral abundance factors (NSAFs). It must be emphasized

Ming Zhou and Timothy Veenstra (eds.), *Proteomics for Biomarker Discovery: Methods and Protocols*, Methods in Molecular Biology, vol. 1002, DOI 10.1007/978-1-62703-360-2_17, © Springer Science+Business Media, LLC 2013

that this is just one of many such techniques that can be used; we use this because it is simple, robust, and inexpensive, and relies on sound mathematical principles. Similarly, there are a myriad of possibilities for how to take biological samples of a given cell or tissue type, and transform them into a set of fractionated peptides or proteins suitable for mass spectrometric analysis. We present in this chapter one of the main techniques we use in our laboratory, which is SDS-PAGE (sodium dodecyl sulfate-polyacrylamide gel electrophoresis) fractionation of proteins prior to in-gel trypsin digestion. Again, we use this separation technique because it is simple, inexpensive, and robust. SDS-PAGE fractionation of proteins also has one advantage over many other techniques in that the SDS buffer is an excellent protein-solubilizing agent, especially in comparison to other, milder detergents.

There are two main ways of measuring changes in protein abundance without using metabolic or isotopic labels that involve measuring precursor ion intensity or counting spectra assigned to a particular protein (1–3). Approaches involving precursor ion intensity are based on the well-established analytical principle that the area under a chromatographic elution curve is proportional to the amount of eluting compound. This works very well for relatively simple mixtures but tends to be less accurate as peptide mixtures become increasingly complex. The approach relies on very accurate and reproducible chromatography, as peptide peaks from different chromatographic elution profiles need to be precisely aligned for the analysis to proceed.

The other main class of approach used in label-free quantitation involves counting spectra identified for the peptides in a protein, also known as spectral counting (4). This approach relies on the simple observation that as more of a digested protein is analyzed in a given mass spectrometric system, more peptides belonging to that protein will be identified. Hence, the number of spectra assigned to each protein present in a complex mixture can be used as a measure of relative abundance for each protein individually (5, 6).

A major conceptual advance in this field arose from the observation that the length of a protein affects the number of spectral counts; a longer protein will generate more identifiable peptides than the same molar amount of a smaller protein (3, 4). This will have an adverse effect when counting raw spectra to calculate abundance of a protein. The introduction of normalized spectral abundance factors (NSAFs) (4) provides an improved measure for relative abundance, by factoring the length of the protein into subsequent calculations (3). An NSAF value for a given protein is calculated by dividing the spectral counts (SpC) for a protein by its length (L). This value is then normalized by dividing by the sum of all SpC/L for all proteins identified in a complex mixture (7). NSAF values provide a measure of relative abundance and the ability to compare the abundance of proteins within a sample (4). The dynamic range for NSAF values is approximately 3.6–3.8 orders of magnitude, allowing the measurement of abundance of a wide range of proteins present in a data set (8). When NSAF values are log-transformed they follow a normal distribution, facilitating analysis of statistically significant changes in expression (4). NSAF values have also been shown to have very similar statistical properties to comparable RNA transcript abundance values (8). This statistical comparability is important as it means that software and analysis tools developed for transcriptomics studies can also be applied to NSAF values in proteomic data sets.

It is important to emphasize the need for high-quality protein identification data when generating NSAF values. Protein and peptide data sets need to be filtered to a very low false discovery rate before meaningful NSAF values can be produced; only then can statistical significance be attached to changes in NSAF values of proteins observed in response to changes in a biological system. Also, it is important to optimize other experimental parameters in order to obtain worthwhile results. One detailed study has already demonstrated that the use of correct dynamic exclusion parameters in nanoLC-MS/MS has little or no effect on data quality, while the use of non-optimal dynamic exclusion parameters can cause distortions in quantitation (9).

Another important consideration is how to account for shared peptides between multiple proteins. It has been shown that the best approach is to apply NSAF values based on distributed spectral counts; shared spectral counts were distributed based on the number of additional spectral counts that belonged uniquely to each isoform (10). Other studies have shown that the far simpler approach of distributing multiple copies of a spectral count across shared protein sequences is a reasonably accurate approach to take (11, 12).

One of the early studies using NSAF-based quantitation involved the analysis of nuclear proteins from yeast. Nuclei were isolated and 2,674 proteins were identified and quantified. Low-abundance proteins associated with transcriptional regulation were identified and found to be present at low amounts, as expected. NSAF values have been used in a broad range of studies as a measure of relative abundance of identified proteins. Examples of such projects include peptide IPG-IEF profiling of rat liver membrane proteins (12), subcellular analysis of nuclear proteins in yeast (3), profiling temperature stress responses in rice (13), comparison of evolutionary adaptation of Pachycladon species (14), assembly of a probabilistic human protein interaction network (15), analysis of mouse renal cortex proteins (10), and characterization of the response of Sydney rock oysters to environmental heavy metal stresses (16).

The essential mathematical steps involved in transforming raw protein identification outputs into NSAF values can be performed in, for example, Excel spreadsheets. However, this is a laborious process and is constrained by the limited mathematical analysis tools available. Hence, numerous research groups have created software analysis packages suited to this purpose. One example is PepC, a program that identifies statistically significant differentially expressed proteins based on spectral counting (17). PepC is a Java-based program that can be used as web server module associated with the trans-proteomic pipeline (TPP) (18). The software statistically assesses spectral counting data based on a *G*-test to assess the difference in spectral counts across samples and a *t*-test to assess data reproducibility, but does not perform a data normalization step. Another example is Census, a software tool capable of processing most types of quantitative data, including both labelled and label-free proteomics experiments; the latter can be either area under the curve (AUC) or spectral counting methods (19). Census is able to quantitate data generated by both AUC and spectral counting methods, and employs an approach based on RelEx, an application previously released by the same group (20).

In this chapter we present details of the analysis pipeline we have used in a number of different publications and other ongoing projects. These details include peptide separation and analysis using nanoLC-MS/MS, peptide identification by peptide-to-sequence matching using the XTandem algorithm, quantitation of identified proteins using normalized spectral abundance factors, statistical analysis of proteins differentially expressed between samples using the Scrappy software package, and consideration of how to best extract biologically relevant information from such experiments.

2 Materials

Prepare all solutions using Milli-Q water or equivalent and the highest quality analytical grade reagents. Prepare and store all reagents at room temperature, unless otherwise indicated. All waste disposal regulations should be strictly adhered to when disposing of waste materials.

1. Zorbax C18 chromatography packing material (5 μm particle size: Agilent Technologies).
2. Readw.exe is available for free download from: http://sourceforge.net/projects/sashimi/files/.
3. The XTandem algorithm is available for free download from: http://www.thegpm.org/tandem/instructions.html.
4. A freely available version of the XTandem algorithm known as GPM-XE Tornado, which installs and runs locally on a

Windows PC, is available from https://proteomecommons.org/dataset.jsp?i=74059.

5. The Scrappy program is available as a series of R modules which can be download from: https://proteomecommons.org.

3 Methods

3.1 Shotgun Proteomics

The workflow described below is applicable to any type of label-free shotgun proteomics experiment. We routinely used SDS-PAGE gel slice shotgun experiments and gas phase fractionation, both of which have been described in detail elsewhere (13, 21–24). It is equally applicable to data produced from online or offline MudPIT experiments, peptide IPG-IEF fractionation, filter assisted sample preparation (FASP) (25), or any of the other myriad techniques commonly available.

The required features are that it is a shotgun data set comprising analysis of three biological replicates of at least two samples to be compared. Each of the individual replicate analyses typically contains hundreds of thousands of individual MS/MS spectra. For reasons of both clarity and brevity we have written this procedure focussing on a pairwise example experiment where the aim is a quantitative comparison of control versus stressed samples. For the figures in this chapter, we have used two data points ("control" and "48 cold") taken from a previously published experiment where rice plants were exposed to low temperature over a 4-day time period (24).

It is also possible to do this type of analysis with more than two samples, such as for a developmental time course or a comparison of varying degrees of temperature or water stress (13, 21, 23, 24). This requires different mathematical assumptions and models, and becomes much more difficult when comparing multiple samples without a defined reference point, such as in our study of five different New Zealand geographical isolates of Pachycladon, an endemic plant (22). That type of analysis becomes more about looking for broad trends in large amounts of data; in our case that was greatly facilitated by concurrent microarray and metabolite analysis which provided an information framework.

3.2 NanoLC-MS/MS

1. Sequentially analyze each of the peptide digest fractions using a nanoLC-MS/MS system, employing an LTQ-XL ion-trap mass spectrometer, Surveyor HPLC pump and Surveyor autosampler (Thermo, San Jose, CA).
2. Prepare an approximately 7 cm (100 μm i.d.) reversed phase columns using 100 Å, 5 mM Zorbax C18 resin (Agilent Technologies, CA, USA) in a fused silica capillary with an integrated electrospray tip (see Note 1).

3. Apply a 1.8 kV electrospray voltage to a gold-electrode liquid junction upstream of the C18 column.
4. Load each sample onto the C18 column followed by an initial wash step with buffer A (5 % (v/v) ACN, 0.1 % (v/v) formic acid) for 10 min at 1 μL/min.
5. Elute the peptides from the C18 column with 0–50 % buffer B (95 % (v/v) ACN, 0.1 % (v/v) formic acid) over a 30 min linear gradient min at 500 nL/min followed by 50–95 % buffer B over 5 min at 500 nL/min, and 5 min was with 95 % buffer B prior to column re-equilibration.
6. Direct the column eluate into the nanospray ionization source of the mass spectrometer (see Note 2).
7. Scan the spectra over the range 400–1,500 amu. Automated peak recognition, dynamic exclusion (90 s), and tandem MS of the top six most intense precursor ions at 40 % normalization collision energy were performed using Xcalibur software (Thermo) (see Note 3).

3.3 Protein and Peptide Identification

1. Acquire the set of data files from one experiment in the proprietary .Raw format. These are first converted to .mzxml format using the freeware Readw.exe program.
2. Place the set of .mzxml data files from a given sample into one directory, and peptide-to-spectrum matching is performed using the XTandem algorithm. We use the Global Proteome Machine software (26, 27), which is freely available and runs the XTandem Tornado version. Searching the set of .mzxml files stored in a directory enables the user to choose for a single combined summary output file to be created, in addition to all 16 individual result files (see Note 4).
3. Export the combined protein and peptide identification output file for all 16 gel slices to an Excel spreadsheet. This spreadsheet contains six columns of data, with the headers identifier, log(I), rI, log(e), pI, Mr (kDa), description, and annotated domains. It is necessary to remove the last column (annotated domains) prior to subsequent analysis as it interferes with subsequent data processing. The Excel file is then exported to comma-separated value format, which is then compatible with input into the Scrappy software.

3.4 Spectral Counting Reporting and Analysis Program (Scrappy): Uploading and Analyzing Data

The Scrappy program is an implementation of the R statistical analysis package, run from a simple web interface. It has a limited amount of variable input allowed, but performs a large number of calculations quickly and efficiently. The following steps are required:

1. Upload the csv files of XTandem protein identification outputs as described above. For a simple pairwise comparison of two

biological samples, it is designed to accept three files for each sample, representing three biological replicate analyses.

2. Define category names (e.g., "control" and "stressed") and upload all six files. It is also possible to upload more than three replicates in a category, or three replicates of any number of samples to be compared. For the purposes of this chapter we will mostly constrain it to the simpler version, a pairwise comparison comprising three biological replicates of each.
3. When finished entering the files, select parameters on the following screen, including the following: the minimum number of peptide identifications for a protein within one sample set to be considered a valid protein identification (default value is 5), whether to use untransformed or log-transformed data for the *t*-test analyses (default is log-transformed), the spectral fraction to be added to all counts for multigroup statistical analysis (default is 0.5), and which of the data categories are to be treated as the baseline for numerical comparisons (see Note 5).
4. Start the analysis. The calculations are performed over several minutes, depending on the size and number of the data files. A results folder is generated, with a series of files containing different data analysis results. At the end of the list the user has the option to download the results, with or without the initial data files.
5. The folder of results will contain a number of different analysis outputs, which are described below.

3.5 Spectral Counting Reporting and Analysis Program (Scrappy): Interpreting Results

The results output files can be grouped into five subheadings: data aggregation, data partitioning, data quality metrics, NSAF ratios, and ANOVA and clustering.

1. Data aggregation

   (a) *Output.csv*

   This file contains the combined data set, namely, the full set of reproducibly present proteins (proteins present in all replicates of at least one sample, having a total peptide count > minimum peptide level as set above), their description, spectral counts, logNSAF values, *t*-test statistic, and *p*-value.

   (b) *Up-regulated.csv*

   This file is the subset of the full data set containing only the up-regulated proteins: *p*-value <0.05 and ratio >1. The ratio is the mean of the two average NSAF values for the two groups, with the denominator being the first group in alphabetical order (so if the groups are stress and control, the ratio will be mean NSAF stress/mean NSAF control).

(c) *Down-regulated.csv*

As for up-regulated, but containing only the proteins with p-value <0.05 and ratio <1.

2. Data partitioning

(a) *DataCategoriesBarChart.png*

The data in the complete data set is partitioned based on reproducible presence and absence in the various experimental categories. For a two-group experiment (e.g., stress–control) this will simply show three bars: proteins present reproducibly (namely, present in all replicates) in stress only, proteins present reproducibly in control only, and proteins present in all samples. If the experiment has an arbitrary number of groups, *n*, there can be up to $2n-1$ separate bars. This image can be used to give a quick idea of which combinations of conditions are most prevalent in an experiment.

(b) *DataCategoriesTable*

The data categories table lists the precise numbers for the categories listed in the bar chart.

(c) *DataCategoriesPieChart.png*

This pie chart shows the numbers of proteins present in one group only, two groups, three groups, etc. For a two-group experiment it would only contain two "slices," proteins present in one group only and proteins present in both groups.

3. Data quality metrics

(a) *QQplot.png*

An image is shown in Fig. 1, displaying the quantile–quantile plot of the average logNSAF data for the control samples from the control and 48 h cold stress experiment referred to earlier. Such a plot shows the quantiles of the control category on the *y*-axis against the quantiles of a standard normal distribution (hence zero mean and unit standard deviation) on the *x*-axis. A separate plot is generated for each category. If the data distribution of the sample is relatively normal, then the points will lie approximately on the diagonal. A small departure from the diagonal is acceptable, but a large deviation may show that the data is not acceptable for further statistical analysis.

(b) *ED1densityPlot.png*

This plot is a kernel density plot of the logNSAF data distribution from each replicate overlaid; in essence it is like a set of smoothed histograms, one from each replicate, placed on top of each other. It shows visually whether the data is approximately normal, or indicates if there is any replicate that has a slightly unusual distribution when compared to

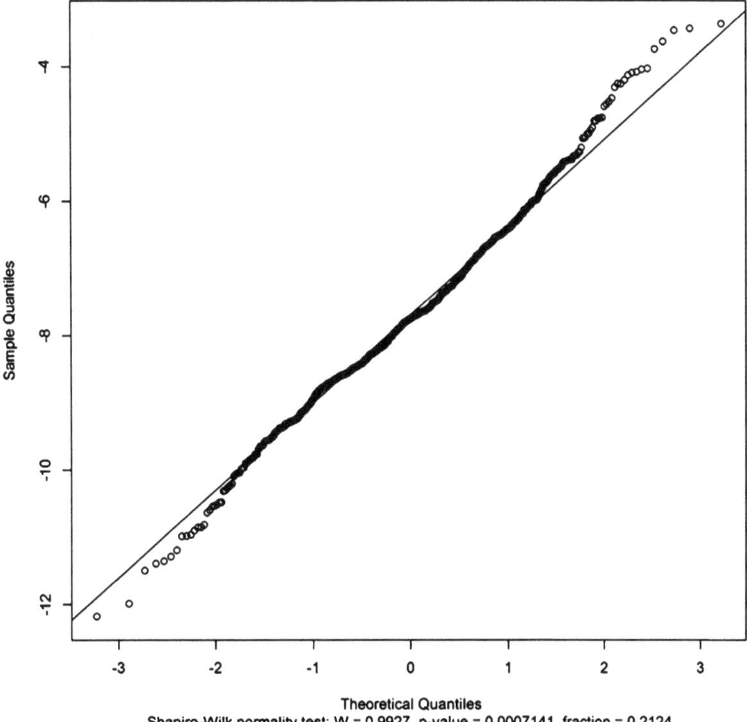

Fig. 1 Q-Q plot

the rest. If any replicate is visually very different from the rest, then the data quality should be examined.

(c) *ED2densityPlot*

This plot is similar to the previous density plot, but performed only for the proteins present reproducibly in all samples. An example is shown in Fig. 2, which includes three replicates each of the control and 48 h cold rice leaf samples referred to earlier. It is clear from this figure that all six samples overlay each other well with no major outliers.

4. Visualizing NSAF ratios

(a) *LogNSAF.png /LogNSAF.svg*

This graphic shows the logNSAF values for identified proteins, presented as logNSAF in the first specified category on the *x*-axis and logNSAF values in the second category on the *y*-axis. The dots are color coded, with light blue circles indicating that the logNSAF values are statistically unchanged between the two categories, while dark blue circles indicate those proteins with statistically significantly different logNSAF values between the two categories. Statistical significance is estimated using a student *t*-test on the log-transformed NSAF values from the original

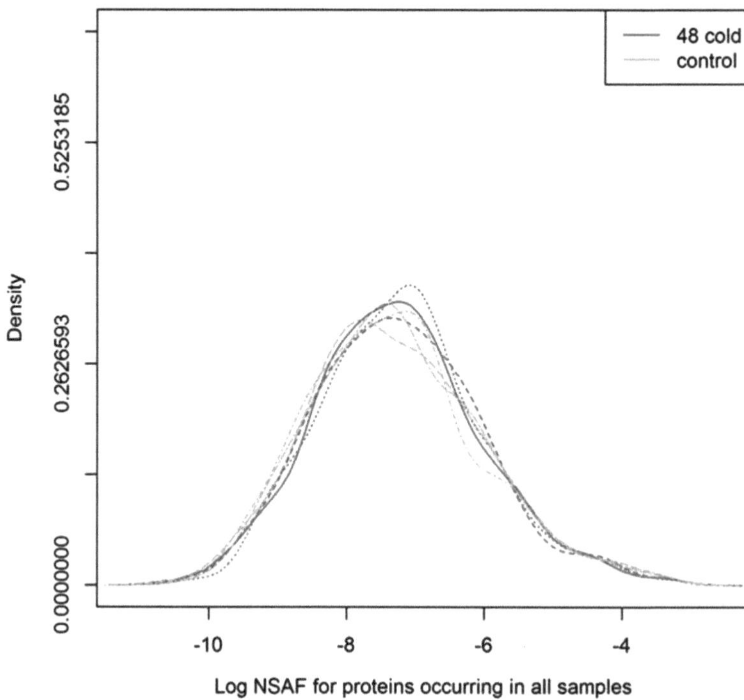

Fig. 2 ED2 density plot

biological triplicate experiments. In most experiments the majority of data points are clustered along the diagonal, with the lower values closer to the origin. The further from the diagonal, the greater the degree of differential expression.

Two different formats of this graph are generated, a png file and an interactive svg file (scalable vector graphics); hovering over the points in the interactive file will show additional information such as the protein identification (see Note 6). An example is shown in Fig. 3, which includes three replicates, each of the control and 48 h cold rice leaf samples referred to earlier.

(b) *RatioChart.png*

This plot shows the ratios of average NSAF in the two groups, ordered in increasing order of the *t*-test statistic. The size of the bars represents the average NSAF ratio for the up-regulated proteins, and the reciprocal (1/ratio) for the down-regulated proteins; therefore high bars show a big difference between groups. The up-regulated proteins are colored green, the down-regulated proteins are colored red, and the proteins showing no statistically significant difference are black. As above, there are two different formats of this image: one plain and one interactive.

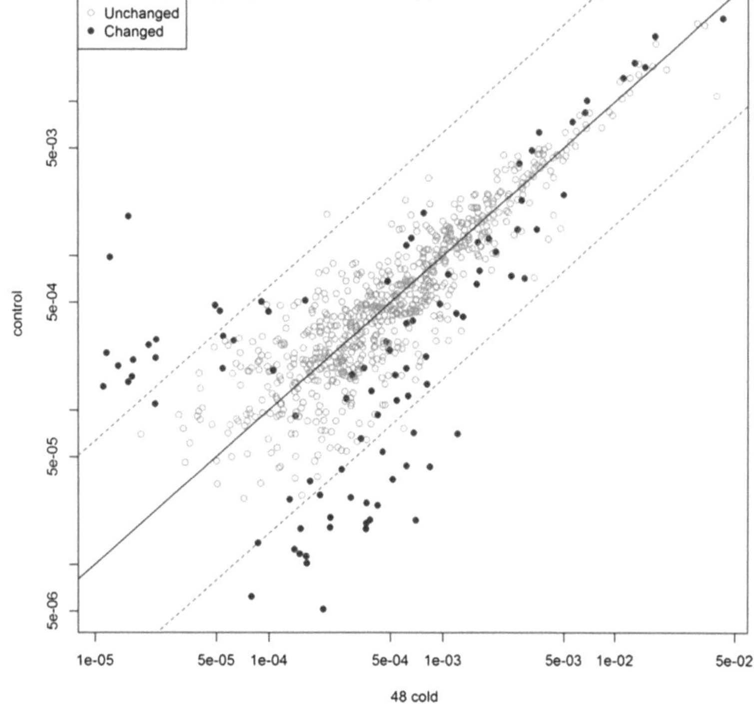

Fig. 3 LogNSAF ratio distribution

5. ANOVA and clustering

   The final category of analysis pertains to multiple group comparisons; hence they are not of immediate interest for a binary comparison. An analysis of variance is run on the subset of proteins present in all samples, separately for each protein. This approach is only of interest for more than two groups; in the two group case this is equivalent to the *t*-test already performed.

   (a) *expressionPatterns.png*

   This file contains the box plots of logNSAF values for the first 20 proteins identified as significantly changing by the analysis of variance (p-value <0.05), showing the pattern of change for those respective proteins. The proteins are ordered in increasing order of the p-value then plotted side by side. In the case of two groups this plot simply shows an up or down pattern. An example is shown in Fig. 4, which includes data from four time points of the rice leaf cold stress study referred to earlier: control, 48, 72, and 96 h cold stress.

   (b) *First20Genes.png*

   This file is very similar to the expression patterns, showing the pattern of change for the first 20 genes in

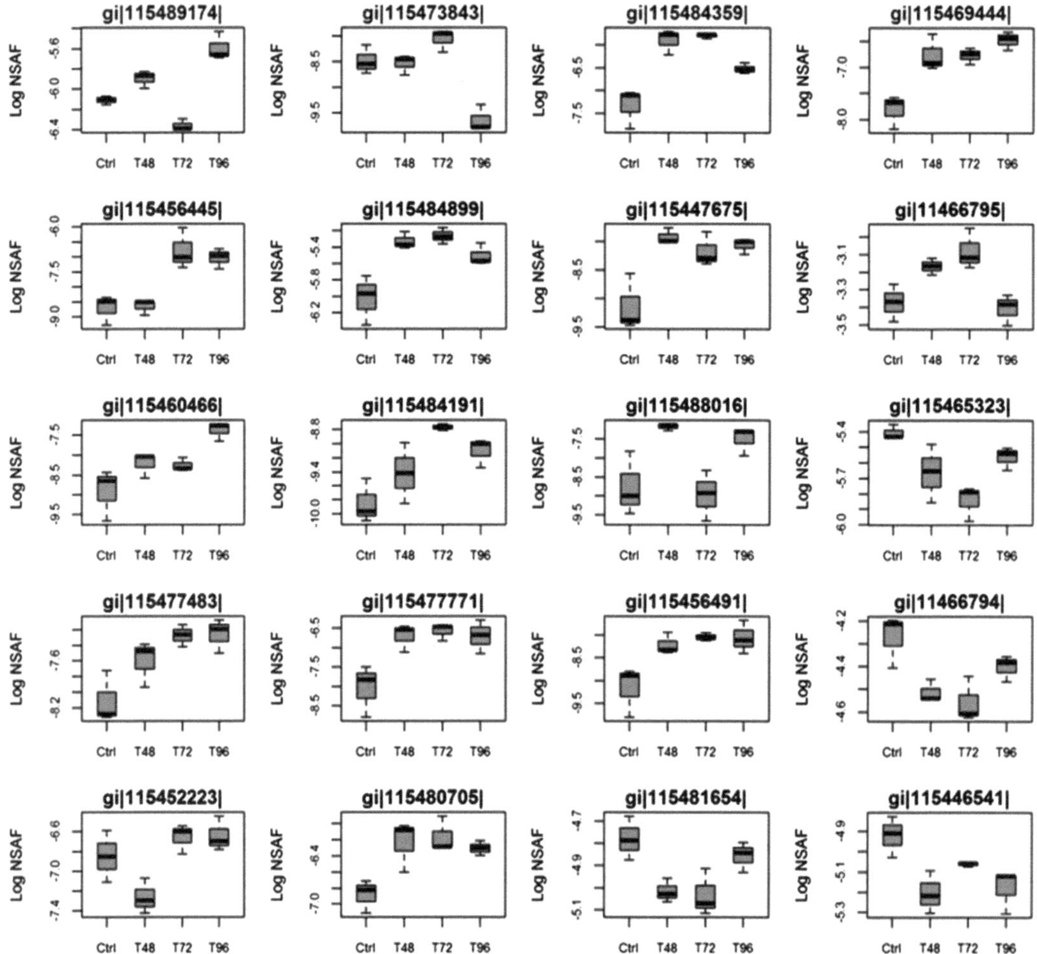

Fig. 4 Expression patterns

order of the ANOVA *p*-value, but in this instance they are plotted on the same scale, so differences in abundance between various proteins are apparent. An example is shown in Fig. 5, which includes data from four time points of the rice leaf cold stress study referred to earlier: control, 48, 72, and 96 h cold stress.

The proteins found to be changing by the ANOVA analysis are clustered on a heatmap and also by hierarchical clustering (complete linkage and Euclidean metric) and using the self-organizing maps algorithm; the resulting clusters are visualized in the images described below.

(c) *CLUST3heatmapCorrDist.png*

A heatmap is generated for all the proteins identified as significantly changing, using a correlation-based distance. A heatmap is a false color image of the logNSAF data;

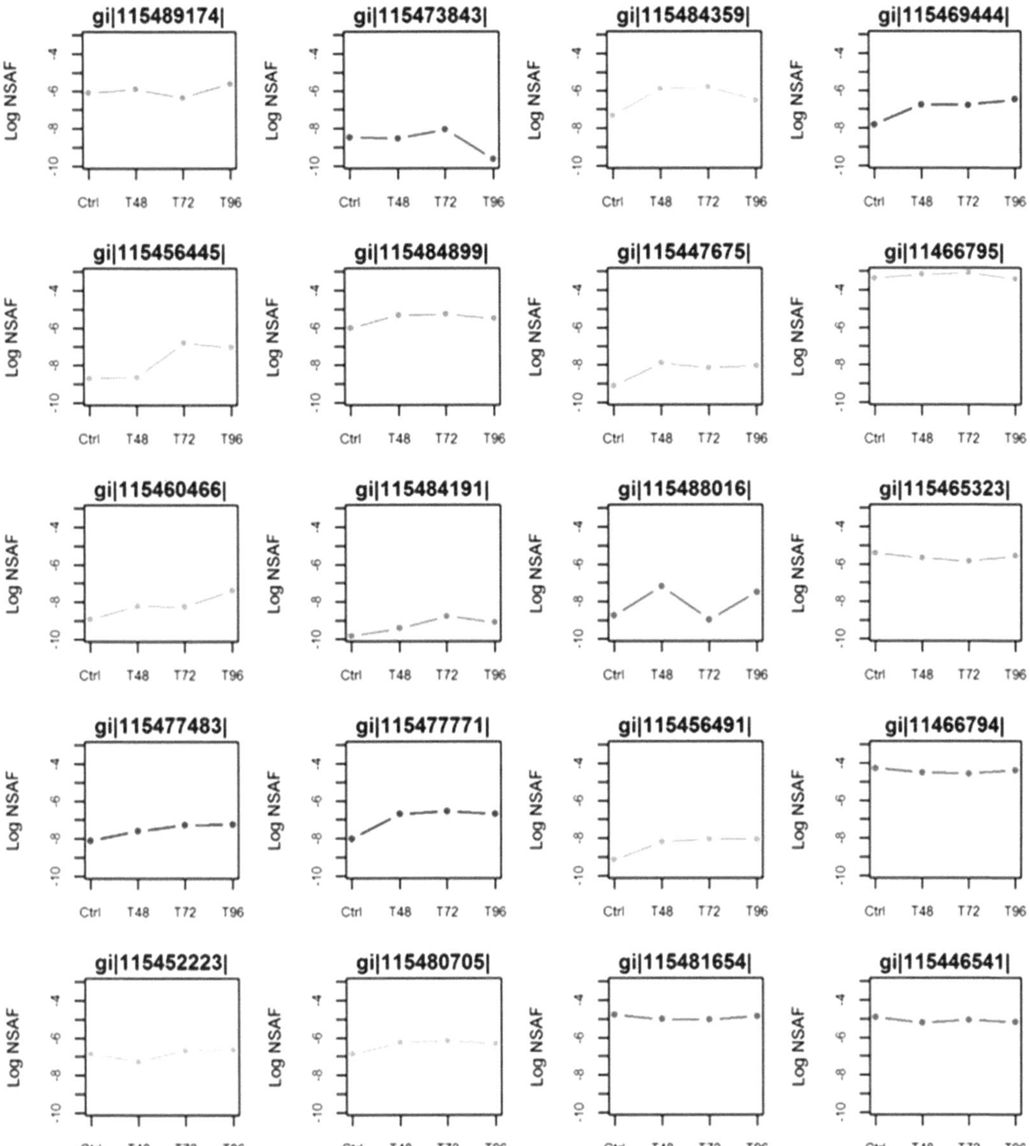

Fig. 5 First 20 genes

higher values appear in darker green. One column corresponds to one sample and one row corresponds to one protein. Rows and columns will be rearranged so that samples and respective proteins with more similar patterns are closer together. Replicate samples should be close together. An example is shown in Fig. 6, which includes data from four time points of the rice leaf cold stress study referred to earlier: control, 48, 72, and 96 h cold stress.

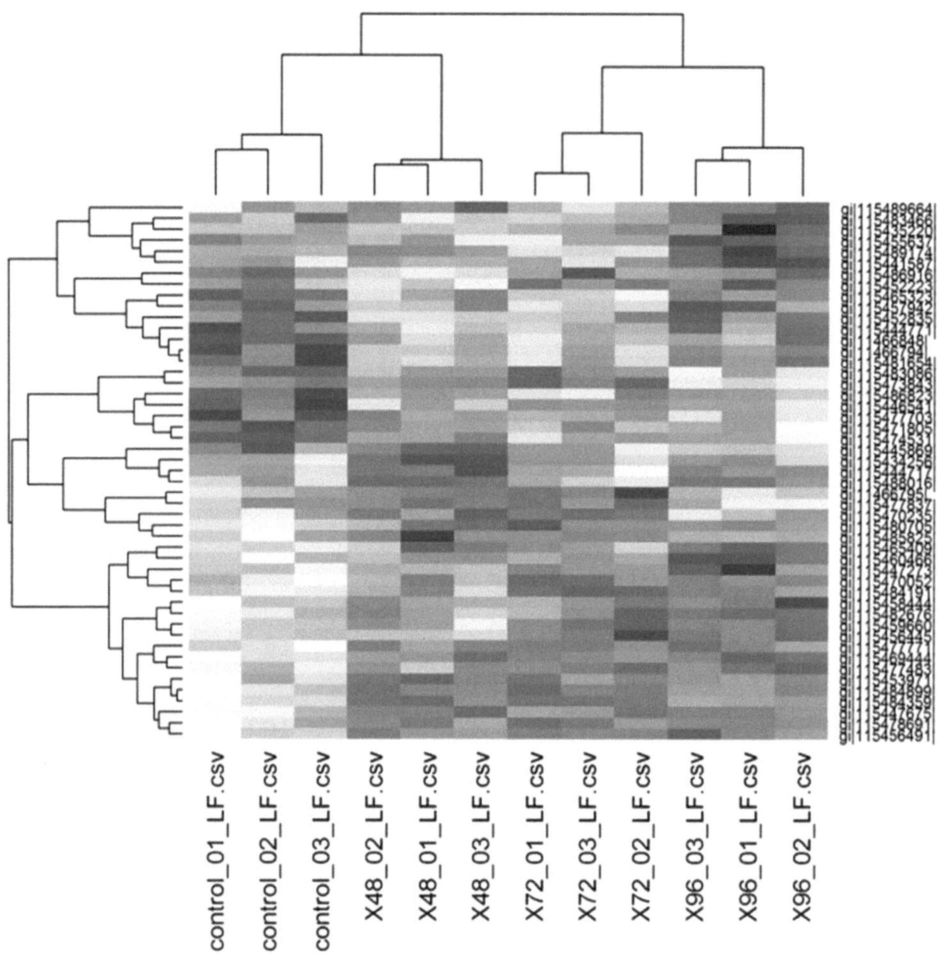

Fig. 6 Clust3 heatmap using a correlation based distance

(d) *CLUST1Genes.png*

This contains the cluster tree (dendrogram) which is the result of hierarchical clustering. There is no automatic choice of the right number of clusters—by cutting the tree at a particular height, one could end up with a certain number of clusters. Though there are ways of assessing the appropriate number of clusters, this is not done automatically for this analysis; instead an arbitrary choice of nine clusters is made, and they are colored with different colors.

(e) *ClusterExpressionProfiles.png*

The average logNSAF values for each experimental group are then plotted separately for each cluster identified in the dendrogram above. The individual protein averages are overlayed in light gray, to get an overall image of the patterns of change in protein abundance for each cluster.

(f) *SOM.png*

The self-organizing map algorithm (as implemented in the R *som* package) is also run for the same data, using an arbitrary *x* and *y* map dimension of 3. This algorithm will result into a partition of the NSAF data on a "map" of $9 = 3 \times 3$ cells, and the means and standard deviations for the average protein expressions in each of samples are visualized on this image. Thus each cell in the graph will show an average trend for the NSAF data in that cell.

(g) *PLGEM*

The power law global error model was shown to give a better estimate of NSAF data variability than that generated from the data alone, and can be used as an alternate to *t*-tests for determining up and down regulation between two conditions. Computationally this approach is made available in the *plgem* R package. In Scrappy it is set up to compare each of the conditions to a baseline, so in the case of two groups it is simply a comparison of, e.g., "stressed" versus "control." For more than two groups it will compare each of them in turn to the selected baseline group, e.g., "control." Unlike the *t*-tests, it is run only on the proteins present reproducibly in both samples (28). The *PLGEMFit.png* image describes the quality of the model fit, and the baselineComparisons.html lists the proteins found to be differentially expressed by this approach.

3.6 Finding Biological Meaning in the Results: Functional Annotation and Enrichment

The main output of Scrappy is the consolidated set of protein quantitation data, along with various subsets of proteins that are highlighted as potentially interesting; for example, proteins up-regulated in a drought-stressed condition, or the various subsets of proteins identified in a complex experiment with many groups. Although this is a richly detailed quantitative data set, extracting biologically interesting information can still be a daunting task.

The first option for proceeding further is via functional enrichment, mapping the identified proteins into categories such as cellular location, biological process, molecular function, or presence in a particular metabolic pathway, and then comparing the sets of interest to see which ones have more of a particular category than would be expected by chance. The tools used here depend very much on the organism studied and the number of groups considered in the experiment. For a binary comparison, such as control and stressed, there are many tools freely available online, with DAVID functional annotation platform (http://david.abcc.ncifcrf.gov/) and Blast2GO (http://www.blast2go.com/b2ghome) amongst the most popular. The researcher can upload lists of up-regulated proteins and find, for example, biological processes or metabolic pathways overrepresented amongst the proteins up-regulated in the stressed condition as compared to the set of proteins at large. Commercial packages such as Ingenuity (http://www.ingenuity.com/) or GeneGO MetaCore (http://www.genego.com/metacore.php) can also provide a rapid and detailed analysis including functions, pathways, and networks, provided the license is purchased and the organism studied is supported in the software. The enrichment analysis may play a dual role: on the one hand it can help focus on a particular list of proteins involved in a biological category of interest, for example focus on signalling proteins up-regulated in the stress condition. On the other, it may help validate the set of proteins identified, provided the biological categories or pathways identified are meaningful in the context of the biology of the experiment.

For an experiment with multiple conditions, such as the five-way comparison of evolutionary adaptation of Pachycladon species mentioned earlier (14), the sheer number of numerical comparisons involved gets rapidly unwieldy: with five separate conditions, there could be ten separate binary comparisons to consider, each yielding up- and down-regulated and unchanged sets of proteins. For such experiments generating many subsets of proteins of interest we use the R-based software package PloGO, developed in our group (29), which allows us to categorize gene ontology information for various batches of proteins at once, compare the relative abundances of various gene ontology categories to a selected reference, and aggregate NSAF quantitation by GO category as well. Again, such an analysis may play multiple roles: firstly, focussing attention on the comparisons of conditions that produce functionally interesting sets of proteins, and secondly giving a list of biological processes or molecular functions that are overrepresented in particular categories of proteins, which can then be cross-correlated with the underlying biological paradigm of the experiment.

4 Notes

1. NanoLC columns can also be purchased from suppliers such as Michrom or New Objective.
2. It is not required to completely eliminate carryover between injections as all the samples in a given set will be combined for subsequent analysis.
3. Our standard experimental design is 16 SDS-PAGE gel slices from each of three biological replicates of a given sample. These samples can be completed in two and half days of mass spectrometric analysis time.

4. We use the "MudPIT combine" option in GPM searches to produce a unified output file from the individual search result files.
5. The default parameters in Scrappy are that a protein must be present in all three replicates with a minimum peptide count of 5. This parameter can be altered if necessary, but seems to work well for most of our experiments.
6. This feature is a very powerful (and popular), as it allows the user to present a single graphic to represent many thousands of data points in a manner that is easy to understand and interpret.

Acknowledgments

The authors acknowledge the funding support from the Macquarie University MQRES scholarship scheme (K.A.N.) and the Australian Research Council (P.A.H.). Aspects of this research were conducted at the Australian Proteome Analysis Facility funded by the Australian Government National Collaborative Research Infrastructure Scheme (NCRIS). P.A.H. acknowledges Robert Black for continued support and encouragement.

References

1. Anderson L, Hunter CL (2006) Quantitative mass spectrometric multiple reaction monitoring assays for major plasma proteins. Mol Cell Proteomics 5:573–588
2. Gao J, Opiteck GJ, Friedrichs MS, Dongre AR, Hefta SA (2003) Changes in the protein expression of yeast as a function of carbon source. J Proteome Res 2:643–649
3. Mosley AL, Florens L, Wen Z, Washburn MP (2009) A label free quantitative proteomic analysis of the Saccharomyces cerevisiae nucleus. J Proteomics 72:110–120
4. Zybailov BL, Florens L, Washburn MP (2007) Quantitative shotgun proteomics using a protease with broad specificity and normalized spectral abundance factors. Mol Biosyst 3:354–360
5. Liu H, Sadygov RG, Yates JR 3rd (2004) A model for random sampling and estimation of relative protein abundance in shotgun proteomics. Anal Chem 76:4193–4201
6. Zhang B, VerBerkmoes NC, Langston MA et al (2006) Detecting differential and correlated protein expression in label-free shotgun proteomics. J Proteome Res 5:2909–2918
7. Zybailov B, Mosley AL, Sardiu ME et al (2006) Statistical analysis of membrane proteome expression changes in *Saccharomyces cerevisiae*. J Proteome Res 5:2339–2347
8. Pavelka N, Fournier ML, Swanson SK, Pelizzola M, Ricciardi-Castagnoli P et al (2008) Statistical similarities between transcriptomics and quantitative shotgun proteomics data. Mol Cell Proteomics 7:631–644
9. Zhang Y, Wen Z, Washburn MP, Florens L (2009) Effect of dynamic exclusion duration on spectral count based quantitative proteomics. Anal Chem 81:6317–6326
10. Zhao Y, Denner L, Haidacher SJ, LeJeune WS, Tilton RG (2008) Comprehensive analysis of the mouse renal cortex using two-dimensional HPLC-tandem mass spectrometry. Proteome Sci 6:15
11. Chick JM, Haynes PA, Bjellqvist B, Baker MS (2008) A combination of immobilised pH gradients improves membrane proteomics. J Proteome Res 7:4974–4981
12. Chick JM, Haynes PA, Molloy MP et al (2008) Characterization of the rat liver membrane proteome using peptide immobilized pH gradient isoelectric focusing. J Proteome Res 7:1036–1045
13. Gammulla CG, Pascovici D, Atwell BJ, Haynes PA (2010) Differential metabolic response of cultured rice (Oryza sativa) cells exposed to high- and low-temperature stress. Proteomics 10:3001–3019

14. Voelckel C, Mirzaei M, Reichelt M et al (2010) Transcript and protein profiling identify candidate gene sets of potential adaptive significance in New Zealand Pachycladon. BMC Evol Biol 10:151
15. Sardiu ME, Cai Y, Jin J et al (2008) Probabilistic assembly of human protein interaction networks from label-free quantitative proteomics. Proc Natl Acad Sci USA 105:1454–1459
16. Muralidharan S, Thompson E, Girch G, Raftos D, Haynes PA (2011) Quantitative proteomics of heavy metal stress responses in Sydney rock oysters. Proteomics 12:906–921
17. Heinecke NL, Pratt BS, Vaisar T, Becker L (2010) PepC: proteomics software for identifying differentially expressed proteins based on spectral counting. Bioinformatics 26:1574–1575
18. Keller A, Eng J, Zhang N, Li X, Aebersold R (2005) A uniform proteomics MS/MS analysis platform utilizing open XML file formats. Mol Syst Biol 1:1–17
19. Park SK, Venable JD, Xu T, Yates JR 3rd (2008) A quantitative analysis software tool for mass spectrometry-based proteomics. Nat Methods 5:319–322
20. MacCoss MJ, Wu CC, Liu H, Sadygov R, Yates JR 3rd (2003) A correlation algorithm for the automated quantitative analysis of shotgun proteomics data. Anal Chem 75: 6912–6921
21. Gammulla CG, Pascovici D, Atwell BJ, Haynes PA (2011) Differential proteomic response of rice (Oryza sativa) leaves exposed to high- and low-temperature stress. Proteomics 11: 2839–2850
22. Mirzaei M, Pascovici D, Keighley T et al (2011) Shotgun proteomic profiling of five species of New Zealand Pachycladon. Proteomics 11:166–171
23. Mirzaei M, Soltani N, Sarhadi E et al (2012) Shotgun proteomic analysis of long-distance drought signaling in rice roots. J Proteome Res 11:348–358
24. Neilson KA, Mariani M, Haynes PA (2011) Quantitative proteomic analysis of cold-responsive proteins in rice. Proteomics 11:1696–1706
25. Wisniewski JR, Zougman A, Nagaraj N, Mann M (2009) Universal sample preparation method for proteome analysis. Nat Methods 6:359–362
26. Craig R, Beavis RC (2003) A method for reducing the time required to match protein sequences with tandem mass spectra. Rapid Commun Mass Spectrom 17:2310–2316
27. Craig R, Beavis RC (2004) TANDEM: matching proteins with tandem mass spectra. Bioinformatics 20:1466–1467
28. Pavelka N, Pelizzola M, Vizzardelli C et al (2004) A power law global error model for the identification of differentially expressed genes in microarray data. BMC Bioinformatics 5:203
29. Pascovici D, Keighley T, Mirzaei M, Haynes PA, Cooke B (2012) PloGO: plotting gene ontology annotation and abundance in multi-condition proteomics experiments. Proteomics 12:406–410

Chapter 18

Employment of Complementary Dissociation Techniques for Body Fluid Characterization and Biomarker Discovery

David M. Good and Dorothea Rutishauser

Abstract

Proteomic analysis of biological fluids has become the de facto method for biomarker discovery over the past half decade. Mass spectrometry, in particular, has emerged as the premier technology to perform such analysis. This shift in the prevailing choice of analytical method is primarily due to the rapid evolution of mass spectrometry technology, with advances in acquisition speed, increased resolving power and mass accuracy, and the development of novel fragmentation methods. The benefits of using one of these new fragmentation methods, electron-transfer dissociation, as a complement to the traditional dissociation technique (i.e., collision-activated dissociation) have been thoroughly illustrated. Detailed here is a method for proteomic analysis of a readily obtainable and often investigated biological fluid, blood plasma, which takes advantage of these complementary dissociation techniques and employs the most recent advances in mass spectrometry technology.

Key words Biomarker, Body fluid, Mass spectrometry, Electron-transfer dissociation

1 Introduction

Biological fluids have become the material of choice for analysis in many modern clinical proteomics studies. This preference is perhaps primarily due to the inherent ease of accessibility to such fluids (relative to tissue biopsy, laser capture microdissection, etc.). In addition, the costs of obtainment (both fiscal and patient health) are in general lower for biological fluids as compared to tissue (1). As a result, the past 5–10 years have seen numerous studies which have illustrated the effectiveness of body fluid investigation for obtaining diagnostic and/or prognostic disease biomarkers (2–5).

Mass spectrometry (MS)-based techniques have become the primary methods for such analyses, often in combination with off- or on-line separations, specifically liquid chromatography and/or electrophoretic methods. Although several studies have shown the utility of having a mass spectrometer as the downstream detector

Ming Zhou and Timothy Veenstra (eds.), *Proteomics for Biomarker Discovery: Methods and Protocols*, Methods in Molecular Biology, vol. 1002, DOI 10.1007/978-1-62703-360-2_18, © Springer Science+Business Media, LLC 2013

of analyte m/z values (peptide mass fingerprinting) (6–8), the current climate in biomarker studies often stipulates the characterization of the observed peptide species. Such sequence analysis is typically achieved by employing a second stage of MS (tandem MS; MS/MS) (9) to fragment selected precursor ions. The most commonly employed method to dissociate precursor ions to date has been the use of collision-activated dissociation (CAD), which has proven its worth over the last two decades, driving development of MS/MS-based proteomics. However, though providing a solid analytical platform for proteomics research, this technique does have limitations. Specifically, CAD has been shown to have a "sweet spot" in terms of dissociation efficiency, performing best for small tryptic peptides; it is less effective for the characterization of larger peptides (~15 residues) and those peptides which contain internal basic residues (10, 11). In addition, the limitations of CAD for analysis of posttranslational modifications (PTMs) have been outlined in detail (12–15), and include both poor sequencing assignment as well as a general inability to localize the site of modification. This is an especially noteworthy limitation, as the role which PTMs play in disease activity has recently garnered a great deal of attention (16–19). Due to the limitations of CAD, and in particular its poor analytical usefulness for the study of PTM-containing analytes, much effort was placed on developing newer technologies which could overcome these restrictions. One result of such research was the development of a fundamentally novel fragmentation technique, electron-transfer dissociation (ETD) (20). In comparison to CAD, ETD offers the ability to characterize PTMs, including not only sequence assignment but also localization of most modifications (21). In addition, and in contrast to CAD, ETD performs better when dissociating larger peptides with multiple internal basic residues, thus providing sample information complementary to that generated by CAD (10). A direct consequence of such complementarity has been the practice of combining both of these techniques within a single analysis to glean the most information from a given sample (22). Furthermore, such broad adoption of this method of combining the two techniques has led to both instrument adaptation (23–25) and the development of "intelligent," automated acquisition methods which allow for the employment of whichever fragmentation method has a greater probability to provide the most information for a given peptide (based upon m/z and charge state) (26, 27).

Here we outline our technique for the analysis of biological fluids, specifically human blood plasma, for biomarker identification and characterization. We provide detailed instructions for sample preparation, as well as for using the complementary fragmentation methods of ETD and CAD in an efficient manner. Finally, we give a brief overview of data analysis, and provide a template for creating data outputs that are compliant with recently published suggestions for the handling of clinical proteomics data (28, 29).

2 Materials

2.1 Sample Prep

1. ProteaseMAX® (Promega, Madison, WI).
2. Bicinchoninic acid (BCA) assay (Pierce/Thermo Fisher Scientific, Rockford, IL).
3. Trypsin (Promega, Madison, WI).
4. C18 Zip-Tips (Millipore, Billerica, MA).
5. Wetting solution (100 % acetonitrile; ACN), washing solution (3 % ACN, 0.1 % trifluoroacetic acid; TFA), elution solution (70 % ACN, 0.1 % TFA).

2.2 Sample Separation and Mass Spectrometry Analysis

1. PicoTip™ capillary column with pre-pulled tip 360 μm × 50 μm (outer diameter × inner diameter; New Objective, Woburn, MA).
2. C8/C18 3.1 μm reversed phase chromatography resin (Dr. Maisch GmbH, Ammerbuch, Germany).
3. Pressure bomb (Thermo Fisher Scientific, Odense, Denmark).
4. HPLC-grade solvents: Solvent A—ACN with 0.1 % FA, solvent B—water with 0.1 % FA.
5. Glass vials (2 mL).
6. Proxeon Easy nano-High Performance Liquid Chromatography Pump (Thermo Fisher Scientific, Odense, Denmark).
7. ETD-enabled Velos-Orbitrap (Thermo Fisher Scientific, Bremen, Germany).
8. Fluoranthene (Sigma-Aldrich, St. Louis, MO).

2.3 Data Analysis

1. Software to provide overview of MS results (i.e., RawMeat; http://vastsci.com/rawmeat/).
2. DTA Generator (30, 31) (freely downloadable at http://www.chem.wisc.edu/~coon/software.php).
3. COMPASS software suite (32) (open-source software; http://www.chem.wisc.edu/~coon/software.php).
4. Database search algorithm; MASCOT (33) (Matrix Science, London, UK).

3 Methods

3.1 Sample Preparation

As the focus here is on analysis of human blood plasma, preparation of this sample does not necessitate any special lysis or preparation steps (such as immunodepletion for serum). However, detailed procedures for the preparation of two other common biomarker-containing media, cells and urine, are provided in Subheading 4 (cells, see Note 1; urine, see Note 2).

1. Though the protein content of plasma samples is relatively consistent, the protein content within each sample should be measured, if the amount of sample allows. We employ a BCA assay for protein content estimation (performed following manufacturer's instructions). If the amount of sample is limited, use the microplate procedure to estimate protein concentration.
2. Prepare sample using ProteaseMAX® surfactant. For each sample, dissolve an aliquot of 10 μg total protein in a mixture of 50 mM ammonium bicarbonate (NH_4HCO_3)/10 % ACN containing 0.1 % ProteaseMAX to a total volume of 80 μL per sample.
3. Incubate samples at 50 °C for 30 min. Use low-intensity shaking to aid solubilization of proteins.
4. Allow samples to cool to room temperature. Sonicate in water bath for 10 min.
5. Centrifuge samples using benchtop centrifuge/microfuge at 20,000 × *g* for 5–10 min. Remove supernatant for further digestion.
6. Digest samples using trypsin (see Note 3).
7. Reduce samples by addition of 25 μL of 20 mM DTT in 50 mM NH_4HCO_3 to a final concentration of 5 mM. Incubate for 30 min at 56 °C.
8. Alkylate via addition of 25 μL of 66 mM iodoacetic acid (IAA) in 50 mM AmBic (final concentration 15 mM). Incubate for 30 min at room temperature, in the dark.
9. Add trypsin in a ratio of 1:30 trypsin:protein (weight:weight). Incubate at 37 °C overnight. Final ProteaseMAX concentration 0.05 %.
10. To inactivate the trypsin, lower the pH of the solution to <2. To achieve this, add concentrated formic acid to a final concentration of 5 %.
11. Centrifuge 10 min at 20,000 × *g*. Remove supernatant to fresh tube and lyophilize. If not using immediately, store at −20 °C until further use.
12. Use Zip-Tips to desalt the sample. Add 20 μL washing buffer to lyophilized sample. Vortex well and let stand at RT for 15 min.
13. To condition the Zip-Tip, first take up 10 μL wetting solution and discard (repeat 3×). Next, draw up 10 μL washing solution and discard (repeat 3×).
14. To bind the sample, introduce 10 μL of sample into the Zip-Tip and discard, keeping Zip-Tip within the same container; repeat this 10× with the Zip-Tip always kept within the original sample.

15. Remove salts and other interfering compounds by rinsing the Zip-Tip with washing solution—take up 10 μL of washing solution and discard to waste (repeat 3×).
16. Elute the cleaned sample by drawing up 10 μL of elution solution and eluting peptides into a fresh tube. Pipette this solution up and down a further three times within the same tube. Repeat this elution step, but using a new tube. Combine the eluents from the two tubes into a single sample.
17. Lyophilize eluent and resuspend in HPLC buffer A, adjusting protein concentration to between 0.1 and 0.5 μg/μL, depending on desired length and depth of analysis to be performed (assuming 5 μg total binding to Zip-Tip). Transfer to autosampler sample vial.

3.2 Sample Separation and Mass Spectrometry Analysis

All methods outlined here have been perfected on a Proxeon Easy nLC coupled to a Thermo Scientific Velos-Orbitrap enabled with ETD. However, the acquisition concepts and many of the parameters may also be carried over for use on similar ETD-enabled, high-resolution instruments.

1. In a first step, pack the microcapillary column with the desired reversed-phase material (C_8 for separation of larger peptides and/or polypeptides and C_{18} for those peptides that are typically formed from tryptic or Lys-C digestion). Create a slurry of the desired packing material in either acetonitrile or methanol within a 2 mL glass sample vial. Add a micromagnetic stir bar to the vial, place the vial within the pressure bomb, and turn on the stirring capability of the bomb.
2. Insert the column into the bomb through the provided opening and pack by applying no more than 50 bar pressure of nitrogen. Pack 7–9 cm of resin into the column.
3. Condition by first flowing solvent A over the column for 15–30 min, followed by 15–30 min solvent B, and a final rinse of solvent A. Load the pressure bomb with 10 fmol/μL bovine serum albumin (BSA) digest and flow over column for 15–30 min. Elute BSA with solvent B and rinse with solvent A prior to use.
4. Connect the microcapillary column to the LC. Equilibrate the column using the "Prepare Column" method on the LC system. During equilibration, the back-pressure of the column should be monitored until it has reached a constant that is below the user- or vendor-defined maximum (typically ~250 bar) (see Note 4).
5. Observe the stability of the electrospray at the desired flow rate (we typically employ 300 nL/min). The spray voltage should be adjusted accordingly until acquiring a spray that is both

stable (no formation of droplets on the tip; indicative of too little voltage) and does not cause arcing (too much voltage applied).

6. Before starting ETD analysis, inspect the fluoranthene anion signal. To do this, change to negative ion mode while the filament is turned on. You should now observe the fluoranthene anion peak should be visible at ~202.1 *m/z*. The injection time should be steady at ~0.1 ms and the signal intensity of the anion should be $\geq 1 \times 10^7$ in centroid view. This peak should be the most intense peak within the spectrum, with a signal to noise ratio of ~10:1 as compared to the peak at 216.
7. Use a data-dependent acquisition method to interrogate the ten most intense precursor ions, as identified from each full MS scan, by both CAD and ETD (two separate sequential events). On the Velos-Orbitrap instrument, this method can be set up using the "Nth order double play" instrument method and enabling a subsequent ETD event. The number of MS^2 events contained in the method should take into account the quantification method being employed; label-free methods should use a lower number of MS^2 events (i.e., 5) while labeling methods such as iTRAQ (Applied Biosystems, Carlsbad, CA) and TMT (34) can have a greater number of MS^2 events between MS^1 scans (10–15 would be typical).
8. The MS^1 scan should be performed at 60,000 resolving power in profile mode over the range of *m/z* 300–2,000. Both "predict ion injection time" and "enable preview mode for FTMS mast scans" should be enabled.
9. We have had the greatest success with all MS^2 scans being performed in the ion trap in profile mode. CAD can be adequately performed using default parameters. For ETD, the default charge state of a precursor ion should be 2, the isolation width should be set to 3, and the activation time should be set to 100 ms. In addition, supplemental activation should be employed (27). Charge state screening should be activated, with monoisotopic precursor recognition and charge state dependent ETD time also employed. Charge state rejection should be used, with both singly charged precursors and unassigned charge states being rejected.
10. Dynamic Exclusion should be employed, with repeat value = 1, repeat duration = 30s, exclusion list size = 500, exclusion duration = 180 s, and exclusion mass width = 0.5 Da low to 1.5 Da high.
11. As an initial quality control, load 100 fmol BSA digest (or cytochrome C, if preferred) and run the desired instrument method. Prior to further testing, this run should be searched using the Data Analysis Protocol, steps 2–4. The desired sequence coverage for BSA is ~80 %. If this coverage is not met, then it is

advisable to rerun this test until a reasonable coverage is acquired.

12. The order of all analyses should be randomized to limit the possible effects of column degradation and changes in instrumental parameters over the course of a run.
13. Insert blank runs to detect carryover between samples where needed. Inject 5–10 μL of solvent B as the blank sample.
14. Quality control: If the experiment is long (>8 samples), it is a good practice to run a standard sample (e.g., BSA digest) before and after the experiment to check column performance.

3.3 Data Analysis

All analysis and data processing should be performed in accordance to the MIAPE guidelines (28) and the suggested guidelines for clinical proteome analysis (29).

1. To gain an initial overview on the runs, the .raw files should be analyzed using RawMeat. Pay specific attention to chromatography reproducibility, checking to ensure there are no gaps in the electrospray or obvious delays in elution of the peptides, which may affect subsequent quantification.
2. Raw data files are converted for subsequent database searching using DTA Generator. The assumed precursor charge state should be set at the following: minimum = 2 and maximum = 8. Precursor filtering and ETD preprocessing should both be enabled.
3. Create a concatenated forward and reversed database using the Database Maker program within COMPASS. Upload this database to the MASCOT server being used.
4. The .mgf files created by DTA Generator are searched against the concatenated, forward, and reversed database using MASCOT. Searches should employ a precursor ion mass window ("Peptide tol.") of 10 ppm and a product ion mass window ("MS/MS tol.") of 0.6 Da, searching in monoisotopic mode. The enzyme should be set accordingly and the maximum number of missed cleavages should be set to 3. Peptide charge should be set to 2+ and 3+, with a fixed modification of carbamidomethyl on cysteines and a variable modification of oxidation on methionines. When searching combined CAD and ETD data, the default MASCOT instrument setting may be employed.

4 Notes

1. Preparation of cells for analysis:
   (a) Prepare lysis buffer: 8 M urea, 75 mM NaCl, 50 mM Tris (pH = 8.0–8.2), one tablet Complete Mini protease inhibitors

cocktail (Roche), and 10 mM sodium pyrophosphate, sufficient for 10 mL total volume.

(b) Perform lysis (on ice): Add lysis buffer to each sample (150 μL cell pellet, about 3:1 lysis buffer:sample). Sonicate using probe 3 × 60 s with 90 s pause (6 s run 3 s pause; AMP 40 %). Vortex, and then centrifuge at 20,000 × *g* for 20 min at 4 °C. Transfer the supernatant to new tube and keep at −20 °C until further analysis.

(c) Digestion: Add DTT to a final concentration of 2 mM. Heat to 37 °C for 45 min and allow to cool to RT. Add IAA to a final concentration of 5 mM, and incubate for 1 h at RT in the dark. Add DTT to a final concentration of 5 mM to quench the remaining IAA, and incubate at RT for 15 min. If using trypsin, dilute with 25–100 mM AmBic until urea concentration is ≤1 M. If necessary, adjust to pH 7.5–8.0. Add enzyme (protein to enzyme ratio between 20:1 and 100:1) and incubate between 5 h to overnight at 37 °C.

(d) Allow to cool to RT and then acidify with glacial acetic acid for larger digestion volumes or a diluted (i.e., 5 %) acetic acid solution for smaller digestion volumes.

2. Preparation of urine for biomarker analysis:

   (a) Use only spontaneously voided, midstream urine samples. Samples should be stored at −80 °C until analysis.

   (b) Thaw a 0.7 mL aliquot of urine immediately prior to use. Dilute with 0.7 mL of 2 M urea, 10 mM NH_4OH containing 0.02 % SDS.

   (c) To remove higher molecular mass polypeptides and proteins (e.g., albumin and immunoglobulin G), filter the sample using Centrisart ultracentrifugation filter devices at 3,000 × *g* until obtaining 1.1 mL of filtrate.

   (d) Apply filtrate onto a PD-10 desalting column and equilibrate with 0.01 % NH_4OH in HPLC-grade water to decrease matrix effects by removing urea, electrolytes, and salts, and also to enrich polypeptides.

   (e) Lyophilize and store at −20 °C until use.

   (f) Resuspend in 50–100 μL HPLC-grade water.

3. If using Lys-C instead of trypsin, dilute with 25–100 mM NH_4HCO_3 until urea concentration is ≤2 M. In addition, digest a minimum of 8 h at 25–30 °C instead of 37 °C.

4. This helps to ensure that the column will not experience an overpressure during loading or running of sample. However, loading too much sample or sample which has not been thoroughly cleaned and/or contains polymers may lead to an overpressure of the column from restricted flow of the solvent

through the column and/or the clogging of the tip. Often, this ultimately results in sample loss, as the columns must typically be thrown away with whatever sample had been loaded onto them.

Acknowledgment

D.M.G. acknowledges support from the Wenner-Gren Foundation.

References

1. Good DM, Thongboonkerd V, Novak J et al (2007) Body fluid proteomics for biomarker discovery: lessons from the past hold the key to success in the future. J Proteome Res 6: 4549–4555
2. Hye A, Lynham S, Thambisetty M et al (2006) Proteome-based plasma biomarkers for Alzheimer's disease. Brain 129:3042–3050
3. Ranganathan S, Williams E, Ganchev P et al (2005) Proteomic profiling of cerebrospinal fluid identifies biomarkers for amyotrophic lateral sclerosis. J Neurochem 95:1461–1471
4. Zhang J, Goodlett DR, Quinn JF et al (2005) Quantitative proteomics of cerebrospinal fluid from patients with Alzheimer disease. J Alzheimers Dis 7:125–133
5. Theodorescu D, Wittke S, Ross MM et al (2006) Discovery and validation of new protein biomarkers for 4 urothelial cancer: a prospective analysis. Lancet Oncol 7:230–240
6. Zimmerli LU, Schiffer E, Zurbig P et al (2008) Urinary proteomic biomarkers on coronary artery disease. Mol Cell Proteomics 7:290–298
7. Lopez MF, Mikulskis A, Kuzdzal S et al (2005) High-resolution serum proteomic profiling of Alzheimer disease samples reveals disease-specific, carrier-protein-bound mass signatures. Clin Chem 51:1946–1954
8. Rossing K, Mischak H, Dakna M et al (2008) Urinary proteomics in diabetes and CKD. J Am Soc Nephrol 19:1283–1290
9. Hunt DF, Yates JR, Shabanowitz J et al (1986) Protein sequencing by tandem mass-spectrometry. Proc Natl Acad Sci USA 83:6233–6237
10. Good DM, Wirtala M, McAlister GC et al (2007) Performance characteristics of electron transfer dissociation mass spectrometry. Mol Cell Proteomics 6:1942–1951
11. Huang YY, Triscari JM, Tseng GC et al (2005) Statistical characterization of the charge state and residue dependence of low-energy CID peptide dissociation patterns. Anal Chem 77:5800–5813
12. Shi SDH, Hemling ME, Carr SA et al (2001) Phosphopeptide/phosphoprotein mapping by electron capture dissociation mass spectrometry. Anal Chem 73:19–22
13. Schroeder MJ, Shabanowitz J, Schwartz JC et al (2004) A neutral loss activation method for improved phosphopeptide sequence analysis by quadrupole ion trap mass spectrometry. Anal Chem 76:3590–3598
14. DeGnore JP, Qin J (1998) Fragmentation of phosphopeptides in an ion trap mass spectrometer. J Am Soc Mass Spectrom 9:1175–1188
15. Wells L, Vosseller K, Cole RN et al (2002) Mapping sites of O-GlcNAc modification using affinity tags for serine and threonine post-translational modifications. Mol Cell Proteomics 1:791–804
16. Renfrow MB, Cooper HJ, Tomana M et al (2005) Determination of aberrant O-glycosylation in the IgA1 hinge region by electron capture dissociation Fourier transform-ion cyclotron resonance mass spectrometry. J Biol Chem 280:19136–19145
17. Gong CX, Liu F, Grundke-Iqbal I et al (2005) Post-translational modifications of tau protein in Alzheimer's disease. J Neural Transm 112:813–838
18. Serrels A, Macpherson IRJ, Evans TRJ et al (2006) Identification of potential biomarkers for measuring inhibition of Src kinase activity in colon cancer cells following treatment with dasatinib. Mol Cancer Ther 5:3014–3022
19. An HJ, Miyamoto S, Lancaster KS et al (2006) Profiling of glycans in serum for the discovery of potential biomarkers for ovarian cancer. J Proteome Res 5:1626–1635
20. Syka JEP, Coon JJ, Schroeder MJ et al (2004) Peptide and protein sequence analysis by electron transfer dissociation mass spectrometry. Proc Natl Acad Sci USA 101:9528–9533
21. Molina H, Horn DM, Tang N et al (2007) Global proteomic profiling of phosphopeptides using electron transfer dissociation tandem mass spectrometry. Proc Natl Acad Sci USA 104:2199–2204
22. Kim MS, Zhong J, Kandasamy K et al (2011) Systematic evaluation of alternating CID and

ETD fragmentation for phosphorylated peptides. Proteomics 11:2568–2572

23. McAlister GC, Berggren WT, Griep-Raming J et al (2008) A proteomics grade electron transfer dissociation-enabled hybrid linear ion trap-orbitrap mass spectrometer. J Proteome Res 7:3127–3136
24. McAlister GC, Phanstiel D, Good DM et al (2007) Implementation of electron-transfer dissociation on a hybrid linear ion trap-orbitrap mass spectrometer. Anal Chem 79:3525–3534
25. Williams DK, McAlister GC, Good DM et al (2007) Dual electrospray ion source for electron-transfer dissociation on a hybrid linear ion trap-orbitrap mass spectrometer. Anal Chem 79:7916–7919
26. Swaney DL, McAlister GC, Coon JJ (2008) Decision tree-driven tandem mass spectrometry for shotgun proteomics. Nat Methods 5:959–964
27. Swaney DL, McAlister GC, Wirtala M et al (2007) Supplemental activation method for high-efficiency electron-transfer dissociation of doubly protonated peptide precursors. Anal Chem 79:477–485
28. Taylor CF, Paton NW, Lilley KS et al (2007) The minimum information about a proteomics experiment (MIAPE). Nat Biotechnol 25: 887–893
29. Mischak H, Apweiler R, Banks RE et al (2007) Clinical proteomics: a need to define the field and to begin to set adequate standards. Proteomics Clin Appl 1:148–156
30. Good DM, Wenger CD, Coon JJ (2010) The effect of interfering ions on search algorithm performance for electron-transfer dissociation data. Proteomics 10:164–167
31. Good DM, Wenger CD, McAlister GC et al (2009) Post-acquisition ETD spectral processing for increased peptide identifications. J Am Soc Mass Spectrom 20:1435–1440
32. Wenger CD, Phanstiel DH, Lee MV et al (2011) COMPASS: a suite of pre- and post-search proteomics software tools for OMSSA. Proteomics 11:1064–1074
33. Perkins DN, Pappin DJ, Creasy DM et al (1999) Probability-based protein identification by searching sequence databases using mass spectrometry data. Electrophoresis 20:3551–3567
34. Thompson A, Schafer J, Kuhn K et al (2003) Tandem mass tags: a novel quantification strategy for comparative analysis of complex protein mixtures by MS/MS. Anal Chem 75: 1895–1904

Chapter 19

Phosphopeptide Microarrays for Comparative Proteomic Profiling of Cellular Lysates

Liqian Gao, Hongyan Sun, Mahesh Uttamchandani, and Shao Q. Yao

Abstract

Protein phosphorylation is one of the most important and well-studied posttranslational modifications. Aberrant phosphorylation causes a wide spectrum of diseases, including cancers. As a result, many of the proteins involved in these pathways are seen as vital drug targets and biomarkers in treatment and diagnosis. The availability of broad-based platforms that identify changes across cellular states is critical in understanding unique disease characteristics and changes at the proteomic level. To highlight how microarrays can be applied in this regard, we describe here a comparative proteomic profiling method using two-color sample labeling and application on phosphopeptide microarrays, followed by a pull-down strategy and MS-based protein identification. This strategy has been applied to uncover candidate biomarkers in breast cancer and colon cancer cell lines. Apart from the synthesis of the phosphopeptide libraries and growth/isolation of cellular lysates, the protocol takes approximately 15 days to complete, once key steps have been optimized, and can be readily extended to other similarly complex biological specimens/samples.

Key words Phosphopeptide microarrays, Biomarkers, Protein phosphorylation, Comparative proteomic profiling

1 Introduction

Microarrays are a well-established screening platform, with a diverse spectrum of applications across many different biomolecule types. Microarrays comprise an assortment of molecules or ligands that are gridded in high density across a planar surface in micrometer-sized spots (1). This compact format reduces the quantities of usually precious samples and ligands required, significantly increasing experimental data yield and throughput. Akin to DNA microarrays, where DNA probes facilitate comparisons across transcriptomes (2), we have developed a method to likewise channel microarray throughput towards comparisons across proteomes (3). Conceivably, this is confounded with the range of molecules available as immobilized ligands on the microarrays, each possessing differing

Ming Zhou and Timothy Veenstra (eds.), *Proteomics for Biomarker Discovery: Methods and Protocols*, Methods in Molecular Biology, vol. 1002, DOI 10.1007/978-1-62703-360-2_19,

binding affinities to protein targets. This is rather unlike DNA sequences where binding and hybridization events are predictable and clearly defined. We thus elected to experiment with a set of phosphopeptide sequences derived from human proteins, to comparatively screen patterns of interactions across different immortalized human cell lines.

Reversible phosphorylation occurs in a significant proportion of the expressed proteome and plays an important role in the regulation of protein function. Up to a third of all proteins are phosphorylated during their lifetimes, and the intricate control over these phosphorylation events and their effects on cellular processes remains a challenge to globally characterize. Kinases and phosphatases are key enzymes that work in opposition to synchronize the phosphorylation states of proteins, mediating key pathways in cell signaling, apoptosis, and differentiation (4–6). Aberrant phosphorylation has been linked to many cancers, which makes the proteins involved in these pathways key targets in cancer research. Cancer cells are also known to exhibit elevated protein phosphorylation levels (7, 8).

Existing strategies for proteomic studies include mass spectrometry with affinity tags (9, 10) and two-dimensional gel electrophoresis (11, 12). These strategies comparatively quantify protein differences across complex samples and have valuably served the proteomics field (13). However, these differences in abundance do not always underscore differences in protein activities (14). We thus sought to leverage on binding interactions on microarrays as a means of distinguishing proteomes based on functional differences (Fig. 1) (3, 15, 16). The experiments were performed in a dual-color microarray format, where protein/proteome samples are cross labeled with two spectrally distinct fluorescent dyes (e.g., Cy3 and Cy5, which are false colored green and red, respectively) (17, 18). A duplicate, reciprocal microarray is also run alongside with dyes swapped, to ensure no dye bias and reproducibility across samples. Spots with equal binding representations from both proteomes on the ligand on the array will hence appear yellow. On the other hand, spots with uneven binding across samples would appear red or green, identifying them as hits capable of differentiating the samples (Fig. 1a).

Our strategy then takes these peptide hits forward towards identifying the proteins within the samples responsible for the differences observed (Fig. 1b). Pulldowns are performed with the hits against the same proteome samples, followed by cutting out bands on fluorescent gels, which are subjected to MS deconvolution, identifying protein candidates. Finally, these candidates are validated as true protein hits, or biomarkers, through Western blots and secondary binding experiments on the microarrays. This workflow provides a valuable strategy to perform functional comparisons across proteomes. We illustrate the strategy here using an

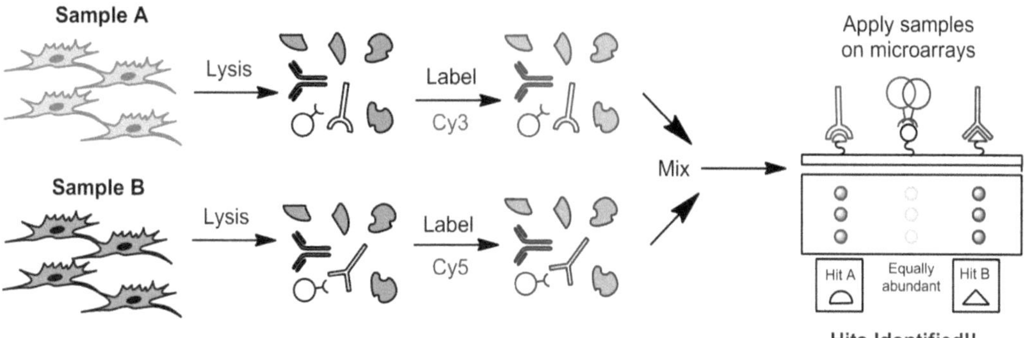

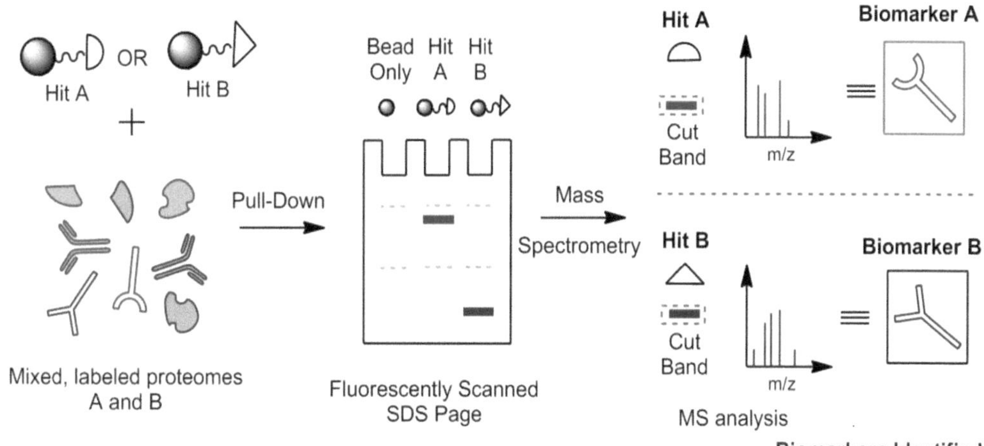

Fig. 1 Workflow of comparative proteomic profiling of mammalian cell lysates and biomarker discovery. (**a**) Microarray application—the phosphopeptide-spotted array is screened with mammalian cell lysates labeled with fluorescent dyes; (**b**) the peptide hit identified was used to perform pulldowns from which biomarkers were identified through mass spectrometry

example of a phosphopeptide library of human peptides and several human-derived cell lines. We also start by optimizing the microarray workflow with a SH2 domain protein, GRB2, which binds phosphotyrosine-containing peptides, for quality control, before describing methods of applying cellular lysates or other complex samples for which the binding profiles would be generally unpredictable.

2 Materials

2.1 Preparation of Avidin Slides

1. Piranha solution (H_2SO_4:H_2O_2 = 7:3).
2. *N*-Hydroxybenzotriazole (HOBt; GL Biochem).

3. APTES solution: 380 mL 100 % ethanol, 8 mL H_2O, and 12 mL aminopropyltriethoxysilane (Sigma-Aldrich).
4. Carboxylic acid activation solution: 400 mL of dimethylformamide (DMF, HPLC grade), 12 g succinic anhydride, and 18 mL of 1 M $Na_2B_4O_7$ (pH = 9.0).
5. NHS activation solution: 15 mL of DMF, 565 mg *O*-benzotriazole-*N*,*N*,*N*′,*N*′-tetramethyl-uronium-hexafluoro-phosphate (HBTU), 173 mg *N*-hydroxysuccinimide, and 550 μL of *N*-ethyldiisopropylamine (DIEA).
6. Avidin solution: 2 mg avidin (for 30 slides), 1.5 mL of Milli-Q water, and 30 μL of 0.5 M $NaHCO_3$, pH 9.0.
7. Quenching solution: 2 mM aspartic acid in 0.5 M $NaHCO_3$ (pH = 9.0).

2.2 Microarray Spotting

1. Phosphopeptide library stock solutions.
2. Phosphate-buffered saline (PBS) (pH = 7.4).
3. 384-well polypropylene microarray plate (Genetix).

2.3 SH2 Domain Protein Expression and Purification

1. GRB2 SH2 domain commercial bacterial construct in expression host (Open Biosystems, Thermo Scientific).
2. Nickel-NTA agarose resin (Qiagen, Cat. No. 1018244).
3. His-tag purification lysis buffer (50 mM Tris–HCl base, 300 mM NaCl, pH = 8.0).
4. His-tag purification wash buffer (50 mM Tris–HCl base, 300 mM NaCl, 20 mM imidazole, pH = 8.0).
5. His-tag purification elution buffer (50 mM Tris–HCl, 300 mM NaCl, 200 mM imidazole, pH = 8.0).

2.4 Protein Labeling, Incubations, and Scanning

1. Cy3 and Cy5 *N*-hydroxysuccinimide ester (Amersham, GE Healthcare, USA).
2. G-25 spin column (Amersham, GE Healthcare, USA).
3. Incubation buffer: 0.05 % Tween® 20 in TBS.
4. 15 % SDS-PAGE gels.

2.5 Pull-Down Experiments

1. MCF7, HEK293T, SW480, and SW620 cell pellets.
2. Fetal calf serum (Hyclone, Cat. No. 1677-006) and penicillin/streptomycin (100 times concentrated) (Invitrogen).
3. Bradford protein assay (Bio-Rad).
4. 12 % SDS-PAGE gels.
5. Blocking buffer: 4 % (w/v) BSA in 0.05 % Tween® 20 in TBS buffer.
6. Bio Trace™ PVDF (polyvinylidene fluoride) transfer membrane (Pall Corporation, Product No. 66543).

7. Antibody for LDH-B, PRDX1, and PHB (Santa Cruz Biotechnology).
8. NeutrAvidin agarose beads (Thermo Scientific).
9. SuperSignal West Pico kit (Pierce).

2.6 Equipment

1. Adhesive film (ABgene, Cat. No. AB-0558).
2. Arrayer (ESI SMA™) (Ontario, Canada) fitted with Stealth Micro Spotting pins (Telechem International, cat. ID SMP7B).
3. Circular shaker.
4. Chromatography column (1.5 cm diameter × 10 cm length).
5. 22 mm × 60 mm coverslips.
6. 22 mm × 22 mm coverslips.
7. 15 and 50 mL centrifuge tubes.
8. Desiccator/dry storage box.
9. Humid incubation chamber.
10. Kodak Medical X-ray Processor 2000 (Carestream Health).
11. LTQ FT ultra mass spectrometer (Thermo Electron, Germany).
12. Microcon YM-3 centrifugal filter.
13. MicroSpin™ G-25 columns (Cat. No. 27-5325-01, Amersham Biosciences, GE Healthcare).
14. Microplate shaker.
15. 75 mm × 25 mm × 1 mm microscope glass slides.
16. 1.5 mL reaction Eppendorf tubes.
17. Slide-A-Lyzer Dialysis Cassettes 7 K MWCO, 0.1–0.5 mL capacity (Thermo Scientific, Cat. No. 0066375).
18. Slide staining rack.
19. Slide staining jar.
20. Sonics Vibra-Cell sonicator (ITS Science & Medical Pte Ltd.).
21. Sterile 22 G needle and 1 mL syringe.
22. 14 mL sterile PP tube (Greiner bio-one, Cat. No. 187261).
23. Tecan Launch LS Reloaded Microarray Scanner (Tecan Trading AG, Switzerland).
24. Typhoon™ 9410 Variable Mode Imager (Typhoon fluorescence gel scanner) (GE Healthcare).
25. UV/visible spectrophotometer.
26. 96-well 1 mL polypropylene stock plates.
27. 96-well polypropylene stock plates.
28. 384-well polypropylene microarray plate.

2.7 Software

1. Array-Pro Analyzer software (Tecan Trading AG, Switzerland).
2. GraphPad Prism 5 software (GraphPad, San Diego, USA).
3. Microsoft Excel.
4. Typhoon scanner control software (operating software for typhoon fluorescence gel scanner) (GE Healthcare).
5. ImageQuant TL (GE Healthcare).

3 Methods

Microarrays have generally been applied in high-throughput screens with individual proteins. Herein, we apply a targeted phosphopeptide microarray towards the differentiation of complex mammalian cell lysates. A library of phosphopeptides were synthesized by solid-phase peptide synthesis and were analyzed by LC-MS, which confirmed that the library members were pure (>85 %). The peptide sequences immobilized on the microarrays were selected from important signal pathway proteins involved in signal transduction. For details on these synthetic methods, the reader is referred to our previously published protocols (19, 20). A total of 176 phosphopeptides with a centrally positioned phosphotyrosine residue were spotted onto the avidin-coated glass slides to generate the corresponding phosphopeptide microarray.

After incubating with pure SH2 domain protein, the fluorescence intensity measured corresponds to the level of protein bound to the respective phosphopeptides spots on the arrays, indicating its affinity to the particular peptide sequence. Next, by adopting a dual-color labeling and application strategy for comparing multiple cellular lysates in a single slide, various microarray "hits" were identified that differentiated cancer and normal proteomes. These phosphopeptide hits were then applied in pulling down their corresponding interacting partners within the cancer samples, revealing potential biomarkers dysregulated in the cancer state. Of the biomarkers successfully validated, LDH-B and PRDX1 were identified in MCF7, a breast cancer cell line, while PHB was uncovered against SW620, a colon cancer cell line. Our workflow thus demonstrates how phosphopeptide microarrays may be applied for the broader comparisons of cancer states and biomarker discovery.

3.1 Microarray Fabrication

1. Fabricate or purchase amine reactive slides (e.g., *N*-hydroxysuccinimide NHS, or epoxy-derivatized slides). Plain glass slides may be derivatized into NHS or epoxy surfaces, through procedures described in earlier protocols (21).
2. Label the top right-hand corner of the activated slide surface, and number the slides with a permanent marker.

3. Coat slides with avidin by first preparing a solution of 1 mg/mL avidin in 10 mM $NaHCO_3$ (pH = 9.0) and then incubating 50 μL of the avidin solution on the surface of amine reactive slides for 30 min under a 60 × 22 mm coverslip.
4. After the reaction, quench the unreacted NHS groups with a solution of 2 mM aspartic acid in a 0.5 M $NaHCO_3$ buffer (pH = 9.0) for 30 min, followed by brief repeated rinses with water.
5. Dry the slides under a stream of nitrogen gas, and store at 4 °C until required for spotting. Prepare peptide solutions in PBS to a concentration of approximately 1 mM. A 1 mM concentration of peptide would saturate the spot completely and ensure that adequate peptides are immobilized (see Note 1).
6. Samples should be prepared in 384-well polypropylene microarray plate, compatible with the microarray spotter.
7. Seal the spotting plate with adhesive film and store it at −20 °C. Thaw when needed.
8. Briefly wash the avidin-coated slides with deionized H_2O, and then quickly dry with nitrogen gas.
9. Spot the biotinylated peptide library (with samples spotted in adjacent duplicates) to the surface of the avidin-coated slides using the ESI SMA arrayer. Let the freshly spotted slides sit inside the spotter for another 2–4 h post-spotting to provide sufficient time for immobilization, and then store the spotted slides at 4 °C until required for use.
10. Rinse the slides with water and shaking for at least 10 min to remove unbound peptides.
11. To ensure the good quality of any given batch of spotted slides (such as consistency between the different subgrids and all the phosphopeptides properly immobilized on the surface of the slides), perform the Pro-Q™ assay as follows:
    (a) Dry the slides under a stream of nitrogen gas.
    (b) Apply 1 mL Pro-Q™ Diamond dye for 1 h in a humid incubation chamber under 60 × 22 mm coverslips at room temperature.
    (c) Destain the slides with 25 mL solution of 20 % acetonitrile in sodium acetate (50 mM, pH = 4) for 0.5 h, and then visualize the slides with the Tecan Launch LS Reloaded Microarray Scanner. Typical Pro-Q™ slide results are shown in Fig. 2.

3.2 Expression and Purification of Positive Control SH2 Domain Protein

1. Grow stocks of BL21DE3 *E. coli* culture positively transformed with GRB2 constructs on LB-agar plates with the relevant antibiotic (kanamycin).

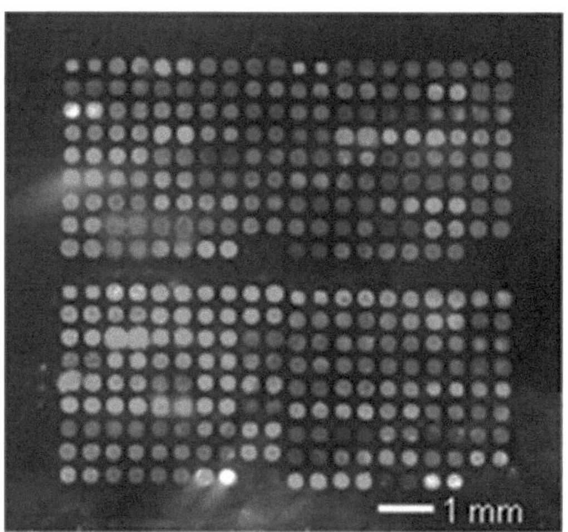

Fig. 2 Pro-Q stained slide images. The slides comprise 176 phosphopeptide library printed in duplicate (spots *left* and *right* represent duplicated samples) using SMP7B pins. Spots were approximately 350 μm in size

2. Inoculate a single bacterial colony from the agar plate into a 14 mL PP sterile tube containing 3–5 mL LB-kanamycin media. Grow the culture overnight at 37 °C in an orbital shaker with constant shaking at 230 rpm.
3. Dilute the overnight culture 1:100 into 200 mL fresh LB-kanamycin (50 μg/mL) media in a conical flask. Incubate at 37 °C with constant shaking at 230 rpm till OD_{600} reaches 0.7 (this takes approximately 2–3 h).
4. Aspirate a 10 mL culture of uninduced cells to be stored as a negative control for protein expression. The optimum absorbance (OD_{600}) for induction is 0.6–0.8. Check the absorbance periodically. If the cells are overgrown, repeat steps 3–4.
5. Add IPTG to the remaining culture to a final concentration of 0.1 mM to induce protein expression. Incubate the culture overnight (12–20 h) at 15 °C with constant shaking at 230 rpm.
6. Collect a 100 mL sample of culture (induced cells), spin, discard the supernatant, and store at −20 °C when use. Cell pellets can be stored at −20 °C for up to a month without any significant deterioration of proteins. However, long-term storage (more than 3 months) of pellets may result in a decrease in protein activities.
7. Resuspend the cell pellet in 3–10 mL ice-cold (4–8 °C) His-tag lysis buffer, and apply vortexing if necessary.
8. Lyse cells by sonication on ice at 30 % amplitude in a sonicator with five 5 s bursts and a 10 s cooling interval between each burst.

9. Clarify the lysate by centrifugation at 16,100 × *g* for 30 min at 4 °C. Transfer the supernatant on ice to a fresh Eppendorf tube.
10. Pack a chromatography column (1.5 cm diameter × 10 cm length) with suitable amount of Ni-NTA resin (150 μL of bead volume may be used for 100 mL protein bacterial culture according to the expression levels). Wash the resin with lysis buffer for 3–5 times (5–10 mL each time) before loading lysate.
11. Incubate and shake the mixture of the beads and lysates at 4 °C for 45 min to 1 h.
12. To maintain the activity of the proteins, we usually carry out the experiments for this step at 4 °C.
13. Slowly flow through the lysates after incubation, and wash the column with His-tag wash buffer for 3–5 times (5–10 mL each time). The columns are designed for low-pressure applications (<1 atm). When used in conjunction with a pump, any disruption in the flow must be avoided as it may cause a rapid increase in back pressure. If this happens, immediately switch off the pump and check gel bed for compression.
14. Elute the SH2 domains protein with 0.1–0.3 mL of His-tag elution buffer for 2–4 times. If possible, incubate the elution buffer with the beads for 2–5 min, and then collect all the fractions of the eluent, measure the protein concentration by Bradford assay, and combine the fractions that contain the highest amounts of SH2 domain proteins (usually the first few fractions).
15. Run the samples 15 % SDS gel to confirm the purity of the expressed GRB2 protein (shown in Fig. 3a).

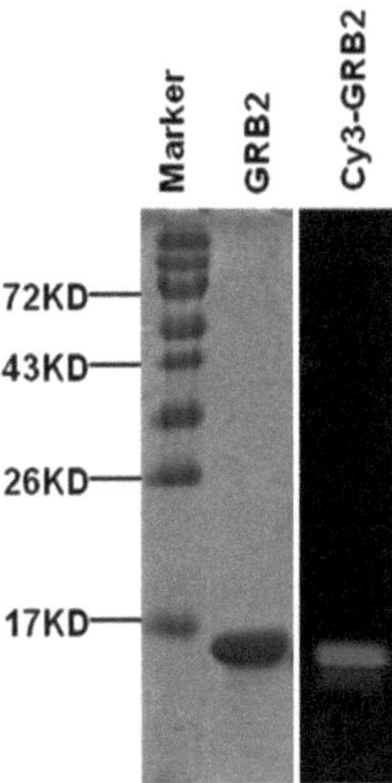

Fig. 3 SH2 domain expression and purification, in this example of GRB2. The *left* image is Coomassie blue-stained gels of purified proteins and on the *left* is the fluorescent gel results of GRB2 Cy3 labeled. Expected GRB2 MW = 12 kDa

3.3 Protein Labeling

1. Perform buffer exchange of SH2 domain proteins from the His-tag buffer to PBS buffer using the dialysis cassette (7 kD cutoff). The dialysis assay for the buffer exchange is performed overnight (about 12 h) at 4 °C. After that, isolate the protein samples and store frozen at −20 °C until needed for use (see Note 2).
2. Label 50 μg GRB2 SH2 domain protein in 50 μL solution of 0.1 M $NaHCO_3$ (pH = 9.0) with 1.25 μL of Cy3- or Cy5-conjugated *N*-hydroxysuccinimide esters in a 1.5 mL tube, and the final protein concentration is 1 mg/mL. After gentle shaking, place the tube on ice and let the mixture react for 1 h. Keep proteins on ice at all times. This allows preservation of activity. It might be necessary to optimize the amount of dye used (see Note 3).
3. Quench the unreacted dye with a tenfold molar excess of hydroxylamine (pH = 8.5) or using 5 mM Tris–HCl (pH = 8.0) on ice, for a further 1 h.
4. Remove unbound dye by performing the buffer exchange overnight into TBS buffer by using the dialysis cassette same as step 1.
5. To confirm the GRB2 SH2 domain protein is successfully labeled, run it on SDS-PAGE and scan it in the fluorescence channel of the Typhoon scanner. The fluorescent gel result is shown in Fig. 3. It is important to implement this step rigorously to ensure that the proteins are evenly labeled in the respective fluorescence channel for following up experiments.

3.4 Pure Protein Screening on Microarrays as Quality Control

1. Block the slide surface with 1 % BSA in TBS for 1 h.
2. Use denatured SH2 domain proteins (heating at 95 °C for 10 min followed by plunging into ice) as well as other model proteins (e.g., albumin or renin) as negative controls.
3. Prepare the labeled protein (active, denatured proteins, and other negative controls) in TBS. Proteins are typically applied at 1 μM. Typical application volumes are at 75 μL, which may be applied directly on the grid which are marked using the permanent marker pen to form rectangular frames, without the need for a coverslip (see Note 4).
4. Incubate different concentrations (0.06–1.0 μM) of the labeled protein solution individually onto the subgrids, in a humid chamber for 0.5–1 h. Care must be taken that no air bubbles are trapped between the coverslips and the slide. These voids would generate false negatives. Besides, to get good and consistent fingerprint results with even backgrounds, care should be taken to apply the labeled proteins solution uniformly across the surfaces of the slides.

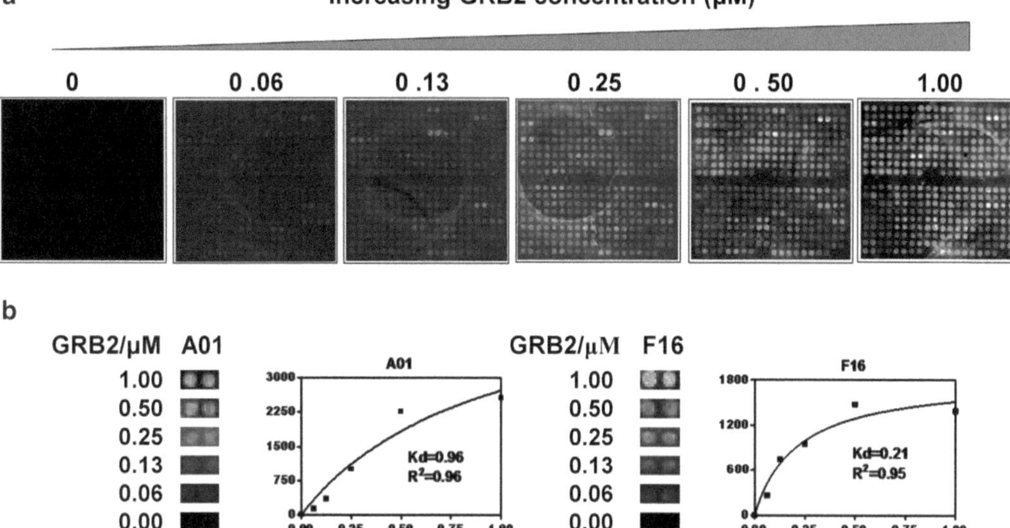

Fig. 4 (**a**) Concentration-dependent fingerprints of the phosphopeptide microarrays with the labeled GRB2 SH2 domain proteins at different concentration; (**b**) 2 representative examples of top binding peptides, A01 (AEKPF*p*YVNVEF) and F16 (SADHL*p*YVNVSE) with 0.96 and 0.21 μM apparent K_D values, respectively

5. After incubation, gently and quickly wash the slides with deionized H_2O to remove all the labeled proteins solution, and then wash with TBST (TBS containing 0.05 % Tween 20). Washing intensity and duration may be optimized as required. We typically use 1–3 short 5 min washes which produces optimal signal to noise.
6. Rinse slides with distilled water, and then quickly dry the slides for nitrogen gas and scan them using microarray scanner.
7. Analyze array images. Confirm the presence of positive control binding on peptides. There should only be signals with active proteins and no (or negligible) signals with negative control proteins or when the proteins are heat denatured.
8. Extract raw intensity values of the spots using microarray software.
9. Perform background subtraction and normalization. Average duplicate spots within the same slide. After that, plot concentration-dependent curves to calculate the corresponding K_D value using GraphPad (Fig. 4).

3.5 Comparative Proteomic Profiling of Mammalian Cell Lysates

1. Grow MCF7 (noninvasive breast cancer line) and HEK293T (normal human embryonic kidney line) cells in growth medium (DMEM supplemented with 10 % fetal bovine serum (FBS), 100 mg/L streptomycin, and 100 IU/mL penicillin) on T-75 culture plates.

2. Maintain cells in a humidified atmosphere of 5 % CO_2 at 37 °C, and then collect the cell pellet when cells reach a ~90 % confluence. Detach cells by trypsin digestion, wash by PBS for three times, centrifuge, and store at −20 °C until required for use.
3. Thaw the cell pellets on ice with PBS buffer added (typically resuspend one T-75 cell pellet with 200–400 μL PBS buffer). Thaw the cell pellet for at least 15 min on ice.
4. When thawed completely, perform the cell lysis by 30–50 passages through a 22 G needle using a 1 mL syringe, and then centrifuge samples at 16,100 × *g* at 4 °C for 30 min. Transfer the clear supernatant to a chilled Eppendorf tube. Determine protein concentration by standard Bradford assays (Bio-Rad).
5. Label 50 μg lysate proteome samples in a final 50 μL solution of 0.1 M $NaHCO_3$ (pH = 9.0) containing 1.25 μL of Cy3- or Cy5-conjugated *N*-hydroxysuccinimide esters (Monofunctional Cy5/Cy3 dye). The final protein concentration is 1 mg/mL. After gentle shaking to make the solution even, put the tube on ice and let the mixture react for 1 h. Adopt labeling procedures as described in steps 2–4 in Subheading 3.3. The amount of the dyes used could be optimized to avoid high dye background when necessary (see Note 3).
6. Run the SDS gel to confirm that the lysates are evenly and successfully labeled (Fig. 5).
7. Reconstitute the labeled proteomes in final TBS buffer with volume of 100 μL (pH = 7.4) to adjust the final application concentration to 0.5 mg/mL.

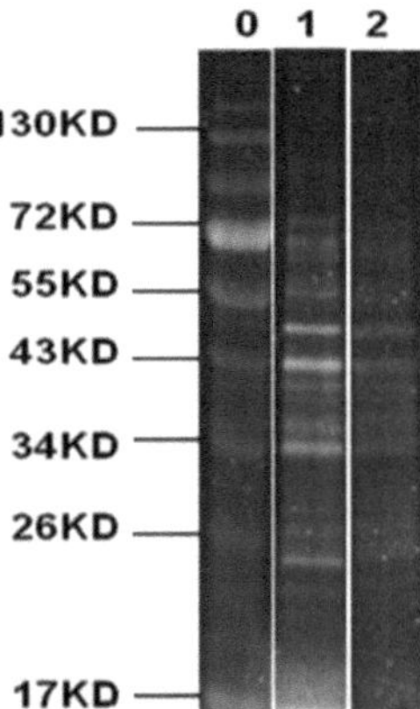

Fig. 5 Cy3 and Cy5 labeled HEK293T cells. In a standard lysate microarray experiment, apply the labeled lysates (0.5 mg/mL; 40 μL) under coverslip to the 2 subgrid arrays (10 μg lysates/subgrid). For the dual-color reciprocal microarray experiments, apply a mixture of Cy3 (or Cy5)-labeled proteome A and Cy5 (or Cy3)-labeled proteome B to two rinsed slides (Reproduced in part from ref. 3 with permission from The Royal Society of Chemistry)

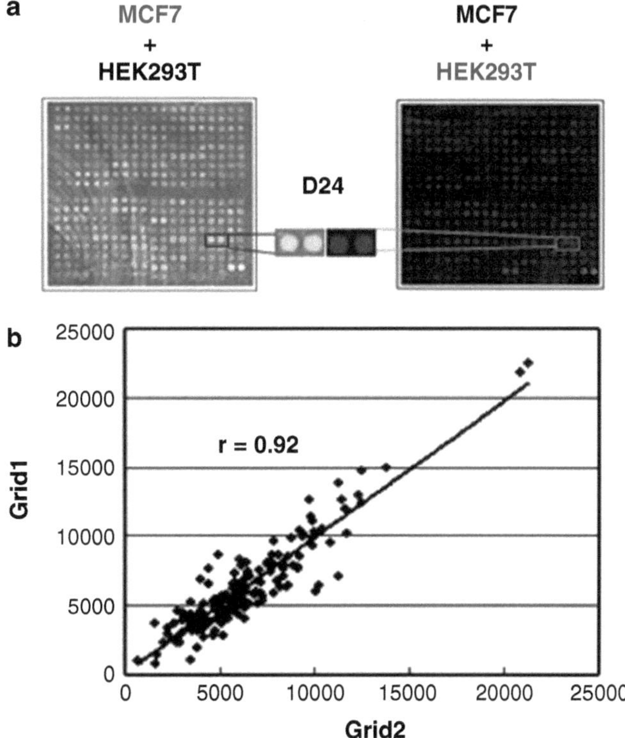

Fig. 6 (**a**) Dual-color reciprocal fingerprinting of MCF7 vs. HEK293T. (**b**) Data quality analysis for reproducibility across the Cy3 and Cy5 channels using Pearson coefficient. Hit molecule D24 (YFMTE*p*YVATRW) was identified (Reproduced in part from ref. 3 with permission from The Royal Society of Chemistry)

8. Apply the proteins to the slides and perform incubation of the samples in a humidified box/chamber for 1 h at room temperature (see Note 5).
9. Perform a dye swap experiment to duplicate the experiments, with the dyes swapped across the test and control samples. This allows result duplication and filtering of any dye bias across two independent slides.
10. Remove the coverslip gently and perform rinses with TBST (TBS containing 0.05 % Tween 20), typically 3 washes, each with 10 min of gentle shaking.
11. Scan slides using a Tecan Launch LS Reloaded Microarray Scanner (Fig. 6).
12. Perform quantitative data analysis to value the data quality and identify the hit molecule that reciprocates across the two slides for follow-up pulldown experiment for biomarker discovery.

3.6 Dual-Color Pulldown

1. Firstly incubate NeutrAvidin™ agarose beads with selected biotinylated peptide probes (final concentration 10 μM) at room temperature for 1 h, and then wash the beads three times with TBS buffer to remove the excess unbound peptide probes.
2. Label cellular lysates with a ratio of 20 μL dye for 2.5 mg lysates, following steps 2–3 in Subheading 3.3, as previously described.
3. After washing the beads, add the cellular lysates (2.5 mg) containing phosphatase inhibitor (to a final concentration 2 mM Na_3VO_4), and incubate at room temperature for 1 h with gentle rocking.
4. Centrifuge NeutrAvidin™ agarose beads for 5 min at 4,000 × *g* at room temperature, and then wash the beads with TBST (0.05 % Tween® 20 in TBS buffer) for 5–10 times.
5. Perform the negative control following steps 2–4 (same pull-down procedures except using beads without the peptide probes).
6. Take about 80 μL beads (about one third of the beads used), resuspend the beads with 1× SDS loading dye buffer, and then run the SDS-PAGE gel after heating at 95 °C for 10–15 min, followed by fluorescent scan and silver stain (shown in Fig. 7a).
7. Carefully cut out only the desired bands in fluorescent gel, and then perform an in-gel trypsin digestion following published protocols (22).

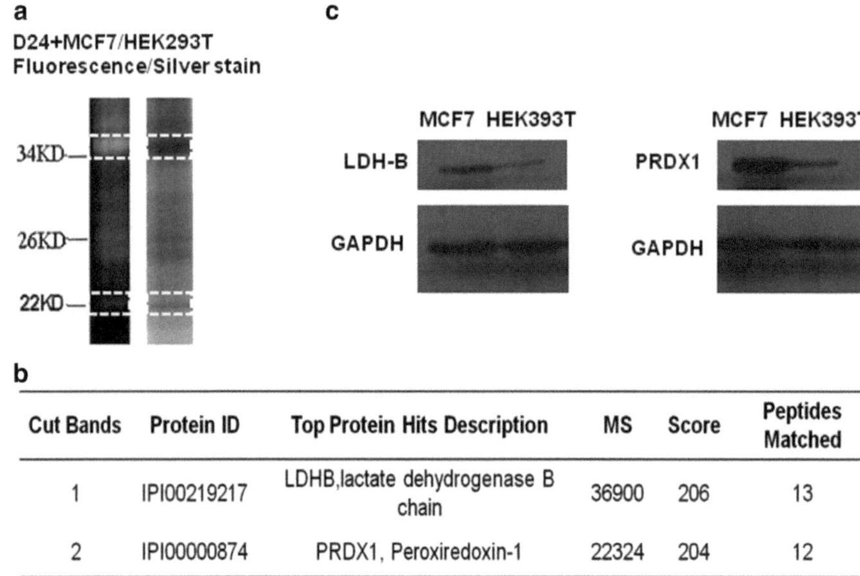

Cut Bands	Protein ID	Top Protein Hits Description	MS	Score	Peptides Matched
1	IPI00219217	LDHB,lactate dehydrogenase B chain	36900	206	13
2	IPI00000874	PRDX1, Peroxiredoxin-1	22324	204	12

Fig. 7 (**a**) Fluorescent gel and silver stain gel results of the dual-color pulldown, (**b**) MS results, and (**c**) Western blotting to validate the different expression levels of the biomarker candidates between the cancer and normal cells (Reproduced in part from ref. 3 with permission from The Royal Society of Chemistry)

3.7 Mass Spectrometric Analysis and MS Identification

1. Extract MS spectra in dta format using the Extract_Msn (version 4.0) program found in Bioworks Browser 3.3 (Thermo Electron, Bremen, Germany) from the raw data of LTQ FT ultra.
2. Convert these dta files into MASCOT generic file format. Intensity values and fragment ion m/z ratios are not manipulated.
3. Use this data to obtain protein identities by searching against the corresponding database by means of an in-house MASCOT server (version 2.2.03) (Matrix Science, Boston, MA). Mass tolerance for peptide precursors is 10 ppm, and mass tolerance for fragment ions is 0.8 Da. Only proteins with a MOWSE score higher than 39, corresponding to $p < 0.05$, are considered significant.
4. Export the peptide/protein list(s) obtained to html file. The top ranked proteins are putative biomarkers, differentially expressed across the test samples. Accordingly, we identify lactate dehydrogenase B (LDH-B) and peroxiredoxin-1 (PRDX1) in MCF7 lysates, as shown in Fig. 7b.

3.8 Validation of Identified Biomarker Candidates by Western Blotting

1. To confirm that these biomarkers are truly differentially expressed, load and run equal amount of lysate proteomes from the cancer and normal cells (30 μg) on SDS-PAGE. Perform a transfer from the SDS gel onto PVDF membrane using the Western blotting transfer buffer (3.03 g/L Tris–HCl, 14.4 g/L glycine, 10 % (v:v) methanol) at 100 V for 1 h at room temperature or 30 V for overnight at 4 °C in cold room.
2. Block the transferred PVDF membrane for 1 h at room temperature or overnight at 4 °C in blocking buffer (comprising 4 % BSA in TBS containing 0.1 % Tween 20).
3. After blocking, incubate the PVDF membrane with primary antibody for 1 h at room temperature in primary antibody buffer (3 % BSA in TBS containing 0.1 % Tween 20). Dilute the primary antibody to a concentration that produces detectable signal in the Western blot.
4. After primary antibody incubation, wash the PVDF membrane three times for 5–10 min each in TBST buffer (TBS containing 0.1 % Tween 20).
5. After wash, incubate the PVDF membrane with HRP-conjugated secondary antibody for 1 h at room temperature in the secondary antibody buffer.
6. After secondary antibody incubation, wash the PVDF membrane three times for 10–15 min with TBST buffer (TBS containing 0.1 % Tween 20).
7. Detect/develop the PVDF signal by chemiluminescence using the SuperSignal West Pico/Dura kit (Fig. 7c).

3.9 Validation of the Biomarkers Identified on Microarrays

1. To confirm that the biomarkers indeed do interact with the hits on the microarray, label each antibody for each respective biomarker identified with Cy3 monofunctional dye: firstly mix 0.3 μL Cy3 dye, with 4–8 μg of antibody in a final 50 μL volume solution of 0.1 M $NaHCO_3$ (pH = 9.0), and then let the mixture solution react for 1 h on ice. After reaction, quench the unreacted dye with 5 μL hydroxylamine solution for an additional 1 h.
2. Run the samples through a G-25 column to remove salts and unreacted dye (see Note 5).
3. After preparation of labeled antibodies, wash the glass slides spotted with peptide library for 10 min with deionized H_2O.
4. Block the slides with 1 % BSA/TBS for 1 h to maximally reduce nonspecific binding.
5. After blocking, incubate the slides with 30 μg MCF7 cell lysates in TBS buffer per subgrid for 1 h.
6. After incubation, briefly and quickly wash the slides with deionized H_2O.
7. After wash, incubate the slides with Cy3-labeled antibodies for another 1 h under 22 mm × 22 mm coverslips. Care must be taken that no air bubbles are trapped between the coverslip and the slide. Prolonged incubation beyond 1 h could cause high background.
8. After antibody incubation, wash the slides between two and three times with deionized H_2O, for about 2 min each time (see Note 6).
9. Dry the slides in a clean fume hood or with nitrogen gas, and then scan the slides with the microarray scanner. Typical results are shown in Fig. 8.

In summary, a microarray-based approach has been described for high-throughput proteomic profiling of mammalian cell lysates. This strategy can be applied in wider biomarker discovery

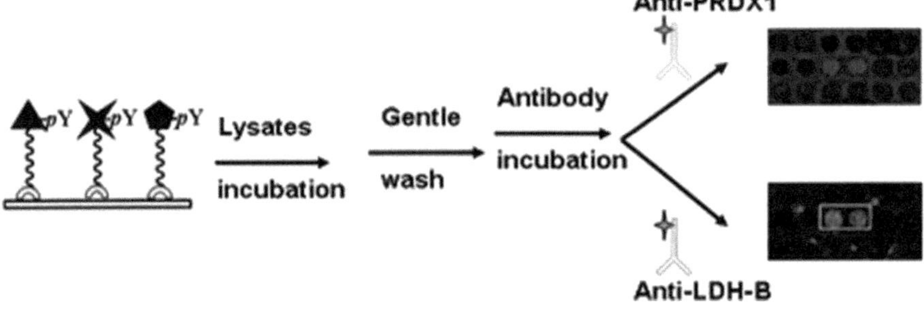

Fig. 8 Antibody detection for the interactions between peptide probes of D24 (YFMTE*p*YVATRW) and the corresponding proteins (biomarker candidates) PRDX1 (boxed in *top array*) and LDH-B (boxed in *lower array*) on the microarray

applications, as it allows optimal use of precious and often limited samples. Low quantities of proteins are required for the initial screens. Once hits are identified, more proteins are required to perform the pulldown assay, for the process of identification of putative biomarkers. Here we have highlighted how selected phosphopeptide hits were able to identify putative cancer biomarker candidates for MCF7 (breast cancer) cell lines, respectively (Fig. 7). These results were validated by Western blots (Fig. 7c) and microarray screens with labeled antibodies (Fig. 8). Overall, this approach enables quick and efficient differentiation of complex cellular proteomes. We hence anticipate that this dual-color microarray strategy can be applied more broadly using large or focused peptide libraries as a vital tool for comparative proteomic profiling.

4 Notes

1. If necessary, lower spotting concentrations may be used, but we would not typically use spotting concentrations lower than 100 μM. We find that up to a 50 % DMSO mixture with PBS is well tolerated as a spotting buffer by the avidin proteins on the slides and should not be exceeded as this will risk denaturing the avidin layer on the slide surface.
2. Occasionally, the SH2 domains proteins may be lost after buffer exchange. The possible reason is that the proteins precipitate or dialysis cassettes were damaged when performing the buffer exchange. To address this, use a more gradual change of buffers or use buffering conditions that minimize protein precipitation.
3. The use of too much dye could lead to excess dye background and reduced protein binding activity. This may be resolved with decreased dyes used and ensuring the proteins are still detectably labeled on fluorescence gel imaging.
4. It is possible that the solutions applied across the grids merge together. The reason is that marker grids drawn onto the slides were partially rinsed off after blocked by BSA or too much solution volume was used. To solve this problem, redraw the subgrid frame using marker pen or decrease the volume of sample solution applied.
5. Too long incubation times could contribute to high signals and high backgrounds that may not be washed away. If this is the case, decrease the incubation time needed or wash the slides with harsher conditions (e.g., using increased Tween concentrations). If too low signals are detected, higher concentrations of proteins samples may be applied, with less stringent washing conditions.

6. It is possible that there may be no signal initially upon antibody application. The reason could be that the antibodies lost activity during the labeling and preparation steps. To solve this problem, confirm that fluorescently labeled antibodies still successfully labeled the target proteins in Western blots.

Acknowledgments

The authors acknowledge funding support from Ministry of Education (R-143-000-394-112), the Agency for Science, Technology and Research (R-143-000-391-305), and DSO National Laboratories.

References

1. Foong YM, Fu J, Yao SQ, Uttamchandani M (2012) Current advances in peptide and small molecule microarray technologies. Curr Opin Chem Biol 16:234–262
2. Afshari CA, Nuwaysir EF, Barrett JC (1999) Application of complementary DNA microarray technology to carcinogen identification, toxicology, and drug safety evaluation. Cancer Res 59:4759–4760
3. Gao L, Uttamchandani M, Yao SQ (2012) Comparative proteomic profiling of mammalian cell lysates using phosphopeptide microarrays. Chem Commun 48:2240–2242
4. Cohen P (2002) Protein kinases—the major drug targets of the twenty-first century? Nat Rev Drug Discov 1:309–315
5. Johnson TO, Ermolieff J, Jirousek MR (2002) Protein tyrosine phosphatase 1B inhibitors for diabetes. Nat Rev Drug Discov 1:696–709
6. Yaffe MB (2002) Phosphotyrosine-binding domains in signal transduction. Nat Rev Mol Cell Biol 3:177–186
7. Sefton BM, Hunter T, Ball EH, Singer SJ (1981) Vinculin: a cytoskeletal target of the transforming protein of Rous sarcoma virus. Cell 24:165–174
8. Hochgrafe F, Zhang L, O'Toole SA et al (2010) Tyrosine phosphorylation profiling reveals the signaling network characteristics of Basal breast cancer cells. Cancer Res 70:9391–9401
9. Sabido E, Selevsek N, Aebersold R (2011) Mass spectrometry-based proteomics for systems biology. Curr Opin Biotechnol 23. doi:10.1016/j.copbio.2011.11.014
10. Bodenmiller B, Aebersold R (2010) Quantitative analysis of protein phosphorylation on a system-wide scale by mass spectrometry-based proteomics. Methods Enzymol 470:317–334
11. Arruda SC, Barbosa Hde S, Azevedo RA, Arruda MA (2011) Two-dimensional difference gel electrophoresis applied for analytical proteomics: fundamentals and applications to the study of plant proteomics. Analyst 136:4119–4126
12. Valledor L, Jorrin J (2011) Back to the basics: maximizing the information obtained by quantitative two dimensional gel electrophoresis analyses by an appropriate experimental design and statistical analyses. J Proteomics 74:1–18
13. Kolch W, Pitt A (2010) Functional proteomics to dissect tyrosine kinase signalling pathways in cancer. Nat Rev Cancer 10:618–629
14. Uttamchandani M, Lu CH, Yao SQ (2009) Next generation chemical proteomic tools for rapid enzyme profiling. Acc Chem Res 42:1183–1192
15. Wu H, Ge J, Yang PY, Wang J, Uttamchandani M, Yao SQ (2011) A peptide aldehyde microarray for high-throughput profiling of cellular events. J Am Chem Soc 133:1946–1954
16. Shi H, Uttamchandani M, Yao SQ (2011) Applying small molecule microarrays and resulting affinity probe cocktails for proteome profiling of mammalian cell lysates. Chem Asian J 6:2803–2815
17. Uttamchandani M, Lee WL, Wang J, Yao SQ (2007) Quantitative inhibitor fingerprinting of metalloproteases using small molecule microarrays. J Am Chem Soc 129:13110–13117
18. Kattah MG, Alemi GR, Thibault DL, Balboni I, Utz PJ (2006) A new two-color Fab labeling method for autoantigen protein microarrays. Nat Methods 3:745–751
19. Lee WL, Li J, Uttamchandani M, Sun H, Yao SQ (2007) Inhibitor fingerprinting of metalloproteases using microplate and microarray

platforms: an enabling technology in Catalomics. Nat Protoc 2:2126–2138
20. Sun H, Lu CH, Shi H, Gao L, Yao SQ (2008) Peptide microarrays for high-throughput studies of Ser/Thr phosphatases. Nat Protoc 3:1485–1493
21. Chattopadhaya S, Tan LP, Yao SQ (2006) Strategies for site-specific protein biotinylation using in vitro, in vivo and cell-free systems: toward functional protein arrays. Nat Protoc 1:2386–2398
22. Tannu NS, Hemby SE (2006) Two-dimensional fluorescence difference gel electrophoresis for comparative proteomics profiling. Nat Protoc 1:1732–1742

Chapter 20

Tissue Preparation for MALDI-MS Imaging of Protein and Peptides

Simona Colantonio and Roberta M. Smith

Abstract

Matrix-assisted laser desorption/ionization (MALDI) imaging is rapidly gaining importance in the biomarkers field because it is able to detect several analytes at the same time and to assign to each one of them not only an *m/z* value but also spatial distribution. Here we present the detailed description of sample preparation for protein and peptide MALDI imaging assays. This chapter describes the microtomy performed in a cryostat to produce tissue slides mounted onto glass slides suitable for MALDI. Sample preparation will include matrix coating procedure with a sensor-controlled aerosol. Finally, we will show some examples of how data can be visualized to suit the purposes of the research.

Key words Mass spectrometry, MALDI, Imaging, Mouse, Protein, Peptide

1 Introduction

Imaging mass spectrometry (IMS) emerged only few years ago (1) as a rapidly evolving technique that allows direct mapping of chemical species on a tissue slide. Its increasing popularity is due to the chemical specificity of mass spectrometry combined with the possibility to detect several analytes at the same time and to obtain for each *m/z* value information about the *x*- and *y*-coordinates in the plane. The main ionization mechanisms that can be used in IMS are matrix-assisted laser desorption/ionization mass spectrometric (MALDI-MS) and secondary ion mass spectrometry (SIMS) (2). They all offer different capabilities in terms of spatial resolution and molecular ion mass range, but MALDI is particularly applicable for proteomics imaging purposes because of its wide mass range that allows molecular weight detection of intact proteins. MALDI can achieve a spatial resolution of 10 μm, while SIMS has an ultrahigh resolution (50 nm) that rapidly looses sensitivity when the *m/z* ration increases above 500. Desorption electrospray ionization (DESI) has also been indicated as another ionization technique

Ming Zhou and Timothy Veenstra (eds.), *Proteomics for Biomarker Discovery: Methods and Protocols*, Methods in Molecular Biology, vol. 1002, DOI 10.1007/978-1-62703-360-2_20, © Springer Science+Business Media, LLC 2013

feasible for protein/peptide imaging studies; however, it still presents limitations concerning resolution and it is mostly used for lipid analysis (3).

Unlike traditional proteomic techniques that cause loss of spatial information due to extraction and purification, MALDI imaging has the advantage of the simultaneous acquisition of the molecular weight of proteins and peptides present on the tissue slide and maintaining their spatial distribution. Compared to classical immunohistochemistry, MALDI imaging does not require a priori knowledge of the analyzed proteins, and it does not involve the use of antibodies that may be unable to distinguish between multiple isoforms of the same protein (4).

MALDI imaging has been recognized for its enormous potential as a tool in cancer biomarker discovery. Many scientists have concentrated their efforts in improving the reproducibility of the analysis through rigorous sample preparation and developing reliable bioinformatic tools capable of analyzing large data sets, particularly for the use of MALDI imaging in clinical studies. In particular, MALDI imaging seems to be the most promising approach to investigate the molecular changes that occur in normal cells surrounding cancer cells. Some clinical applications of the techniques have been reported so far; however, the lack of protocols for sample collection, sample treatment and data analysis, and integration with other analytical platforms in clinics needs to be properly addressed (5).

The increasing interest in MALDI imaging has stimulated technical development in instrumentation to decrease laser focal diameter without losing sensitivity. As a future prospective, MALDI imaging will reach subcellular spatial resolution (<3 μm), allowing to discover details at the molecular levels that are still buried in the tissue. Research focuses also on matrix deposition techniques to reduce analyte displacement, in order to guarantee the most accurate and reliable sample analysis.

Matrix deposition is crucial to obtain high-quality data. Matrix deposition should not cause dispersion or dislocation of the analytes during the extraction or co-crystallization process that occurs during the coating procedure (6). The size of the analyte-matrix crystals should not exceed in size the imaging resolution. Homogenous and reproducible matrix distribution can be achieved with several methods, including ink-jet printers, spray coating, and sublimation, and new efforts are trying to reduce the time of sample preparation and to prepare multiple tissue slides at the same time, without interfering with matrix uniformity and formation of small size crystals (7).

IMS experiments in proteomics can be classified in two categories: imaging and profiling. Profiling is a targeted analysis in which small areas of interest are compared (e.g., cancer vs. healthy tissue), while imaging focuses on the overall distribution of specific molecules in the region of interest. The profiling studies are typically directed by histology and require sophisticated bioinformatic analysis (8). The imaging approach correlates molecular distribution with histological feature (1). Both approaches require the same sample preparation and data acquisition, although the data analysis is dependent on the experimental objective.

One of the aspects of MALDI imaging that needs to be further developed is the direct identification of biomarkers in tissue. On-tissue trypsin digestion could be one approach, and so far, many efforts have been made in this direction, both on frozen tissue section (9) or formalin-fixed paraffin-embedded tissue sections (10, 11). A direct identification on the tissue approach could be an ideal solution (12); however, beyond the MALDI experiment, time-consuming traditional techniques that involve extraction and purification are still indispensable tools for reliable protein identifications (13, 14).

This chapter describes in details MALDI experiments for both protein and peptide imaging, from the sample preparation, from the tissue sectioning and coating to the data acquisition. This chapter also shows examples of how data can be visualized in a MALDI imaging experiment, addressing some technical issues that should be avoided in every step of the experiment performance.

2 Materials

2.1 Tissue Sectioning

1. Cryostat (Leica CM 1950).
2. 1″ square of cork.
3. Embedding medium for frozen tissue specimens (OCT, Sakura, Tissue-Tek).
4. Indium-tin oxide glass slides coated (Bruker Daltonics, Billerica, MA, USA).
5. Digital multimeter M-1000C (Elenco® Electronics, Inc., Wheeling, IL).

2.2 Tissue Slide Coating

1. Sinapic acid (SA), matrix substance for MALDI-MS.
2. α-Cyano-4-hydroxycinnamic acid (CHCA), matrix substance for MALDI-MS.
3. Protein and peptide calibration standards (Bruker Daltonics, Billerica, MA, USA).
4. Glass coverslips (VWR, Batavia, IL).
5. ImagePrep (Bruker Daltonics, Billerica, MA, USA).
6. Sonicator.

2.3 Data Collection and Analysis

7. Scanner (suggested 800 dpi resolution or higher capability).
8. MTP Slide-Adapter II.
9. Bruker Ultraflex III TOF/TOF (Bruker Daltonics, Billerica, MA, USA) with smartbeam Nd:YAG laser.

3 Methods

The methods described herein are for use with molecular imaging of protein and peptides. Figure 1 gives a general overview of a MALDI imaging experiment. The methods explain tissue sample preparation and matrix fixation. Data acquisition and analysis will be briefly presented as well.

3.1 Microtomy

1. Remove brain (or any other target organ of interest) quickly after animal is euthanized (see Note 1).
2. Cut brain in half sagittally and place each half cut side down on a precooled (previously submerged in liquid nitrogen) 1″

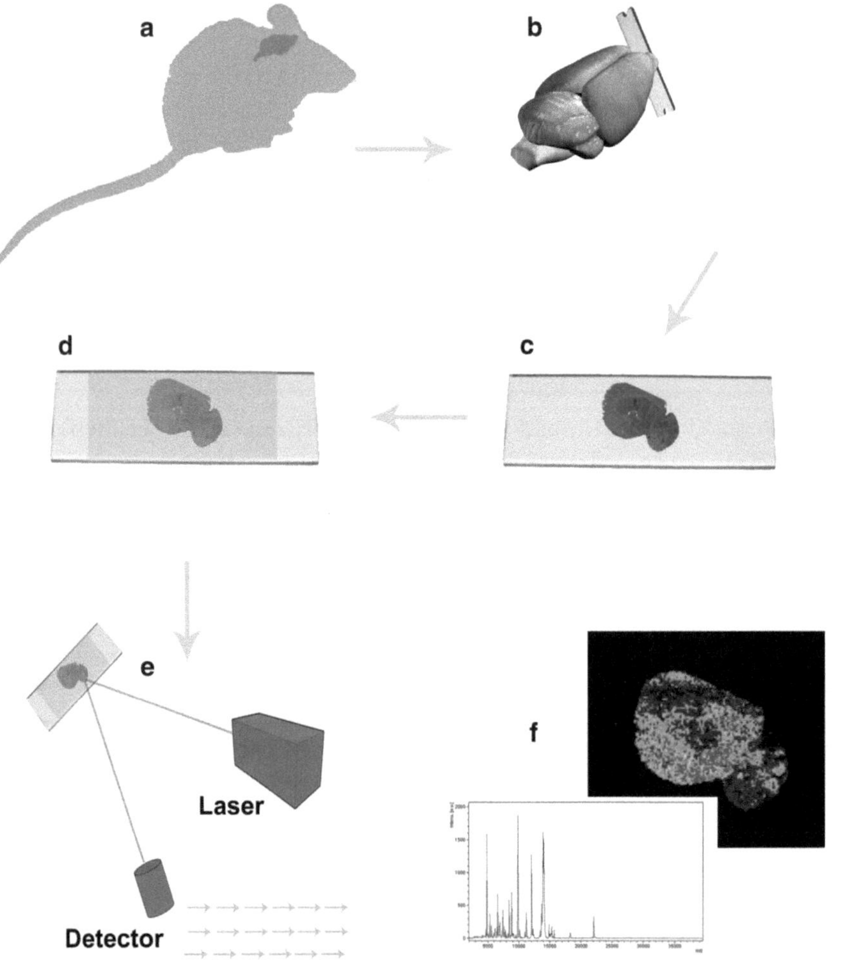

Fig. 1 General flow for a MALDI imaging experiment. The organ of interest is removed from the euthanized animal (**a**). Cryosection of the organ (**b**) provides thin tissue slides that are thaw-mounted on the ITO-treated surface of a glass slide (**c**). Alcohol washes are optional. The slides are coated with the appropriated matrix solution (**d**), to be analyzed in a MALDI instrument (**e**). For each raster spot, a spectrum is acquired to build images that reflect the molecular ion's distribution (**f**)

square of cork with just sufficient OCT to adhere it to the cork. Take care to keep the OCT material away from the surface to be sectioned (see Note 2).

3. Quickly place cork square in an aluminum foil pocket (tissue side up), seal and submerge in liquid nitrogen.
4. Once frozen, store foil pocket at −80 °C until cut.
5. Place the following in cryostat for 45 min prior to cutting: tissue blocks, MALDI imaging glass slides, Superfrost Plus slides, an artist's brush, and fine forceps.
6. Rough cut tissue until entire brain is present in the sections.
7. Cut two consecutive sections (10 μm thickness) and gently tease the two sections onto the ITO-coated surface (see Note 3) of a precooled MALDI slide with the brush and forceps, leaving space between the sections.
8. Flatten the tissue section as much as possible on the slide and then place the slide on the warm palm of your gloved hand (tissue side up) inside the cryostat (see Note 4).
9. Number slides and transfer immediately to a rack inside a desiccator.
10. Cut and place a third consecutive section on a Superfrost Plus slide in the same fashion and number it (desiccation is not necessary). This will be used for H&E staining (for comparison purposes).
11. Repeat numbers 7–10 as many times as necessary until the desired number of sections has been reached.
12. Desiccate MALDI slides under vacuum for 30 min.
13. Remove two slides from the desiccator and place them consecutively in 70, 90, and 95 % HPLC grade ethanol for 30 s each. Label slides with the fixation time (30 s) and return them to the desiccator (see Note 5). This set of slides will be used for protein imaging purposes in linear mode experiments (data shown in Figs. 2 and 3).
14. Repeat step 13 with the two additional slides but reduce the time in each alcohol to 15 s. Label the slides with the fixation time (15 s) and return them to the desiccator (see Note 5). This set of slides will be used for peptide imaging purposes in reflectron mode experiments (Figs. 4 and 5).
15. Desiccate the slides under vacuum for an additional 10 min.
16. Place fixed slides in cardboard slide holders according to the fixation time, wrap in pre-labeled aluminum foil, place inside a plastic bag, and freeze at −80 °C until ready to use.
17. Stain H&E slides.

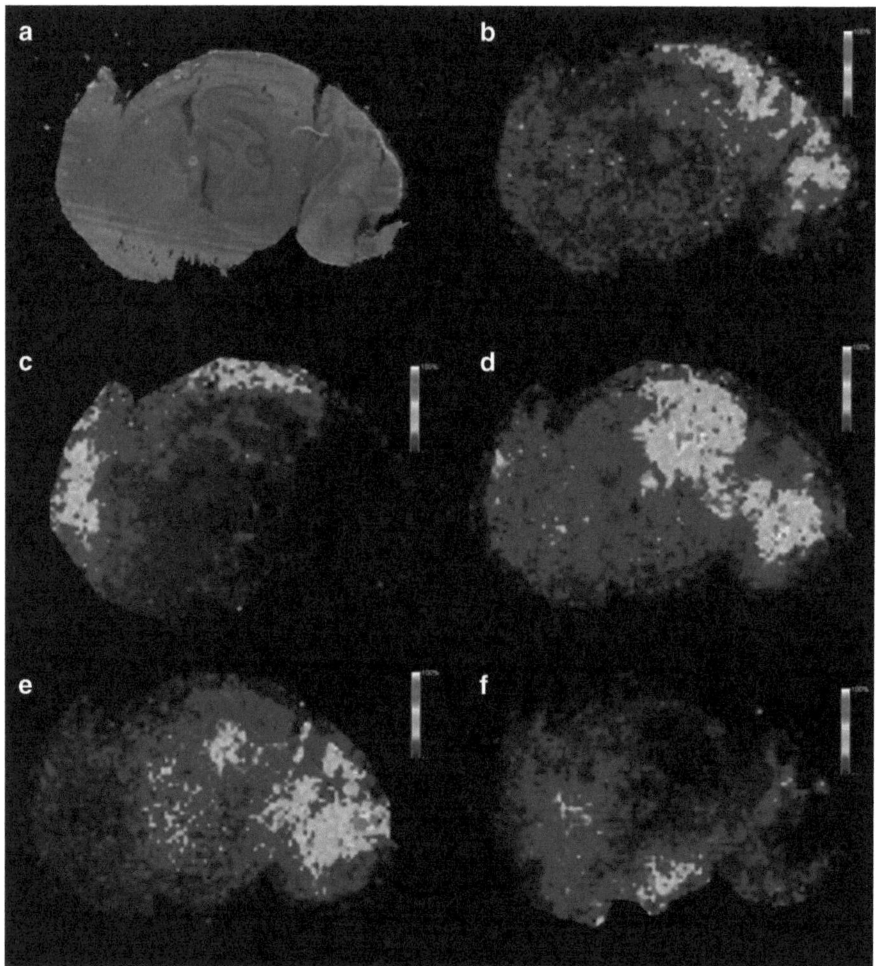

Fig. 2 Gradient color representation of five different ions (linear mode acquisition). (**a**) Optical image of sagittal mouse brain section; (**b**) distribution ion 5,342.522 ± 0.25 %; (**c**) distribution ion 7,432.365 ± 0.25 %; (**d**) distribution ion 8,878.842 ± 0.25 %; (**e**) 9,811.284 ± 0.25 %; (**f**) distribution ion 18,323.8 ± 0.25 %

3.2 Matrix Deposition

1. For protein imaging, prepare a fresh solution of SA 10 mg/mL with a final volume of 10 mL in acetonitrile/0.2 % TFA, 60/40. Weight accurately 100 mg of sinapic acid in a 15 mL plastic tube and add 6 mL of acetonitrile and 4 mL of water with 0.2 % TFA. Vortex the matrix solution for at least 30 s and then sonicate it for 30 min (see Note 6). For peptide imaging studies, prepare a solution of CHCA 7 mg/mL with a final volume of 10 mL in acetonitrile/0.2 % TFA, 50/50. Weight out 70 mg of CHCA and dissolve it 5 mL of acetonitrile and 5 mL of 0.2 % TFA. Vortex and sonicate the CHCA solution as described above for the SA solution.

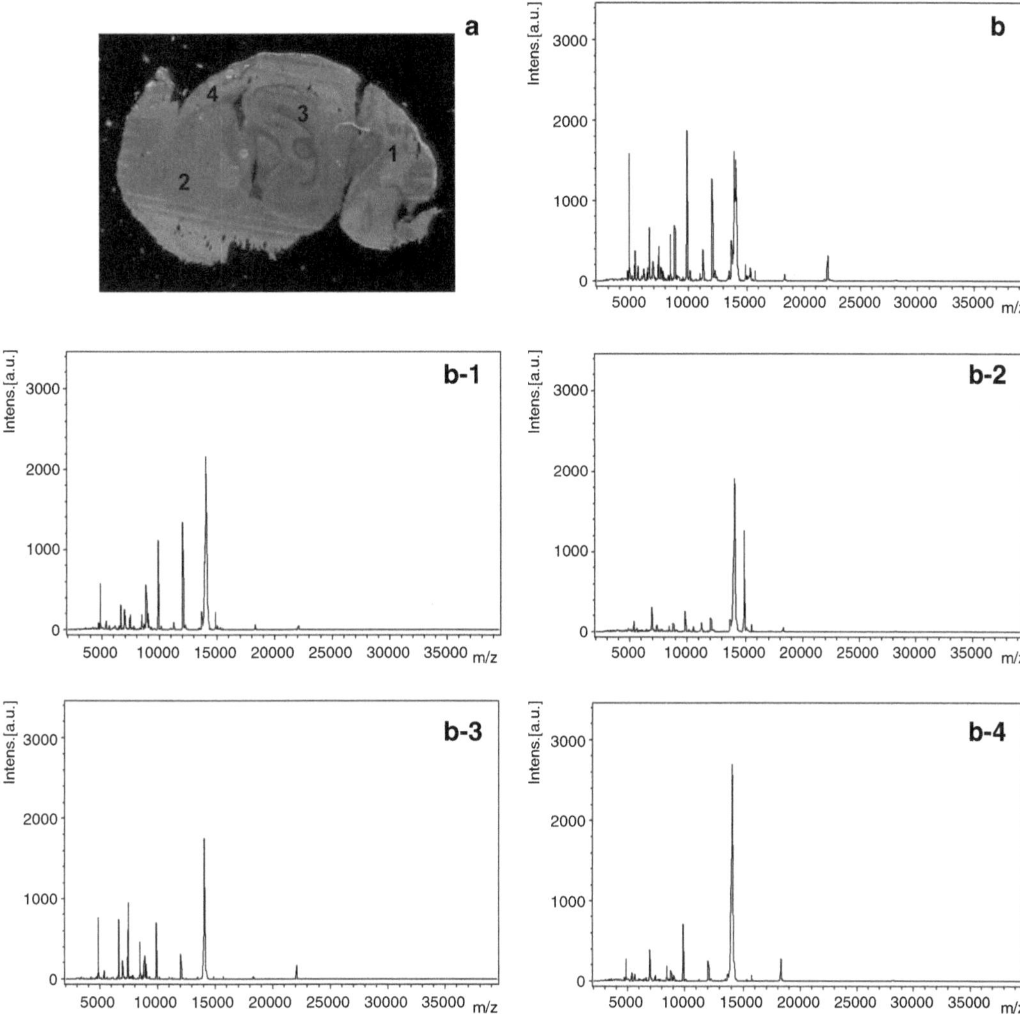

Fig. 3 Comparison of average spectrum from different regions of interest (linear mode acquisition). (**a**) Optical image of sagittal mouse brain section. Four different regions selected: (1) cerebellum, (2) striatum, (3) hippocampus, and (4) cerebral cortex. (**b**) Overall average spectrum of tissue section. (**b-1**) Cerebellum region average spectrum. (**b-2**) Striatum region average spectrum. (**b-3**) Hippocampus region average spectrum. (**b-4**) Cortex average spectrum

2. Allow the frozen tissue slide to thaw in vacuo (see Note 7). It should take a few minutes.
3. Mark three distinctive white spots on the glass slide to be analyzed, and mount the slide into the MTP Slide-Adapter II slide holder. Scan the slide in the holder and save the image as a "*.jpeg" format (see Note 8).

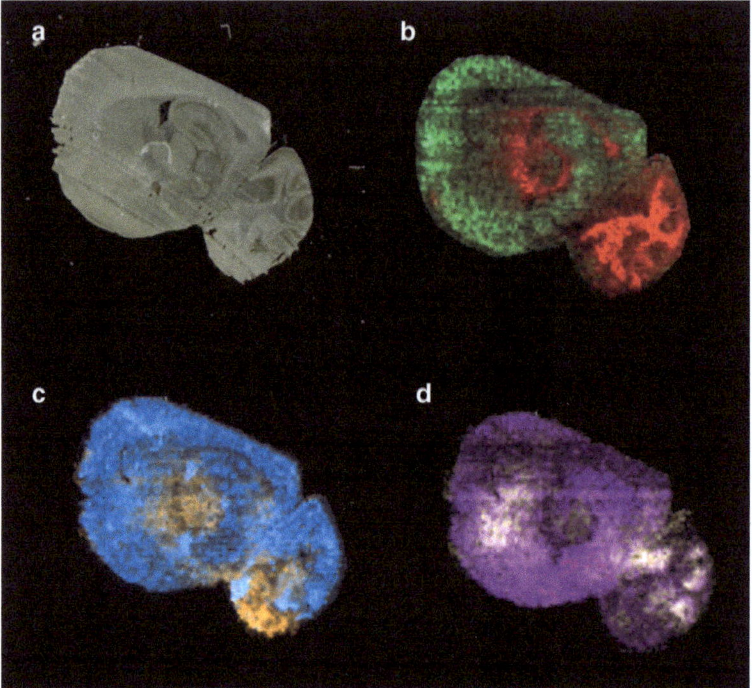

Fig. 4 Relative distribution of molecular ions (reflectron mode acquisition). (**a**) Optical image of sagittal mouse brain. (**b**) *Green* corresponds to ion 1,292.77 ± 0.25 %, while *red* represents ion 789.6595 ± 0.25 %. (**c**) Color *blue* shows the distribution of ion 617.9791 ± 0.25 %, when ion 1,872.461 ± 0.25 % is represented in *orange*. (**d**) *Purple* and *yellow* are chosen to represent, respectively, ion 5,006.548 ± 0.25 % and ion 1,757.404 ± 0.25 %

4. Release the slide from the holder and cover the sides of the glass with a layer of Parafilm (see Note 9).
5. Clean ImagePrep spray head with methanol using the manufacture automated procedure, and repeat the operation if necessary.
6. Load the matrix solution into the matrix reservoir of ImagePrep, and test the efficiency of the spray (see Note 10).
7. Introduce the glass slide into the ImagePrep chamber assuring that the optical device that detects the thickness of the matrix layer is not covered by the tissue and it is positioned in the middle of the slide. Carefully place a small glass coverslip on the top of the glass slide in correspondence with the optical detector (see Note 11). Start the coating using the protocol suggested by the manufacturer for SA (or CHCA) deposition. In both cases the process lasts usually between 60 and 80 min.

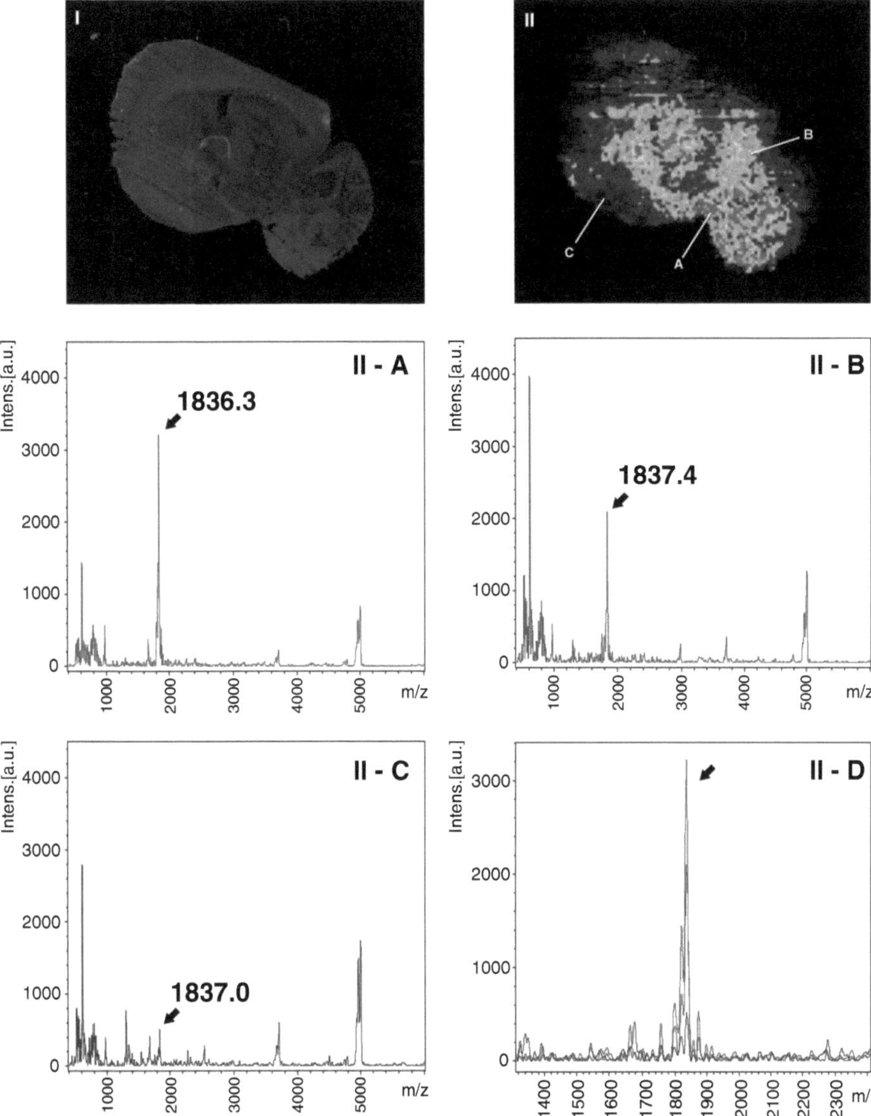

Fig. 5 Extraction and elaboration of single spectra from region of interest (reflectron mode acquisition). (**I**) Optical image of sagittal mouse brain section. (**II**) Distribution of ion 1,836.6 ± 0.25 %. (**II-A**) Spectrum at coordinates X097 and Y140 (high intensity signal). (**II-B**) Spectrum at coordinates X085 and Y127 (medium intensity signal). (**II-C**) X062 and Y096 (low intensity signal). (**II-D**) Overlapping of the three spectra in the region of interest

8. When the coating process is complete, remove the sample from the ImagePrep and proceed to another one of two cycles of methanol washes to remove matrix residues from the spray head.
9. Remove the coverslip from the slide surface.

10. Mix 1 μL of matrix solution used in the coating process and 1 μL of protein (or peptide) standard. Spot 1 μL of this mixture on the surface of the glass slide not coated with matrix.
11. Remove the Parafilm from the side of the glass slide and load the slide on the target holder. The tissue slide is now ready to be loaded into the mass spectrometer.

3.3 Data Acquisition

1. Select an acquisition method (see Note 12).
2. Calibrate the instrument using in the desired mass range and save the newly calibrated method.
3. Test the laser power to be used during the imaging experiment on the tissue edges and establish the laser power range to be assigned in the automatic acquisition process.
4. Define imaging sequence properties, including spatial resolution and measurement regions. Start the automatic acquisition (see Note 13).
5. When the acquisition is completed, data will be automatically loaded into FlexImaging, and it is possible to visualize the localization and the relative intensity of selected *m/z* values.

3.4 Data Analysis

It is necessary to assure that data are normalized. Some software offers several options for data normalization; however, normalization on the total ion current is usually appropriate for most purposes. Data can be visualized in various ways, according to the investigator's need. Some basic examples are presented in this last section.

In Fig. 2, we show an example of protein expression mapping using a color gradient. This type of data visualization is usually very effective, and in many cases, it is able to emphasize dramatic differences in expression of a protein or peptide across the tissue slide. In our example, several protein distributions are reported in comparison with the optical image of the same tissue slide that it has been used to generate the data set.

If the focus of the study is the comparison of different histological regions, it is possible to extract an average spectrum from different tissue region and compare them as reported in Fig. 3. In this case, we have four regions of interest corresponding to four different anatomic portions of the mouse brain, and we display the average spectra from each region in comparison with the average spectra from the whole tissue section.

Figure 4 shows another approach that indicates the relative localization of two or more species. To have a better understanding and a more effective visual, it is mandatory to choose contrasting color as presented in our example. To each pixel corresponds a spectrum, indicated by *x*- and *y*-coordinates. Each spectrum can be extracted and individually annotated as shown in Fig. 5.

4 Notes

1. Animals are previously euthanized by CO_2 inhalation as described in the *US Public Health Service Policy on Humane Care and Use of Animals* (2000); the *Guide for the Care and Use of Laboratory Animals* (1996); and the *US Government Principles for Utilization and Care of Vertebrate Animals Used in Testing, Research, and Training* (1985).
2. Optimal cutting temperature (OCT) contains polymers that can be detected in MALDI. The polymer signal can be strong enough to suppress the analyte signal; therefore, where possible, the use of OCT should be avoided or reduced as much as possible.
3. This step allows the section to thaw to the slide (thaw-mounted) and prevents condensation.
4. To assure the correct sample preparation, it highly recommended measuring the resistance value with a digital multimeter or similar apparatus. Set the resistance range of the digital multimeter at 200 kΩ and connect the two test leads with two different points of one side of the glass slide. The ITO-coated side should show a reading of 0.2–0.3, while the not conductive side will show no reading.
5. Alcohol-based washes assure lipid removal and improve protein and peptide detections. Washes seem to be correlated also with matrix deposition regularity and tissue integrity from the histological point of view. For peptide analysis, the time of the alcohol rinse needs to be reduced to avoid losing the analyte itself.
6. To achieve a homogeneous matrix solution, it is preferable to add acetonitrile first, briefly vortex the solution, and then add 0.2 % TFA water. Sonication also greatly contributes to this aim. Solution should be visually inspected before use, to look for insoluble particles that should not be present.
7. Thawing the tissue rather than in atmospheric pressure in vacuo should reduce oxidation phenomena.
8. The three teaching point will be used in FlexImaging for spatial references. Extreme caution should be used during this step to avoid disturbing the surface of the tissue during the scanning process.
9. The ITO glass slide's sides need to be covered during the matrix deposition process to assure the best possible contact between the holder and the surface of the slide during the imaging experiment. Alternatively, the matrix layer can be removed from the sides afterwards with the use of methanol and a wipe or cotton-tipped applicator.

10. It is always recommended to assure cleanness of the spray head repeating methanol washes in the sensor-controlled aerosol system ImagePrep. Spray efficiency test instead should be mandatory, and adjustments should be performed starting from a spray power of 100 % with an offset of "+0" and a spray power of 10 % with an offset of "+8." Both operations are extremely simple and quick, also easily accessible from the command prompt of ImagePrep.
11. Glass coverslip on the top of the surface of the glass slide has been found to improve coating performance and reproducibility.
12. Protein imaging experiments are performed in linear mode and the typical range is 2,000–40,000 Da, while peptide imaging experiments are usually in reflectron mode with a range of 400–6,000 Da.
13. The acquisition time depends on the chosen laser repetition rate, the spatial resolution, and the tissue size. Time rapidly increases with lower raster size. Reasonable resolution should be chosen based on the nature of the experiment and the capability of the instrument. As an example, on the Bruker Ultraflex III MALDI TOF/TOF, data acquisition from a mouse brain slide in linear mode with a medium laser diameter (50 μm) 100 Hz and a raster of 100 μm can take up to 10 h.

References

1. Stoeckli M, Chaurand P, Hallahan DE, Caprioli RM (2001) Imaging mass spectrometry: a new technology for the analysis of protein expression in mammalian tissues. Nat Med 7:493–496
2. McDonnell LA, Heeren RM (2007) Imaging mass spectrometry. Mass Spectrom Rev 26:606–643
3. Ifa DR, Wiseman JM, Song Q, Cooks RG (2007) Development of capabilities for imaging mass spectrometry under ambient conditions with desorption electrospray ionization (DESI). Int J Mass Spectrom 259:8–15
4. Seeley EH, Schwamborn K, Caprioli RM (2011) Imaging of intact tissue sections: moving beyond the microscope. J Biol Chem 286:25459–25466
5. McDonnell LA, Corthals GL, Willems SM, van Remoortere A, van Zeijl RJ, Deelder AM (2010) Peptide and protein imaging mass spectrometry in cancer research. J Proteomics 73:1921–1944
6. Chaurand P, Schwartz SA, Caprioli RM (2002) Imaging mass spectrometry: a new tool to investigate the spatial organization of peptides and proteins in mammalian tissue sections. Curr Opin Chem Biol 6:676–681
7. Mounfield WP 3rd, Garrett TJ (2012) Automated MALDI matrix coating system for multiple tissue samples for imaging mass spectrometry. J Am Soc Mass Spectrom 23: 563–569
8. Cornett DS, Mobley JA, Dias EC et al (2006) A novel histology-directed strategy for MALDI-MS tissue profiling that improves throughput and cellular specificity in human breast cancer. Mol Cell Proteomics 5:1975–1983
9. Groseclose MR, Andersson M, Hardesty WM, Caprioli RM (2007) Identification of proteins directly from tissue: in situ tryptic digestions coupled with imaging mass spectrometry. J Mass Spectrom 42:254–262
10. Lemaire R, Desmons A, Tabet JC, Day R, Salzet M, Fournier I (2007) Direct analysis and MALDI imaging of formalin-fixed, paraffin-embedded tissue sections. J Proteome Res 6:1295–1305
11. Stauber J, Lemaire R, Franck J et al (2008) MALDI imaging of formalin-fixed paraffin-embedded tissues: application to model animals

of Parkinson disease for biomarker hunting. J Proteome Res 7:969–978

12. Gemoll T, Roblick UJ, Habermann JK (2011) MALDI mass spectrometry imaging in oncology. Mol Med Report 4:1045–1051
13. Kim HK, Reyzer ML, Choi IJ et al (2010) Gastric cancer-specific protein profile identified using endoscopic biopsy samples via MALDI mass spectrometry. J Proteome Res 9:4123–4130
14. Lee HS, Park JW, Chertov O et al (2012) Matrix-assisted laser desorption/ionization mass spectrometry reveals decreased calcylcin expression in small cell lung cancer. Pathol Int 62:28–35

Chapter 21

Plant Proteogenomics: From Protein Extraction to Improved Gene Predictions

Brett Chapman, Natalie Castellana, Alex Apffel, Ryan Ghan, Grant R. Cramer, Matthew Bellgard, Paul A. Haynes, and Steven C. Van Sluyter

Abstract

Historically many genome annotation strategies have lacked experimental evidence at the protein level, which and have instead relied heavily on ab initio gene prediction tools, which consequently resulted in many incorrectly annotated genomic sequences. Proteogenomics aims to address these issues using mass spectrometry (MS)-based proteomics, genomic mapping, and providing statistical significance measures such as false discovery rates (FDRs) to validate the mapped peptides. Presented here is a tool capable of meeting this goal, the UCSD proteogenomic pipeline, which maps peptide-spectrum matches (PSMs) to the genome using the Inspect MS/MS database search tool and assigns a statistical significance to the match using a target-decoy search approach to assign estimated FDRs. This pipeline also provides the option of using a more reliable approach to proteogenomics by determining the precise false-positive rates (FPRs) and *p*-values of each PSM by calculating their spectral probabilities and rescoring each PSM accordingly. In addition to the protein prediction challenges in the rapidly growing number of sequenced plant genomes, it is difficult to extract high-quality protein samples from many plant species. For that reason, this chapter contains methods for protein extraction and trypsin digestion that reliably produce samples suitable for proteogenomic analysis.

Key words Proteogenomics, Proteomics, Peptide identification, Protein extraction, Trypsin digest, False discovery rate, False-positive rate, Posterior error probability, Local false discovery rate, p-Value, q-Value, Inspect, MS-GF

1 Introduction

Many annotation strategies in the past have not been based on peptide level evidence and instead relied on ab initio gene prediction tools, homology searches, and EST or RNA-Seq mapping. These strategies relied on a set of relatively stringent parameters, such as gene size, and applying a set of narrow principles across a variety of different organisms. In reality gene size can vary significantly, and precisely identifying the boundaries of transcription

Ming Zhou and Timothy Veenstra (eds.), *Proteomics for Biomarker Discovery: Methods and Protocols*, Methods in Molecular Biology, vol. 1002, DOI 10.1007/978-1-62703-360-2_21,

and translation remains challenging. Consequently many annotated genomic sequences now contain erroneously predicted genes or alternatively genes that have been missed entirely, undefined exon borders and missed small exons, small open reading frames (smORFs), unpredicted frame translations, or undetermined interlaced reading frames (1). For example 9–44 % of genes in some bacteria and archaeal genomes have erroneously predicted start/stop codons (2, 3), and the start codons of 60 % of the genes in a 143 prokaryotic genome comparison were revealed to be annotated in error (4).

Over the last few years, a new strategy, called proteogenomics, which was first coined in 2004 by Jaffe et al. (5), has emerged that uses tandem mass spectrometry (MS/MS)-based proteomics (6) to assist with genomic annotation. MS/MS-based proteomics is based on the separation of peptides (derived from trypsin digestion of proteins) using high-performance liquid chromatography (HPLC) and their subsequent identification using a mass spectrometer. The separation and identification of the peptides is collectively referred to as LC-MS/MS.

This method alone often results in relatively low coverage of the proteome. A variation of this approach, which has been utilized more widely and has seen vast improvements in the form of high-throughput and greater proteome coverage over the last decade, is shotgun proteomics, also referred to as 2D or multidimensional LC-MS/MS. Shotgun proteomics is similar in concept to the shotgun sequencing technique used in genomics. A whole proteome sample is enzymatically digested, and the resulting peptides can then be separated using two or more dimensions of orthogonal fractionation. One example is multidimensional HPLC, where the peptides are fractionated by charge on an HPLC column composed of strong cation exchange (SCX) resin and the fractions subsequently separated on C_{18} reverse phase (RP) material (7). The resulting peptides that elute off the column are ionized (usually by electrospray ionization or ESI) and directly analyzed by the mass spectrometer where their mass over charge ratios (m/z) and intensities are measured by the detector. The precursor ion masses during the first stage of MS (MS^1 or MS) and resulting fragmented ions during the second stage of MS (MS^2 or MS/MS) are determined so that the MS/MS spectra can be matched to known peptide sequences through the use of database search algorithms.

In contrast to most discovery-mode proteomics experiments that use the annotated proteome as a database, proteogenomics takes advantage of peptide-spectrum matches (PSMs) against an expanded database of potential protein sequences associated with genomic locations. Matches to proteogenomic databases reveal regions of the genome that were previously believed to contain nonprotein coding elements. The peptides may reveal that the region is in fact translated and can bring to light novel annotation events, such as the discovery of novel genes, the reversed reading frames, the determination of start and stop sites of genes or programmed frame shifts, the verification of splice variants and validation of gene functions, and the hypothetical open reading frames (8–10).

Proteogenomics is fast becoming a viable approach to vastly improve upon genomic annotation, with many annotation attempts in recent years being conducted alongside proteomics at the initial stages (11–13) instead of correcting the genomic annotations at a later date. Through this approach, a more thorough genomic annotation can be conducted from the very beginning, reducing the time spent annotating and the need for expensive manual curation. The following sections outline the challenges faced by proteogenomics and provide a suitable proteogenomic pipeline to address these challenges. More generally, sample preparation for peptide LC-MS/MS from many plant tissues is challenging. We present extraction and trypsin digestion protocols that have successfully been applied to different tissues of several plant species.

2 Materials

2.1 Protein Extraction from Plant Tissues and Digestion (See Note 1)

1. Hurkman Extraction Buffer (14) prepare fresh before use.
2. Protease Inhibitor Cocktail Tablets (Roche 11836170001).
3. TRIS-saturated phenol pH 7.9 (store at 4 °C, see Note 2).

2.2 TFE-FASP: Filter Aided Sample Preparation with Trifluoroethanol

1. 50 % TFE, 0.1 M ammonium bicarbonate (ABC/ambic) containing freshly prepared 10 % 0.5 M iodoacetamide (IAM). If the solid IAM or stock solution is not colorless, then extract with 100 μl chloroform by vortexing and centrifuging. Use the top aqueous layer for 0.5 M IAM.
2. 0.5 μg/μl Lys-C (Lysyl Endopeptidase, Wako Pure Chemical Industries). Store in aliquots at −80 °C.
3. 1 μg/μl sequencing grade modified trypsin (Promega) in 20 μl resuspension buffer (50 mM acetic acid).
4. 0.5 ml Ultrafiltration devices (Millipore Amicon Ultra 10 or 30 k).

3 Methods

3.1 Protein Extraction from Plant Tissues and Digestion

High-quality and efficient protein extraction can be difficult sometimes due to the difficulty of grinding samples for extraction and interfering substances in the sample (15). For example, plant materials such as grape berries can be large and contain hard seeds if frozen immediately, prior to crushing the tissue. Subsequent crushing in a blender or grinder without warming the sample can be challenging (e.g., whole berries in dry ice in a commercial-grade blender). Furthermore, phenolic compounds and a high abundance of polysaccharides and sugars can interfere with clean separation of the proteins. Additional reagents are needed in the extraction cocktail to prevent protein modifications such as oxidative reactions. The following detailed protein extraction protocols represent current extraction procedures used in our labs. The results presented later in this chapter are from grape berry skin proteins. Protein yields are commonly 0.2–1 mg per g tissue and vary depending on tissue type.

1. Prepare fresh Hurkman extraction buffer in 50 ml aliquots. PMSF, Protease Inhibitor Tablets (1 tablet/50 ml buffer), and 2-mercaptoethanol are added immediately before use.
2. Place ~4.0 g of frozen, finely ground skin tissue into 50 ml Falcon tubes. Quickly add 10 ml of Hurkman extraction buffer to each tube. Allow material to dissolve in buffer by inverting and vortexing tubes. Frozen tissue will appear to "gel" and take a while to go into solution. After material is fully dissolved, vortex for 30 s.
3. Let stand at 4 °C (on ice) for 10 min.
4. Add 10 ml TRIS-saturated Phenol pH 7.9 to each Falcon tube. The TRIS and Phenol will be separated into two phases (use the lower phase). Handle the bottle gently and do not mix the phases. Lower the pipette tip through the upper TRIS phase to collect phenol.
5. Vortex for 30 s.
6. Let stand 30 min at 4 °C (or on ice), inverting tubes every 10 min.
7. Centrifuge the tubes at 3,650 × *g* and 4 °C for 30 min.
8. Carefully remove the upper phenol phase (~7 ml) and place into a new 50 ml Falcon tube.
9. Add an equal volume of Hurkman extraction buffer to the volume of phenol (~7 ml) in each of the 50 ml Falcon tubes.
10. Vortex for 30 s.
11. Let stand for 30 min at 4 °C (or on ice), inverting every 10 min.
12. Centrifuge tubes at 3,650 × *g* and 4 °C for 30 min.
13. Prepare 0.1 M methanol-ammonium acetate during the 30 min centrifugation in preparation for the next step, if not previously prepared.
14. Carefully remove the upper phenol phase (~5 ml) and place into a new 50 ml Falcon tube.
15. Add 5 volumes (~25 ml) cold 0.1 M methanol-ammonium acetate to each Falcon tube.

16. Vortex for 30 s.
17. Let stand overnight at –20 °C.
18. Centrifuge 50 ml Falcon tubes at 3,650 × *g* and 4 °C for 30 min.
19. Discard supernatant from each 50 ml Falcon tube. This waste contains phenol and should be discarded appropriately.
20. Add 5 ml of cold 0.1 M methanol-ammonium acetate to each Falcon tube.
21. Vortex for 30 s.
22. Let stand for 1 h at –20 °C, inverting tubes every 10 min.
23. Centrifuge tubes at 3,650 × *g* and 4 °C for 30 min.
24. Discard supernatant from each 50 ml Falcon tube.
25. Add 1 ml cold acetone to each Falcon tube and transfer pellets to 2 ml eppendorf tubes for cold acetone wash. Add an additional 500 μl to the 50 ml tubes to wash remaining protein down and transfer to the respective 2 ml eppendorf tubes.
26. Vortex for 30 s.
27. Let stand for 1 h at –20 °C, inverting tubes every 10 min.
28. Centrifuge 2 ml eppendorf tubes at 19,000 × *g* and 4 °C for 5 min.
29. Repeat step 5 two more times.
30. Discard acetone supernatant from each of the 2 ml tubes and add a final volume of 500 μl cold acetone to the pellets.
31. Place tubes in a –20 °C freezer until ready for peptide digestions.
32. This is a good stopping point if samples are to be shipped at ambient temperature. For shipping, the acetone should be replaced with 10 mM sodium metabisulfite, 0.1 % 2-mercaptoethanol in 90 % methanol.

3.2 TFE-FASP: Filter-Aided Sample Preparation with Trifluoroethanol

This protein method is based on the ultrafiltration (UF) device-assisted methods of Manza et al. (16) and Wisniewski et al. (17). Protein in solution is placed in the top of a UF device and washed several times by centrifugation and refilling the top reservoir with buffer. The washes are discarded, and the protein retentate, in the top reservoir, is digested with Lys-C and then trypsin. The difference between our method and the previous methods is that we use trifluoroethanol as a denaturant to increase digestion efficiency. At the end of the digestions, the resulting peptides are small enough to pass through the UF membrane and are collected in the flow through by centrifugation. This method produces clean samples for LC-MS/MS (see Note 3).

Protein amounts are determined by evenly suspending precipitated proteins and removing a subsample. The subsample is spun down in a chilled centrifuge, the supernatant removed, and the protein dissolved in SDS-PAGE sample buffer. Protein can be assayed by detergent compatible commercial kits such as the EZQ Protein Quantitation kit from Invitrogen or a bicinchoninic acid (BCA) assay-based kit (see Notes 4 and 5).

1. Wash 0.25 mg protein pellets in 500 μl methanol twice (collect protein by centrifugation, discard methanol). Dissolve pellets in 200 μl solution A (50 % TFE, 0.1 M ammonium bicarbonate, 50 mM DTT) and heat to 75 °C in heating block for 15 min and sonicate if necessary. A higher volume of solution A is acceptable if necessary, but it is best not to exceed the capacity of the ultrafiltration devices. In cases where samples are very difficult to get into solution, it is possible to dissolve protein insoluble in solution A with a small amount of 8 M guanidine HCl and combine that with the rest of the protein in solution A.
2. Add dissolved proteins to 0.5 ml Amicon UF device; centrifuge at 14,000 × *g*, 45 min or until retentate volume is ~20 μl.
3. Add 100 μl (50 % TFE, 0.1 M ammonium bicarbonate, 50 mM iodoacetamide), put in thermomixer at 600 rpm (room temp) for 1 min, then 1 h in the dark, then centrifuge at 14,000 × *g* for 45 min.
4. Add 200 μl (50 % TFE, 0.1 M ammonium bicarbonate), mix 1 min, centrifuge at 14,000 × *g*, 45 min or until retentate <20 μl. Discard flow through.
5. Repeat step 4 four more times. Discard flow through as necessary.
6. Add 45 μl (50 % TFE, 0.1 M ammonium bicarbonate) and 5 (1 μg/μl Lys-C in 50 mM acetic acid). Put on thermomixer 5 min.
7. Incubate UF devices in tube box containing damp paper towel overnight at 25–30 °C.
8. Add 350 μl (20 % acetonitrile, 50 mM ammonium bicarbonate), then 2.5 μl (1 μg/μl trypsin in 50 mM acetic acid). Mix carefully with thermomixer (start out at low rpm—the tops of the UF devices will leak if the sample reaches them while mixing), 1 min. Incubate 4 h at 37 °C. Stop incubation with all samples with 10 μl 50 % formic acid.
9. Rinse new UF device receptacles with 100 μl (50 % ACN, 2 % formic acid) and discard rinses. Replace the used receptacles on the UF devices with the rinsed receptacles. Centrifuge 14,000 × *g*, 45 min. Save the used receptacles for future digests.
10. Add 150 μl (50 % ACN, 2 % formic acid) to the top of the UF device; thermomix at 600 rpm, 1 min. Centrifuge 14,000 × *g*, 45 min.
11. Repeat step 10.
12. Remove filters from UF devices and transfer flow through (filtrates) to 1.5 ml tubes. Dry down flow through with a speed vac to near dryness.

13. Add 10 μl 50 % formic acid and 2.5 μl of 50 % TFE (without ammonium bicarbonate) to dried peptides. Vortex 15 min. Add water to 60 μl (do not just add 60 μl water), vortex, store frozen. Alternatively, if the LC-MS/MS system uses a trap column so that bound peptides can be washed off-line before loading onto the analytical column, then the peptides can be diluted with 0.1 % trifluoroacetic acid and stored frozen.

3.3 Mass Spectrometry

Multidimensional MS methods are beyond the scope of this chapter but could include SCX-RP, peptide isoelectric focusing (IEF) (18) followed by reversed-phase LC-MS/MS, gas-phase fractionation (GPF) (19), or data-independent acquisition (e.g., PACIFIC) (20). In the example presented in this chapter, three GPF fractions were used spanning a precursor range of m/z 400–2,000. Reversed-phase separations of 220 min were performed at 500 nl/min over 220 min by an Easy nLC II (Thermo) system with 0.1×300 mm 3 μm Magic C18AQ (Bruker-Michrom) analytical column and 0.22×25 mm vented trap packed with the same material. The mass spectrometer used here was an LTQ Velos Pro (Thermo) with Advance Captive Spray Source (Bruker-Michrom).

3.4 Scoring and Validation of Peptide-Spectrum Matches (See Note 6)

Historically, protein database search algorithms have used PSM scores, which vary depending on which algorithm is implemented, based on either the correlation between the theoretical MS/MS spectra and the observed MS/MS spectra (21), the number of ions matching a peptide sequence (22), *k*-similarity statistics (23), and calculated probability factors (24). The scores are ranked, and with many algorithms, the top-ranking match is chosen as the best match. Each top PSM is then assigned a significance value, in the form of a *p*-value (probability of being incorrect), which is calculated based on an estimate of the false-positive rate (FPR). The FPR is determined from the fraction of all incorrect PSMs over a chosen PSM score threshold, over the number of all incorrect PSMs. A *p*-value cutoff is then applied across all peptide matches, which is generally around a *p*-value of 0.01–0.05. There is almost always a FPR greater than zero, which is a result of the overlap of correct PSMs above a particular score threshold with all incorrect PSMs. The PSM scores do not correlate well with their FPRs, since the search algorithms usually assign similar scores to the topmost ranking PSMs, where the estimation of their FPRs is more inaccurate. In an attempt to reduce the number of false positives, many search algorithms employ different heuristic measures, such as considering the difference between the top-scoring PSM and the second best-scoring PSM, the peptide charge state of each PSM, the peak intensities of the spectra matching a particular peptide, the fraction of b ions within the matching spectra, and many other attributes of the PSM to further improve the confidence of the match (25).

This overall approach towards the validation of peptide matches was found to be flawed in a 2006 HUPO study, which found many inconsistencies with peptide identifications, due to differences between experiments, mass spectrometers, and laboratories, as each researcher applied his or her own set of heuristic rules and score cutoffs (26). Since then, standard practices have been put in place to ensure consistency of results within the proteomics community, as well as including a statistical measure to accompany each peptide match. In the past decade, many methods have been developed to assign statistical significance to peptide-spectrum matches. One popular statistical measure is the false discovery rate (FDR). The FDR of a collection of peptide-spectrum matches can be estimated by using a decoy database (27), which is typically reversed or randomized sequences from the target database. While the FDR is a useful measure for a collection of spectra, recent efforts have sought a statistical measure for assessing the significance of each individual peptide-spectrum match with the *q*-value (28) and the posterior error probability (PEP) (29, 30), or local FDR (lFDR) as coined by Efron et al. (31). The false-positive rate is a measure of the quality of an individual peptide-spectrum match, and in recent years a number of efficient tools for directly computing the FPR have been proposed (32, 33). It is important to note that the FDR can be computed from the FPR, avoiding the many pitfalls of the target-decoy approach (34). While the FPR has been used as a score for computing the FDR using the target-decoy approach, there have been no proteogenomic studies to date, which rely solely on the FPR, even though a precise FPR for each PSM is crucial for any study requiring a high reliability for individual PSMs, such as with proteogenomics and with the identification of rare posttranslational modifications (PTMs) (34). In light of this, there has been one recent study which, along with the use of a decoy database incorporated the precise calculation of the FPR and *p*-values in place of using the PEP/lFDR for each PSM, uses the scoring algorithm MS-Generating Function (MS-GF) (33), to annotate 46 different bacterial and archaea species using a proteogenomic approach to genome annotation (35).

3.5 Proteogenomics Pipeline

The proteogenomic pipeline outlined below was developed at the University of California, San Diego, by Castellana et al. (36) and relies on a database search tool called Inspect (37). The pipeline works by performing a search against the known proteome, a six-frame translated genome and an exon splice-graph of the genome. Each PSM is mapped to the genome, and rigorous statistical measures are used to determine the significance of each of the genomic annotation events found.

3.5.1 Data Formats

As part of the proteogenomic pipeline, the Inspect search tool uses the mzXML data format but also can accept formats mgf, mzData, ms2, and dta. If these formats are not at hand and only raw data formats such as .wiff are available, a number of MS data format conversion tools can be used depending on the user's needs. A list of available MS tools, including conversion tools, is provided by the Seattle Proteome Center and can be found at http://tools.proteomecenter.org/wiki/.

3.5.2 Databases

A number of databases should be searched against to ensure maximum coverage of the spectra against all likely peptide sequence matches.

A list of common proteogenomic databases are the following:

- A six-frame translation or/and exon splice-graph of the genome (required).
- The sequence from the known proteome (required).
- Contaminant peptides (e.g., keratin, trypsin) (recommended).
- Additional peptides to likely match the spectra (e.g., other proteome contaminants) (optional).

3.5.3 Six-Frame Translations

A six-frame translation of the genome sequence is required to be able to find all possible annotation events, particularly outside the boundaries of known genes and exons where novel genes and exons may reside, which have not been found with previous annotation attempts. The six-frame translation of the genome is prepared using the script SixFrameBuilder.jar supplied by the proteogenomics pipeline. The script reads all possible stop-to-stop ORFs of a minimum specified size (default 40 amino acids) across all six reading frames of the genome and outputs the sequences to a fasta file. The type of codon usage can also be chosen by selecting from a number of genetic code tables. Each fasta sequence entry header is of the form Sequence name@Start@End@Strand, which is important when mapping peptides to genomic positions. Using a six-frame translation results in an inflated false-positive rate, mentioned in Subheading 2.1, which is a concern, but is addressed adequately through the use of stringent significance thresholds, at the peptide, protein, and annotation event level, outlined in Subheadings 2 and 2.1.

An alternative approach to six-frame translations is through the use of large protein datasets, which avoids the ambiguity of assessing putative ORFs from the six-frame translation. Although this is a valid approach, it can leave gaps where the proteomic coverage of the protein database may be lacking. A six-frame translation considers every possible ORF, resulting in a more comprehensive assessment of potential protein coding genes.

Usage:

Required:

- r (FILE) FASTA file containing nucleotide sequence(s) to translate
- w (FILE) FASTA file as six-frame translation output

Optional:

- c (NUM) Specifies the genetic code table to use (default: 0)
- s (NUM) Max size of sequences per file (default: 100 MB)
- t (NUM) Minimum ORF size (default: 40 amino acids)
- z Prepare and shuffle each DB (default: do not prep and shuffle)

Example:
java -jar SixFrameBuilder.jar -r Genome.fasta -w Genome_6frame.fasta -c 8 -t 0

3.5.4 Database Formatting and Indexing

All fasta sequence files need to be formatted and indexed to a .trie database format, which can be done using the PrepDB.py script provided by Inspect. The script creates a .trie and .index file from the fasta file. Both .trie and .index need to stay together within the same directory.

Example:

```
PrepDB.py FASTA Genome_6frame.fasta
Output: Genome_6frame.trie
Genome_6frame.index
```

3.5.5 Exon Splice-Graphs

Splice-junctions can be identified when searching a protein database using tandem mass spectra. The identified peptides then get mapped to the genome to validate splicing-junctions (6, 38, 39). The majority of protein databases, such as Swiss-Prot (40), have poor representation of splicing variants due to the removal of redundant sequences, making this approach not ideal for the detection of alternatively spliced peptides. Another approach more suited to the detection of alternative splicing is to search against databases of Expressed Sequence Tags (ESTs), which are derived from a large proportion of experimental evidence in the form of mature RNA (mRNA), which have introns already removed, making them ideal for the identification of splice-junctions (41). A major problem with this strategy is the fact that EST databases are notoriously large, containing a large number of errors and redundant sequences. The exon splice-graph, first pioneered by Edwards et al. (42), is based upon a de Bruijn graph (43) and circumvents the problem of highly redundant and error-prone databases by representing the sequences of the exons or cDNA fragments as a de Bruijn graph, which significantly compresses the size of the database and removes redundancy.

The exon splice-graph is created from the genomes or gene prediction GFF file(s) and underlying sequences in .trie format using the BuildMS2DB.jar script provided by the proteogenomic pipeline. Multiple gene predictions in GFF format can be supplied as well as multiple region interval tag lines from within the GFF file, such as lines with "CDS" and any other tags that indicate an exon region or the expression of a protein. The type of codon usage can also be chosen, by selecting from a number of genetic code tables. The script outputs a .ms2db file containing the exon splice-graph.

Usage:

Required:

- r (FILE) GFF file built from the sequences in the sequence file given (may be specified multiple times)
- s (FILE) The FASTA or trie file version of the sequence file(if . trie then a corresponding .index file is assumed to exist)
- w (FILE) The output file

Optional:

- c (NUM) Genetic code table to use for translation (default: 0)
- t (STRING) The interval type to use (e.g., CDS, Exon, Match), case-insensitive (may be specified multiple times, default: CDS)
- d Run in Debug Mode
- f Interpret the "frame" as the "phase." The frame is the number of nucleotides to skip before the start of the codon. The phase is the position of the first nucleotide in the codon (0,1,2).
- g (NUM1-NUM2) Add introns between compatible exons if they are within this base pair range (NUM1-NUM2) (inclusive). (Default: no introns are added)
- m (NUM) Max number of artificially added introns for each exon. (Default: 10 if intron range is specified with -g)

Example:

java -jar BuildMS2DB.jar -r genePrediction1.gff -r genePrediction2.gff -s Genome.trie -w Genome.ms2db -t CDS -c 8

3.5.6 Decoy Databases

A decoy database, mentioned in Subheading 2, is a random or reversed sequence database of the target database, used to estimate the FDR of a group of PSMs against the target database. Estimating the FDR through a decoy database is widely accepted as standard practice within the proteomics community, although as mentioned in Subheading 2, a precise FDR and FPR is more practical and useful to the field of proteogenomics, and a few algorithms have been developed to address this issue. A decoy database in .trie format can be created using the ShuffleDB.py script provided by Inspect. The script has the option of either reversing or shuffling the sequences to create the decoy sequences. The script outputs another .trie file of the database with included decoy sequences.

Usage:

- r (Trie file name): Path to input database
- w (FileName): Path to output database
- s: If set, proteins will be REVERSED. Default behavior is to SHUFFLE.
- b: If set, ONLY the scrambled proteins are written out. Default behavior is to write both forward and scrambled proteins.
- p: In shuffled mode, avoid repeating peptides of length 8 or more in the shuffled database. Treat I and L as identical and don't treat Q and K as identical. The script requires a little longer time to run, as some "bad words" (repeated 8mers) will still be seen for repetitive records.
- t: Number of shuffled/reversed copies to write out (defaults to 1)

Examples:

For shuffled sequences:

ShuffleDB.py -r Genome_6frame.trie -w Genome_6frame_shuffle.trie -p

Or

For reversed sequences:

ShuffleDB.py -r Genome_6frame.trie -w Genome_6frame_reverse.trie -s

Additionally, if using an exon splice-graph, a decoy version can also be created. This can be done using the MS2DBShuffler.jar script provided from the proteogenomic pipeline. The script outputs another .ms2db file containing the exon splice-graph with included decoy sequences.

Example:

java -jar MS2DBShuffler.jar -r spliceGraph.ms2db -w spliceGraph_wDecoy.ms2db

3.5.7 Tandem Mass Spectrometry Database Search Tool: Inspect

Inspect (37) is a database search tool used to interpret peptide tandem mass spectra, based on a hybrid approach to MS/MS database searching, which was initially developed by Mann and Wilm in 1994 (24). The hybrid approach combines de novo sequencing and database searching. A de novo algorithm first generates a list of partial peptide sequences (44–48) or short sequence tags (37) of roughly 3–5 residues in length which are then used in a mutation-tolerant database search (49) against a protein database, which can allow for one or more mismatches, caused by sequencing errors, mutations, or PTM peptides. By searching the database and only matching to peptides that contain these short sequence tags, the search space is significantly reduced, an advantage towards improved search times and a reduction in the FDR, which is convenient when searching against large databases such as six-frame translations of genomes. Additionally, Inspect is the only known MS/MS database search tool that can search against an exon splice-graph, making it suitable for proteogenomics.

The proteome of any organism, tissue or cell type, is dynamic and so being able to accommodate for mutations, or PTM peptides, which potentially have no records in conventional databases, becomes an advantage when aiming to annotating a genome and/or proteome.

To run Inspect, an input text file needs to be created first containing commands for Inspect to follow in the form of (Command),(Value).

Commands that Inspect follows are:

- Spectra, (File name)—a spectrum file to search, such as a . mzXML,.mgf, .mzData, .ms2, .dta.
- Instrument, (Type)—The options are ESI-ION-TRAP (default), QTOF, and FT-HYBRID.
- db, (File name)—a database (.trie or .ms2db exon splice-graph file) to search.
- SequenceFile, (File name)—the name of a single FASTA-format protein database to search (if not using PrepDB.py to create a database for the "db" command).
- Protease, (Name)—The protease name, either "Trypsin," "None," or "Chymotrypsin." If a protease is specified, then non-tryptic termini are penalized.
- mod, (Mass), (Residues), (Type), (Name)—Specifies an amino acid modification. The delta mass (in Daltons) and the affected amino acids are required. Valid values for "type" are "fix," "c-terminal," "n-terminal," and "opt" (the default).
- Examples: mod, +57, C, fix—Fixed modification carbamidomethylation.
- mod, 80, STY, opt, phosphorylation—Optional modification phosphorylation.
- mod, 16, M—Oxidation of methionine.
- mod, 43, *, nterminal—N-terminal carbamylation.
- mods, (Count)—Number of allowed PTMs on a single peptide (1 or higher) (1 or 2 recommended).

Further instructions on the use of Inspect can be found at http://proteomics.ucsd.edu/InspectDocs/index.html.
An example Inspect input text file:

```
spectra,MySpectra.mzXML
instrument,FT-Hybrid
protease,Trypsin
DB,mySixFrameTranslation_wDecoy.trie
DB,cRAP_wDecoy.trie
DB,mySpliceGraph_wDecoy.ms2db
DB,myProteome_wDecoy.trie
# Protecting group on cysteine:
mod,57,C,fix
mod,16,M
mods,2
```

Additionally, Inspect can report errors during processing (-e parameter), and the resources directory (if not running from within its own path) can be specified (-r parameter).

Example:

inspect -i ../input/InspectInput.txt -o InspectOutput.txt -e InspectError.txt -r /InsPecT/20120109/data/

Description of Inspect output formatting:

- SpectrumFile—The file searched.
- Scan#—The scan number within the file; this value is 0 for .dta files; For MGF files, the scan# is equivalent to the SpecIndex but is 0-based numbering.
- Annotation—Peptide annotation, with prefix and suffix and (non-fixed) modifications indicated. Example: K. DFSQIDNAP+16EER.E.
- Protein—The name of the protein this peptide comes from. (Protein names are stored in the .index file corresponding to the database .trie file.)
- Charge—Precursor charge. If "multicharge" is set or if no charge is specified in the source file, Inspect attempts to estimate the charge.
- MQScore—A database independent Match Quality Score of the peptide-spectrum match, which looks at non-database features, such as intensity and fraction of b and y ions found.
- Length—The length of the matched peptide in amino acids.
- TotalPRMScore—Summed score for break points (between amino acids), based upon a Bayesian network modeling fragmentation propensities.
- MedianPRMScore—The median score for break points.
- FractionY—The fraction of charge 1 y ions detected.

- FractionB—The fraction of charge 1 b ions detected.
- Intensity—Fraction of high-intensity peaks which are b or y fragments. For a length-n peptide, the top n*3 peaks are considered.
- NTT—Number of tryptic termini (or Unused, if no protease was specified). Note that the N- and C-terminus of a protein are both considered to be valid termini.
- InspectFDR—This is the FDR of all matches with F-Score equal to or greater than this match. Since Inspect knows nothing about a decoy database, it is often best to run ComputeFDR.jar (available from the MS-GFDB package) (50) or InspectFDRComputer.jar (available from the proteogenomic pipeline) to compute an empirical FDR.
- F-Score—The database-dependent quality score of the peptide-spectrum match. A linear combination of MQScore and DeltaScore.
- DeltaScore—The difference between the MQScore of this match and the best alternative match is database dependent and becomes less useful with much larger databases.
- DeltaScoreOther—The difference between the MQScore of this match and the best alternative from a different locus. To see the difference between this and the previous column, consider a search that finds similar matches of the form "M+16MALGEER" and "MM+16ALGEER." In such a case, the DeltaScore would be very small, but DeltaScoreOther might still be large.
- RecordNumber—Index of the protein record in the database.
- DBFilePos—Byte-position of this match within the database.
- SpecFilePos—The byte offset of the spectrum in the spectrum file. Useful for passing to the "Label" script provided by Inspect.
- PrecursorMZ—The precursor m/z given in the spectrum file.
- PrecursorError—The difference (in m/z units) between the precursor m/z given in the file and the theoretical m/z of the identified peptide.
- SpecIndex—This is a one-based number of the index of the spectrum in the original spectrum file. Only MS2+ spectra are counted.

3.5.8 Precise Calculation of Individual FPRs and Rescoring of PSMs

Estimating FDRs using decoy databases is an imprecise tool to use in the field of proteogenomics, where the calculation of precise FPRs is vastly more applicable in assigning confidence to individual PSMs and hence annotation events. By calculating the precise FPR of individual PSMs and rescoring accordingly, a higher level of stringency to genomic annotations can be obtained. The tool

MS-GF performs this function by calculating the precise FPR and *p*-value, referred to as spectral probabilities, of each individual PSM. MS-GF computes generating functions in combinatorics to calculate the spectral energy and spectral probability (*p*-values) properties of the spectra and its matching peptide (33).

Usage:

– i Results File

– d Spectra Directory

– (-o Output File Name) (Default: standard out)

– (-m Fragmentation Method 0/1) (0: CID (default), 1: ETD, 2: CID/ETD pairs)

– (-e Enzyme 0/1/2/3/4/5/6/7) (0: No enzyme, 1: Trypsin (default), 2: Chymotrypsin, 3: Lys-C, 4: Lys-N, 5: Glu-C, 6: Arg-C, 7: Asp-N)

– (-fixMod 0/1/2) (0: No Cysteine Protection, 1: Carbamidomethyl C (default), 2: Carboxymethyl C)

– (-aaSet AASetFileName (default: standard amino acids))

– (-x 0/1) (0: all (default), 1: One Per Spectra)

– (-p Spectral Probability Threshold) (Default: 1)

– (-param Scoring Parameter File)

– (-addScore 0/1) (0: don't add MSGFScore (default), 1: add MSGFScore)

Example:

java –jar MSGF.jar -i InspectOutput.txt -d massSpectraFileDirectoy/ -o InspectOutput.msgfRescore.txt

3.5.9 Evaluating Significance Thresholds

Inspect calculates a *q*-value (FDR adjusted *p*-value) to determine the minimum FDR cutoff for significant PSMs. An E-value or FDR value threshold can be applied across all these significant PSMs. The script InspectFDRComputer.jar, provided by the proteogenomic pipeline, applies a cutoff based on different score thresholds, either the F-Score (linear combination of MQScore and DeltaScore) or using MS-GF to determine the false-positive rate (Spectral Probability). A FDR cutoff of 1 % is usually applied in proteogenomics, where decoy hits must be present and should be applied to the local FDRs for each PSM.

Usage:

Required:

– r (FILE) Input file that has been run through MS-GF

– w (FILE) Output file to write filtered results

– a (0/1) Perform an EValue (0) or FDR (1) cutoff

- p (NUM) EValue or InspectFDRComputer cutoff

Optional:

- t (FILE) The searched trie file, if using EValue cutoff (default 10 Mb)
- s (0/1) Score mode (0: use MS-GF, 1: usage F-Score (default))
- l Compute local FDR for each PSM
- q Compute peptideFDR instead of psmFDR
- m (NUM) Max missed cleavages permitted in a reported peptide
- H report the decoy matches as well. The FDR of a decoy match is the same as the FDR for a target match with the same score

Example:
java -jar InspectFDRComputer.jar -r InspectOutput.msgfRescore.txt -w InspectOutput.msgfRescore.FDRcompute.txt -a 1 -p 0.01 -s 0 –l -m 2

3.5.10 Novelty Determination

The peptides are sorted into categories of known, or previously predicted, and novel peptides by comparing to a reference proteome using the DetermineNovelty.jar script, provided by the proteogenomic pipeline. Any peptides matching known protein sequences are labeled as known, and any remaining peptides not matching within these sequences are labeled as novel peptides. The known peptides can then be used to confirm previously predicted proteins and the novel peptides used to determine new annotation events.

Usage:

Required:

- r (FILE) An Inspect results file
- t (FILE) A .trie file of known protein sequences
- w (FILE) An output file name for the "known" peptides
- n (FILE) An output file name for the "novel" peptides

Optional:

- x (FILE) A .trie file of filtered protein sequences
- d Run in debug mode

Example:

java -jar DetermineNovelty.jar -r InspectOutput.msgfRescore.FDRcompute.txt -t myProteome.trie -w InspectOutput.msgfRescore.FDRcompute.known.txt -n InspectOutput.msgfRescore.FDRcompute.novel.txt

3.5.11 Peptide Mapping

Novel peptides are mapped to genomic positions using the six-frame translation of the genome sequence in .trie format and also, if applicable, the exon splice-graph, using the GenomeLocator.jar script, provided by the proteogenomic pipeline. This is done to determine which genomic regions are likely to be translated to peptides. The results are produced in the same format as Inspect output, including the genomic location of each peptide-spectrum match; whether the peptide is unique within the genome, its location count (number of times it maps to genomic locations) and the best local FDR of each peptide-spectrum match.

Usage:

Options:

- r (FILE) Inspect result file to analyze containing novel peptide-spectrum matches
- t (FILE) May be .trie or .ms2db, produced by SixFrameBuilder.jar or BuildMS2DB.jar, respectively
- w (FILE) File to write converted results

Example:

java -jar GenomeLocator.jar -r InspectOutput.msgfRescore.FDRcompute.novel.txt -t mySixFrameTranslation.trie -t mySpliceGraph.ms2db -w InspectOutput.msgfRescore.FDRcompute.novel.genomicLocation.txt

Novel peptides can optionally be validated using the NovelPeptideVerifier.jar script, provided by the proteogenomic pipeline. It searches for novel peptides with isoleucine/leucine substitutions and lysine/pyroglutamate substitutions, which would make any novel peptides ambiguous and no longer novel. This step of the pipeline can be implemented either after the DetermineNovelty.jar script or after the GenomeLocator.jar script, where the type of input (inspect results or genome locations file) can be chosen.

Usage:

- r (DIR/FILE) File or directory containing novel peptides or locations
- w (DIR/FILE) File or directory to write novel peptides to
- t (FILE) Trie file containing known proteome
- p (0/1) The type of file containing novel peptides (0: inspect results (default), 1: locations)
- q Run in a loose mode (also include W->AD)
- d Run in debug mode

Example:

java -jar NovelPeptideVerifier.jar -r InspectOutput.msgfRescore.FDRcompute. novel.genomicLocation.txt -w verified NovelPeptides/ -t myProteome.trie -p 1

Once the GenomeLocator.jar has been run and if chosen, novel peptides verified, the output text file should contain the following columns (in order from left to right):

- Spectrum File—The file containing the best spectrum representative for this peptide. The best spectrum is judged as the spectrum with the lowest localFDR.
- Scan#—The scan number for the best spectrum representative.
- Annotation—The peptide sequence mapped to this location, otherwise the same as the sequence from the Inspect output.
- SpecFilePos—Same as for the Inspect output.
- MQScore—Same as for the Inspect output.
- F-Score—Same as for the Inspect output.
- SpecProb—The precise *p*-value for the peptide-spectrum match (Produced by MS-Generating Function).
- InspectFDR—Same as for the Inspect output. (Not necessary for further downstream processing.)
- Protein—Same as for the Inspect output.
- DBFileName—The database file that was matched.
- SequenceName—Usually, this is the chromosome or BAC name for this location.
- Start—The 0-based start of the peptide location. The Start is always a lower coordinate than the End.
- End—The noninclusive end of the peptide location.
- Strand—The strand of the location.
- SpliceSequence—A representation of the peptide indicating the location of splice-junctions (if any). A colon (":") indicates a splice junction that falls between two codons. A semicolon (";X;") indicates a splice junction that splits a codon, and the split codon produces the amino acid "X."
- Splices—A space-delimited list of the coordinates of the splices (0-based, left-inclusive).
- SpectrumCount—The number of spectra that match this peptide.
- IsUnique—Does this peptide map to only this location?
- IsNovel—Is this peptide novel?
- Charge—The charge of the best spectrum representative for this peptide.

- LocationCount—The total number of genomic locations that this peptide matches.
- BestLocalFDR—The localFDR for the best spectrum representative for the peptide.

The known peptides can also be processed and formatted into a GFF file for later visualization and comparison with the annotation events through a genome browser. The known peptides outputted by the DetermineNovelty.jar script can be formatted into a GFF file using the CreatePeptideGFF.jar script, provided by the proteogenomic pipeline.

Data Format Assumptions:

1. The peptides contain only modifications of the form +NUM, or -NUM. Flanking amino acids are ignored (e.g., only XXXXX is used in A.XXXXX.B)
2. The protein name contains no spaces and is the first contiguous set of characters in the header (e.g., in ">GRMZM2G005939_P01 This is my protein," only GRMZM2G005939_P01 is the protein name).
3. In the GFF file, features of type CDS are considered. The protein for each CDS is determined from the attributes column (e.g., Parent=GRMZM2G005939_P01). If no parent is given, then it considers the Name and finally the ID.

Usage:

- r (FILE/DIR) File or directory containing tab delimited files of peptides
- c (NUM) The number of the columns containing the peptides (0 is the first column)
- p (FILE) FASTA or Trie file containing protein sequences
- g (FILE) The GFF file containing the coordinates of the proteins
- o (FILE) The output GFF file to create
- t (FILE) A file containing conversions from FASTA file protein names to GFF file protein names
- q (NUM) The column of the FASTA gene name in the conversions file (default is 3)
- d (NUM) The column of the GFF gene name in the conversions file (default is 2)
- v Run in verbose mode

Example:

java -jar CreatePeptideGFF.jar -r InspectOutput.msgfRescore.FDRcompute.known.txt -c 2 -p MyProteome.trie -g MyProteome.gff -o InspectOutput.msgfRescore.FDRcompute.known.gff

3.5.12 Interpretation of Peptide Clusters and Determination of Annotation Events (See Note 7)

The peptides and their aligned genomic positions are clustered into groups of colocated peptides, where upon novel annotation events can be determined. The minimum number of peptides per cluster can be set (default of 1 peptide per cluster) and a minimum number of unique peptides per cluster can also be set (default of 0 peptides per cluster). Using the default settings will use all the peptides to determine annotation events, and using a minimum number of peptides per cluster set to 2 and a unique peptide clusters set to 0 would meet the "two peptides per protein rule" which is a commonly used strategy to give confidence to protein identifications, often followed within the proteomics community. In contrast, considering the precise FPRs and *p*-values calculated using MS-GF, and the likely multiple locations a peptide may map to a six-frame translation, a minimum number of 1 peptide per cluster and a minimum number of 1 unique peptide per cluster could be set, using a "One (unique) peptide per protein rule" where every annotation event has at least 1 highly significant peptide and which maps uniquely to a location within the six-frame translation. This approach adds further confidence to that annotation and prevents the loss of annotations with only one peptide for evidence.

The concept of using the single-peptide rule was explored and has recently been described and enhanced by Gupta et al. (51), using MS-GF scores, and can improve upon the number of annotation events discovered, which may not be detected otherwise with limited peptide evidence. Additionally, taking a comparative proteogenomic approach, the single-peptide rule can be further enhanced by finding the same single peptides within other related species, further adding confidence to the annotation event. Although this pipeline does not consider multiple related genomes in a comparative setting and there has not yet been a scoring algorithm for comparative proteogenomics, each peptide from each cluster is outputted to a GFF file containing their genomic positions.

Each annotation event is determined and is assigned an event probability, calculated based on the number of peptide locations and the local FDR, which can be used to apply a cutoff before results are outputted to an event GFF file. The event GFF file contains the genomic positions of each peptide, its event probability, and its event name (e.g., Novel Gene, Reverse Strand, Exon Boundary, Frame Shift, Novel Exon, Novel Splice, Translated UTR, Gene Boundary). The event GFF can be used as evidence towards a gene prediction tool or used in a genome browser tool, such as Gbrowse.

Usage:

Required:

- r (DIR) Directory containing peptide locations. If directory, then all files are loaded.
- w (DIR) Directory to write cluster files and/or GFFs to
- b (NAME) Species name

Optional:

- g (0/1) Write clusters in GFF format, if 0 then only human-readable format, if 1 then write GFF format and human-readable format
- l (NUM) The distance between two adjacent peptides (default: 1,000 bp)
- c (NUM) Minimum number of locations per cluster (default: 1)
- m (NUM) Minimum number of uniquely located peptides per cluster (default: 0)
- k (FILE) Known proteins GFF (Filtered Set)
- s (FILE) Known proteins GFF (Working Set), if only one set of known proteins, use "-k"
- o Overwrite (don't load the locations) of any files in the output directory
- x (NUM:NUM:NUM or NUM) Minimum event probability (give the event cutoffs separately for novel genes, distal events, and proximal events, or just give a single cutoff)
- f (0/1) Human-readable format (0:table or 1:full info)
- e (FILE) Compute events and write them to specified file
- i (DIR) Directory containing known peptides
- t (FILE) Trie file of known proteins
- y (FILE) Write sister cluster!
- a (FILE) File with conversions between GFF transcript name to FASTA protein name
- h (NUM) Column containing the GFF Transcript name in the conversion file (default:2)
- j (NUM) Column containing the FASTA protein name in the conversion file (default:3)
- n (NUM) Column containing the gene name in the conversion file (default: 1)
- p (NAME) Case-insensitive feature name which contains the gene boundaries in the GFF files (default: mrna)

Example:

java -jar GenomeClusterer.jar -r novelGenomicLocation Directory/ -w novelGenomicLocationClusterDirectory/ -b MySpecies -g 1 -l 20000 -c 1 -m 1 -k myProteome.gff -o -x 0.9 -f 0 -e myEvents.txt -t myProteome.trie

3.6 Future Directions

There are still many areas within the field of proteogenomics that require further work to develop a robust and comprehensive annotation pipeline. For example the field of comparative proteogenomics with multiple proteomes and genomes has yet to be implemented into an automated proteogenomic pipeline. In such a pipeline the annotations of multiple genomes could be performed and cross-referenced with each other, adding confidence to annotations particularly in the context of using single peptides per protein for annotation (commonly referred to "one-hit wonders") (52) and assigning improved confidence scores based on the cross species evidence. This comparative approach to assigning confidence to single PSMs, along with calculating the precise FPRs and *p*-values with tools such as MS-GF, will result in a much more confident and comprehensive analysis without an overwhelming need to manually validate each and every one-hit-wonder, as was conducted recently by Christie-Oleza et al. (53) using RT-PCR. In cases where only one proteome is available, simply calculating the precise FPR and *p*-value should be a sufficient level of confirmation to confirm one-hit-wonders, as was demonstrated by Kim et al. (33). In such situations where a single proteome is available, an ortho-proteogenomics approach could be applied, by first correcting annotations in one genome and then finding homologues to genes in other related species and inferring annotations based on comparative genomics (3).

Other interesting scientific issues that an automated comparative proteogenomic pipeline can potentially address are the detection and correlation of multiple common and rare PTMs with cross-referencing between multiple species; such PTMs can be confirmed to exist. Additionally, other applications include resolving sites of RNA-editing, which are difficult to confirm using single MS-based proteomics or genomics methods; site determination of N-terminal Methionine Excision (NME), which are unreliable from single proteomes and genomes; determination of signal peptides, where experimental evidence is limited; and lastly improving upon the predictions of operons within bacterial genomes, in which expression of some proteins can be ambiguous in a single proteome/genome setting (52).

Additionally, an error-tolerant search approach within an automated comparative proteogenomic pipeline could use tools such as short sequence tag algorithms like Inspect and TagRecon (54), to detect single nucleotide polymorphisms (SNPs) or mutant peptides within the genome, which would allow for the comparison between normal and diseased tissues. Such pipelines could potentially expand and accelerate fields such as cancer biomarker discovery.

Approaching the challenges of proteogenomics from a mass spectrometry and proteomics perspective reveals generally inadequate protein coverage. This can be overcome with further fractionation techniques, although much of the protein coverage is lost through mixture spectra in which multiple peptides can be found hidden within single spectra. Through tools such as MixDB (55), ProbIDtree (56), and M-SPLIT (57) incorporated into proteogenomic pipelines, annotation events, previously undetected due to "invisible" peptides, may be uncovered. Additionally, using search tools, such as MS-GFDB (50), which can make use of alternative spectral types such as electron-transfer dissociation (ETD), would result in improved protein coverage.

From a computational perspective, there is also a need to run these proteogenomic pipelines on high-performance computers, due to the resource overheads and time expended on searching large databases with large spectral files. Such pipelines in their native form are geared to run on a single computer, but by splitting up the tasks into smaller chunks and running on a computer cluster, the time to completion can be reduced by many orders of magnitude. Additionally, these pipelines in their native form are not easily accessible to biologists, and allowing them access to graphical workflow environments to run them on will greatly improve their appeal and accessibility to the wider scientific community at large.

4 Notes

1. This protocol is based on Hurkman and Tanaka (14) and an efficient and high-quality extraction protocol published previously (method 4 in Vincent et al. (15)).
2. Phenol should be handled in a fume hood and personal protective equipment such as gloves, eye protection, and a lab coat should be worn. It is highly recommended that users be familiar with proper phenol handling procedures including waste disposal and responses to accidental exposure.
3. All the following steps are for the preparation of 0.25 mg protein. We often digest 0.1 mg and scale down amount of Lys-C and trypsin 2.5-fold to account for the smaller protein amount. In that case the addition volumes of the Lys-C and trypsin stocks stay the same; the enzyme concentrations in the stocks are reduced 2.5-fold.
4. All steps using TFE must be done in a fume hood, including centrifugation steps. Wear gloves, eye protection, and a lab coat when working with TFE. Use a separate waste container in hood for TFE waste.
5. Wisniewski reports similar results for 10 k and 30 k UF devices, i.e., no significant loss of smaller proteins. The 30 k offers the advantage of faster centrifugation steps. In the method below, each centrifugation step for 45 min may be reduced to a shorter time as long as the retentate volume is ~20 μl.
6. MS-based proteomics faces challenges due to error rates from peptide identifications caused by incomplete fragmentation of precursor ions, spectral noise, and isometric peptides (structurally different peptides with the same mass), which are all due to the limitations of the technology. These problems are inflated in the field of proteogenomics where error rates are many times

larger due to the larger proteogenomic databases, such as with six-frame translated genomic databases, containing many putative open reading frames (ORFs), the vast majority of which would be false positives. Also to improve proteome coverage for proteogenomic studies, spectral datasets tend to be very large, containing well over 10 million spectra in the majority of cases, which can also contain low-quality spectra due to limitations in the MS technology mentioned earlier. These problems reduce the sensitivity of the database searches and inflate the false-positive rate (FPR) significantly. Therefore, the need to apply stringent rules and thresholds when defining annotation events, such as the boundaries of genes, the discovery of novel genes, and/or the identification of new gene isoforms, by applying a higher level of filtering at the peptide, protein, and annotation event level becomes absolutely critical.

7. Proteogenomic pipeline annotation example results. The example input spectra files were three mzXML files, converted from Thermo RAW format, from three 220 min LC-MS/MS runs using gas-phase fractionation. The peptide samples were from grape berry skin proteins, and the spectra were matched to a splice-graph constructed with 12 GFF files of gene evidence from CRIBI Genomics (http://genomics.cribi.unipd.it/Grape_Genome). There were 2,107 PSMs to the splice-graph. Twelve hundred thirty-three unique peptides were located within 688 previous protein predictions; 26 peptides were outside previous predictions. The final annotation results of this pipeline yielded a number of novel annotations based on a FDR threshold of 1 %, with local FDRs ranging from 0.25 to 6.4 %, and an overall event probability threshold of 90 %,

Table 1
Summary of annotation events

Event types	Number
Frame shift	1
Translated UTR	3
Exon boundary	1
Novel splice	0
Novel exon	6
Gene boundary	9
Reverse strand	3
Novel gene	6

which was calculated based on the number of peptide locations within the genome and the best local FDR. Each peptide was clustered into an annotation event, based on a colocated peptide distance of 20,000 bp, which was based on the size distribution of genes in the current protein predictions; >95 % of predicted grape genes are <20,000 bp. A summary of the results can be seen in Table 1.

Acknowledgments

The authors acknowledge funding support from the Australian Research Council and the NSF Grape Research Coordination Network. P.A.H. acknowledges Robert Black for continued support and encouragement.

References

1. Windsor AJ, Mitchell-Olds T (2006) Comparative genomics as a tool for gene discovery. Curr Opin Biotechnol 17:161–167
2. Aivaliotis M, Gevaert K, Falb M et al (2007) Large-scale identification of N-terminal peptides in the halophilic archaea Halobacterium salinarum and Natronomonas pharaonis. J Proteome Res 6:2195–2204
3. Gallien S, Perrodou E, Carapito C et al (2009) Ortho-proteogenomics: multiple proteomes investigation through orthology and a new MS-based protocol. Genome Res 19:128–135
4. Nielsen P, Krogh A (2005) Large-scale prokaryotic gene prediction and comparison to genome annotation. Bioinformatics 21:4322–4329
5. Jaffe JD, Berg HC, Church GM (2004) Proteogenomic mapping as a complementary method to perform genome annotation. Proteomics 4:59–77
6. Domon B, Aebersold R (2006) Mass spectrometry and protein analysis. Science 312:212–217
7. Washburn MP, Wolters D, Yates JR 3rd (2001) Large-scale analysis of the yeast proteome by multidimensional protein identification technology. Nat Biotech 19:242–247
8. Mann M, Pandey A (2001) Use of mass spectrometry-derived data to annotate nucleotide and protein sequence databases. Trends Biochem Sci 26:54–61
9. Ansong C, Purvine SO, Adkins JN, Lipton MS, Smith RD (2008) Proteogenomics: needs and roles to be filled by proteomics in genome annotation. Brief Funct Genomic Proteomic 7:50–62
10. Armengaud J (2009) A perfect genome annotation is within reach with the proteomics and genomics alliance. Curr Opin Microbiol 12:292–300
11. de Groot A, Dulermo R, Ortet P et al (2009) Alliance of proteomics and genomics to unravel the specificities of Sahara bacterium Deinococcus deserti. PLoS Genet 5:e1000434
12. Jaffe JD, Stange-Thomann N, Smith C et al (2004) The complete genome and proteome of Mycoplasma mobile. Genome Res 14:1447–1461
13. Zivanovic Y, Armengaud J, Lagorce A et al (2009) Genome analysis and genome-wide proteomics of Thermococcus gammatolerans, the most radioresistant organism known amongst the Archaea. Genome Biol 10:R70
14. Hurkman WJ, Tanaka CK (1986) Solubilization of plant membrane proteins for analysis by two-dimensional gel electrophoresis. Plant Physiol 81:802–806
15. Vincent D, Wheatley MD, Cramer GR (2006) Optimization of protein extraction and solubilization for mature grape berry clusters. Electrophoresis 27:1853–1865
16. Manza LL, Stamer SL, Ham AJ, Codreanu SG, Liebler DC (2005) Sample preparation and digestion for proteomic analyses using spin filters. Proteomics 5:1742–1745
17. Wisniewski JR, Zougman A, Nagaraj N, Mann M (2009) Universal sample preparation method for proteome analysis. Nat Methods 6:359–362
18. Chick JM, Haynes PA, Molloy MP, Bjellqvist B, Baker MS, Len AC (2008) Characterization of the rat liver membrane proteome using peptide immobilized pH gradient isoelectric focusing. J Proteome Res 7:1036–1045

19. Scherl A, Shaffer SA, Taylor GK, Kulasekara HD, Miller SI, Goodlett DR (2008) Genome-specific gas-phase fractionation strategy for improved shotgun proteomic profiling of proteotypic peptides. Anal Chem 80:1182–1191
20. Panchaud A, Scherl A, Shaffer SA et al (2009) Precursor acquisition independent from ion count: how to dive deeper into the proteomics ocean. Anal Chem 81:6481–6488
21. Yates JR 3rd, Eng JK, McCormack AL, Schieltz D (1995) Method to correlate tandem mass spectra of modified peptides to amino acid sequences in the protein database. Anal Chem 67:1426–1436
22. Perkins DN, Pappin DJ, Creasy DM, Cottrell JS (1999) Probability-based protein identification by searching sequence databases using mass spectrometry data. Electrophoresis 20:3551–3567
23. Pevzner PA, Mulyukov Z, Dancik V, Tang CL (2001) Efficiency of database search for identification of mutated and modified proteins via mass spectrometry. Genome Res 11: 290–299
24. Mann M, Wilm M (1994) Error-tolerant identification of peptides in sequence databases by peptide sequence tags. Anal Chem 66:4390–4399
25. Brosch M, Choudhary J (2010) Scoring and validation of tandem MS peptide identification methods. Methods Mol Biol 604:43–53
26. States DJ, Omenn GS, Blackwell TW et al (2006) Challenges in deriving high-confidence protein identifications from data gathered by a HUPO plasma proteome collaborative study. Nat Biotechnol 24:333–338
27. Elias JE, Gygi SP (2007) Target-decoy search strategy for increased confidence in large-scale protein identifications by mass spectrometry. Nat Methods 4:207–214
28. Storey JD, Tibshirani R (2003) Statistical significance for genomewide studies. Proc Natl Acad Sci USA 100:9440–9445
29. Choi H, Ghosh D, Nesvizhskii AI (2008) Statistical validation of peptide identifications in large-scale proteomics using the target-decoy database search strategy and flexible mixture modeling. J Proteome Res 7:286–292
30. Kall L, Storey JD, MacCoss MJ, Noble WS (2008) Posterior error probabilities and false discovery rates: two sides of the same coin. J Proteome Res 7:40–44
31. Efron B, Tibshirani R, Storey JD, Tusher V (2001) Empirical Bayes analysis of a microarray experiment. J Am Stat Assoc 96:1151–1160
32. Alves G, Yu YK (2008) Statistical Characterization of a 1D Random Potential Problem—with applications in score statistics of MS-based peptide sequencing. Physica A 387:6538–6544
33. Kim S, Gupta N, Pevzner PA (2008) Spectral probabilities and generating functions of tandem mass spectra: a strike against decoy databases. J Proteome Res 7:3354–3363
34. Gupta N, Bandeira N, Keich U, Pevzner PA (2011) Target-decoy approach and false discovery rate: when things may go wrong. J Am Soc Mass Spectrom 22:1111–1120
35. Venter E, Smith RD, Payne SH (2011) Proteogenomic analysis of bacteria and archaea: a 46 organism case study. PLoS One 6:e27587
36. Castellana NE, Payne SH, Shen Z, Stanke M, Bafna V, Briggs SP (2008) Discovery and revision of Arabidopsis genes by proteogenomics. Proc Natl Acad Sci USA 105: 21034–21038
37. Tanner S, Shu H, Frank A et al (2005) InsPecT: identification of posttranslationally modified peptides from tandem mass spectra. Anal Chem 77:4626–4639
38. Tanner S, Shen Z, Ng J et al (2007) Improving gene annotation using peptide mass spectrometry. Genome Res 17:231–239
39. Desiere F, Deutsch EW, Nesvizhskii AI et al (2005) Integration with the human genome of peptide sequences obtained by high-throughput mass spectrometry. Genome Biol 6:R9
40. Apweiler R, Bairoch A, Wu CH et al (2004) UniProt: the Universal Protein knowledgebase. Nucleic Acids Res 32:D115–D119
41. Fermin D, Allen BB, Blackwell TW et al (2006) Novel gene and gene model detection using a whole genome open reading frame analysis in proteomics. Genome Biol 7:R35
42. Edwards NJ (2007) Novel peptide identification from tandem mass spectra using ESTs and sequence database compression. Mol Syst Biol 3:102
43. de Bruijn NG, Erdos P (1946) A combinatorial problem. Koninklijke Netherlands: Academe Van Wetenschappen 49:758–764
44. Bern M, Cai Y, Goldberg D (2007) Lookup peaks: a hybrid of de novo sequencing and database search for protein identification by tandem mass spectrometry. Anal Chem 79:1393–1400
45. Frank A, Pevzner P (2005) PepNovo: de novo peptide sequencing via probabilistic network modeling. Anal Chem 77:964–973
46. Kim S, Gupta N, Bandeira N, Pevzner PA (2009) Spectral dictionaries: integrating de novo peptide sequencing with database search of tandem mass spectra. Mol Cell Proteomics 8:53–69
47. Mo L, Dutta D, Wan Y, Chen T (2007) MSNovo: a dynamic programming algorithm for de novo peptide sequencing via tandem mass spectrometry. Anal Chem 79:4870–4878

48. Ma B, Zhang K, Hendrie C et al (2003) PEAKS: powerful software for peptide de novo sequencing by tandem mass spectrometry. Rapid Commun Mass Spectrom 17:2337–2342
49. Pevzner PA, Dancik V, Tang CL (2000) Mutation-tolerant protein identification by mass spectrometry. J Comput Biol 7:777–787
50. Kim S, Mischerikow N, Bandeira N et al (2010) The generating function of CID, ETD, and CID/ETD pairs of tandem mass spectra: applications to database search. Mol Cell Proteomics 9:2840–2852
51. Gupta N, Pevzner PA (2009) False discovery rates of protein identifications: a strike against the two-peptide rule. J Proteome Res 8:4173–4181
52. Gupta N, Benhamida J, Bhargava V et al (2008) Comparative proteogenomics: combining mass spectrometry and comparative genomics to analyze multiple genomes. Genome Res 18:1133–1142
53. Christie-Oleza JA, Miotello G, Armengaud J (2012) High-throughput proteogenomics of Ruegeria pomeroyi: seeding a better genomic annotation for the whole marine Roseobacter clade. BMC Genomics 13:73
54. Dasari S, Chambers MC, Slebos RJ, Zimmerman LJ, Ham AJ, Tabb DL (2010) TagRecon: high-throughput mutation identification through sequence tagging. J Proteome Res 9:1716–1726
55. Wang J, Bourne PE, Bandeira N (2011) Peptide identification by database search of mixture tandem mass spectra. Mol Cell Proteomics 10(M111):010017
56. Zhang N, Li XJ, Ye M, Pan S, Schwikowski B, Aebersold R (2005) ProbIDtree: an automated software program capable of identifying multiple peptides from a single collision-induced dissociation spectrum collected by a tandem mass spectrometer. Proteomics 5: 4096–4106
57. Wang J, Perez-Santiago J, Katz JE, Mallick P, Bandeira N (2010) Peptide identification from mixture tandem mass spectra. Mol Cell Proteomics 9:1476–1485

Chapter 22

Label-Free Differential Analysis of Murine Postsynaptic Densities

Scott P. Goulding, Michael J. MacCoss, and Christine C. Wu

Abstract

This chapter provides detailed methodology for the enrichment and label-free differential analysis of postsynaptic density (PSD) proteins. Methods discussed will include tissue homogenization, subcellular fractionation, protein digestion, and label-free differential analysis after liquid chromatography–tandem mass spectrometry. When combined, these protocols facilitate the identification of receptors and signal transducers that comprise the PSD and provide an optimized workflow for the differential analysis of PSD proteomes. This strategy supports a utility for coupling fractionation with proteomics analysis to enrich for low-abundant proteins in cellular localizations that would otherwise be lost in a global tissue context.

Key words Postsynaptic density, Label-free, Differential analysis, Neuroproteomics, Tandem mass spectrometry, Sucrose gradient centrifugation, Topograph

1 Introduction

An excitatory synapse is characterized by an electron-dense specialization at the postsynaptic membrane, commonly referred to as the postsynaptic density (PSD) (1). Consistent with structural observations, the PSD contains a dense clustering of integral membrane receptors and associated proteins that mediate neuronal signal transduction, synaptic plasticity, and homeostasis. Because perturbations at the PSD have been implicated in a wide range of psychiatric and neurological disorders (2, 3), including schizophrenia, Alzheimer's disease, and alcohol addiction, its purification and characterization is of high clinical importance. For this reason, methodologies discussed in this chapter will provide an optimized workflow for the label-free differential analysis of PSD proteomes.

Since sample complexity is negatively correlated with the ability to detect low-abundant peptides by mass spectrometry (4), we have optimized subcellular enrichment strategies to selectively

Ming Zhou and Timothy Veenstra (eds.), *Proteomics for Biomarker Discovery: Methods and Protocols*, Methods in Molecular Biology, vol. 1002, DOI 10.1007/978-1-62703-360-2_22, © Springer Science+Business Media, LLC 2013

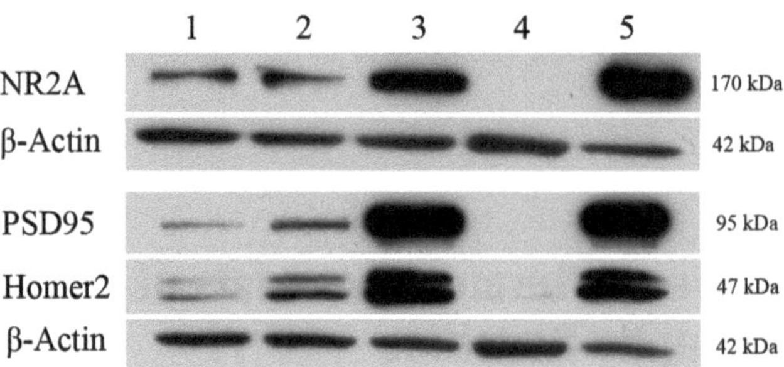

Fig. 1 Immunoblot verification of PSD enrichment by sucrose gradient centrifugation. Western blotting for three PSD proteins (NR2A, PSD95, and Homer2) and a loading control (β-actin) confirms clean isolation of the PSD and separation from the presynaptic fraction. *Lanes* correspond to 10 μg of (1) brain homogenate, (2) synaptosomal fraction, (3) intact-synapse fraction, (4) presynaptic fraction, and (5) PSD fraction

isolate PSD proteins from a whole mouse brain with minimal sample loss (Fig. 1). After manual homogenization, synaptosomes are obtained by sucrose gradient centrifugation and subjected to sequential extraction in increasingly alkaline Tris–Triton X-100 buffers. This PSD isolation strategy relies on the relative insolubility and solubility of the clathrin dense presynaptic junction in acidic and basic environments, respectively (5–12). Following a protein purification method designed to remove Triton X-100 (12), PSD proteins are solubilized in an enzyme-friendly mass spectrometry-compatible surfactant (13). The PSD proteome is then enzymatically digested using modified trypsin endoproteinase, and primary protein structure is determined by liquid chromatography–tandem mass spectrometry (LC–MS/MS). Elevated chromatographic temperatures are employed to enhance the identification of hydrophobic peptides (14, 15), and data analysis by label-free differential analysis software offers a robust and simplistic alternative to stable-isotope labeling neuroproteomic analyses since the use of isotopically labeled internal standards becomes impractical in complex samples that contain thousands of peptides (16, 17).

The methods described in this chapter detail (1) tissue homogenization, (2) subcellular fractionation, (3) protein purification, (4) protein digestion, (5) LC–MS/MS, and (6) label-free differential analysis (Fig. 2).

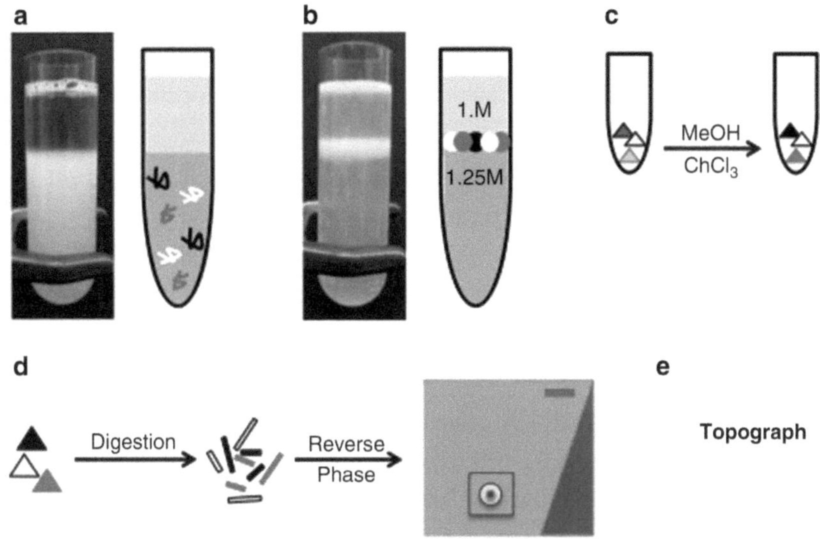

Fig. 2 Workflow of PSD enrichment and differential analysis. (**a**) Brain homogenate in 1.25 M centrifugation buffer (*bottom*) and covered with 1 M overlay buffer (*top*). (**b**) Post ultracentrifugation, the synaptosomal fraction moves to the 1/1.25 M sucrose interface. (**c**) Methanol (MeOH) and chloroform ($ChCl_3$) protein purification of the PSD fraction removes TX-100. (**d**) Proteolytic digestion of the PSD yields tryptic peptides that are separated by reverse-phase chromatography and analyzed by mass spectrometry. (**e**) Chromatographic alignment by Topograph facilitates label-free differential analysis

2 Materials

Prepare all stock solutions using ultrapure water and store at room temperature (unless otherwise noted). To remove contaminates from aqueous solutions, we recommend sterile filtration using a vacuum driver disposable filtration system with a 0.22 μm polyethersulfone membrane.

2.1 Homogenization and Subcellular Fractionation

1. Potter-Elvehjem tissue grinder with PTFE pestle, 15 mL (Wheaton Science Products, Millville, NJ) (see Note 1).
2. Sorvall WX90 Ultracentrifuge (Thermo Scientific, Rockford, IL).
3. AH-629 swinging bucket aluminum rotor, 6 × 36 mL (Thermo Scientific, Rockford, IL).
4. Open-top thickwall polycarbonate ultracentrifuge tubes, 32 mL (Beckman Coulter, Brea, CA).
5. Sorvall MTX 150 Micro-Ultracentrifuge (Thermo Scientific, Rockford, IL).
6. Snap-on cap polyallomer micro-ultracentrifuge tubes, 1.5 mL (Beckman Coulter, Brea, CA).

7. Labquake® tube shaker/rotator (Thermo Scientific, Rockford, IL).
8. 5417R refrigerated microcentrifuge (Eppendorf, Hauppauge, NY).
9. Insulin syringes, 1 cc.
10. Polystyrene serological pipettes, 5 mL and 10 mL.
11. 100× Halt Protease and Phosphatase Inhibitor Cocktail, EDTA-free (Thermo Scientific, Rockford, IL) (see Note 2).
12. Stock solutions: 2 M sucrose (store at 4 °C), 0.1 M $CaCl_2$ (store at 4 °C), 0.1 M $MgCl_2$, 1 M Tris–HCl pH 6, and 10 % (v/v) Triton X-100 (unfiltered) (see Note 3).

2.2 Protein Purification

1. 5417R refrigerated microcentrifuge (Eppendorf, Hauppauge, NY).
2. Misonix XL-2000 ultrasonic liquid homogenizer (Misonix, Farmingdale, NY).
3. Vacuum trap.
4. Benchtop vortex.

2.3 Protein Solubilization and Quantitation

1. 5417R refrigerated microcentrifuge (Eppendorf, Hauppauge, NY).
2. Misonix XL-2000 ultrasonic liquid homogenizer (Misonix, Farmingdale, NY).
3. Epoch Microplate Spectrophotometer (BioTek, Winooski, VT).
4. DC™ Protein Assay (Bio-Rad Laboratories, Hercules, CA).
5. *Rapi*Gest™ SF Surfactant (Waters Corporation, Milford, MA).
6. Bovine serum albumin (BSA).
7. 96-well flat-bottom microplates.
8. Vacuum trap.

2.4 Protein Digestion

1. *Rapi*Gest™ SF Surfactant (Waters Corporation, Milford, MA).
2. Thermomixer® shaker (Eppendorf, Hauppauge, NY).
3. Sequencing grade modified trypsin endoproteinase (Promega Corporation, Madison, WI).
4. 5417R refrigerated microcentrifuge (Eppendorf, Hauppauge, NY).
5. Stock solutions: 0.05 M ammonium bicarbonate (AmBic) pH 7.8, 0.5 M dithiothreitol (DTT) (unfiltered; store aliquoted at −20 °C), 0.5 M iodoacetamide (IAA) (unfiltered; light sensitive; store aliquoted at −20 °C), and 0.1 M $CaCl_2$ (store at 4 °C) (see Note 4).

2.5 Liquid Chromatography and Mass Spectrometry

2.5.1 Reverse-Phase Column

1. P-2000 laser-based micropipette puller (Sutter Instruments, Novato, CA).
2. Flexible fused silica capillary tubing, 75 μm inner diameter (ID) (Polymicro Technologies, Phoenix, AZ).
3. Alcohol lamp.
4. Ceramic scribes.
5. Aqua C18 reverse-phase (Phenomenex, Torrance, CA).
6. Compressed helium.
7. High-pressure packing bomb (built in-house, plans available upon request).

2.5.2 Liquid Chromatography

1. EASY-nLC 1000 liquid chromatography system (Thermo Scientific, Rockford, IL).
2. Heated nanoelectrospray ionization source (built in-house, plans available upon request).
3. Formic acid, 96 %.

2.5.3 Mass Spectrometry

1. Orbitrap Elite Mass Spectrometer (Thermo Scientific, Rockford, IL).

2.5.4 Sample Dilution and Runs

1. Glass autosampler vials with resealing caps, 200 μL (National Scientific, Rockwood, TN).
2. 6-Bovine protein digestion equal molar mix (Bruker–Michrom, Aubrun, CA).
3. Skyline (see https://skyline.gs.washington.edu/labkey/project/home/software/Skyline/begin.view).

2.6 Label-Free Differential Analysis

1. Beowulf LINUX cluster.
2. MakeMS2 (see http://proteome.gs.washington.edu/software/makems2/).
3. SEQUEST (see http://fields.scripps.edu/sequest/).
4. Hardklör (see http://proteome.gs.washington.edu/software/hardklor/).
5. Bullseye (see http://proteome.gs.washington.edu/software/bullseye/).
6. Percolator (see http://noble.gs.washington.edu/proj/percolator/).
7. Topograph (see https://skyline.gs.washington.edu/labkey/project/home/software/Topograph/begin.view?) (see Note 5).

3 Methods

Together, these instructions outline the preparation of one whole snap-frozen adolescent mouse brain (~450 mg), but it is possible to fractionate smaller tissue quantities (~100 mg) without changing buffer volumes. When combined, sample preparation for mass spectrometry analysis will take approximately 3 days to complete (see Note 6). We recommend performing the homogenization, subcellular fractionation, and protein purification on day 1, the protein solubilization and quantitation on day 2, and splitting the protein digestion between days 2 and 3, but we will mention where sample preparation may be paused and the sample stored −80 °C. The length of mass spectrometry data collection will depend on the number of biological and technical replicate injections, while the time required for Topograph analysis will increase with data size. If possible, perform homogenization and fractionation in a refrigerated room and give materials 30 min to equilibrate to 4 °C. Unless otherwise noted, keep sample on ice at all times.

3.1 Homogenization and Subcellular Fractionation

3.1.1 Buffer Preparation

1. Prepare 3 mL of homogenization buffer per brain: for working concentrations of 0.32 M sucrose, 0.0001 M $CaCl_2$, 0.001 M $MgCl_2$, and 1× protease and phosphatase inhibitor cocktail, mix 480 μL 2 M sucrose, 3 μL 0.1 M $CaCl_2$, 30 μL 0.1 M $MgCl_2$, 30 μL 100× protease and phosphatase inhibitor cocktail (add directly before use), and 2.457 mL water in a 15 mL conical tube. Keep on ice.
2. Prepare 17 mL of centrifugation buffer per brain: for a working concentrations of 1.25 M sucrose, 0.0001 M $CaCl_2$ and 1× protease and phosphatase inhibitor cocktail, mix 12.02 mL 2 M sucrose, 17 μL 0.1 M $CaCl_2$, 170 μL 100× protease and phosphatase inhibitor cocktail (add directly before use), and 4.793 mL water in a 50 mL conical tube. Keep on ice.
3. Prepare 10 mL of overlay buffer per brain: for working concentrations of 1 M sucrose and 0.0001 M $CaCl_2$, mix 5 mL 2 M sucrose, 10 μL 0.1 M $CaCl_2$, and 4.99 mL water in a 15 mL conical tube. Keep on ice.
4. Prepare 40 mL of pH 6 Tris–Triton buffer per brain: for working concentrations of 0.02 M Tris–Cl pH 6, 1 % (v/v) Triton X-100, 0.0001 M $CaCl_2$, and 1× protease and phosphatase inhibitor cocktail, mix 880 μL 1 M Tris–Cl pH 6, 4.4 mL 10 % (v/v) Triton X-100, 44 μL 0.1 M $CaCl_2$, 440 μL 100× protease and phosphatase inhibitor cocktail (add directly before use), and 34.236 mL water in a 50 mL conical tube. Keep on ice (see Note 7).
5. Prepare 10 mL of pH 6 Tris–Cl rinse buffer per brain: for working concentrations of 0.02 M Tris–Cl pH 6 and 0.0001 M

$CaCl_2$, mix 220 μL Tris–Cl pH 6, 11 μL 0.1 M $CaCl_2$, and 9.769 mL water in a 15 mL conical tube. Keep on ice.

6. Prepare 1 mL of pH 8 Tris–Triton buffer per brain: for working concentrations of 0.02 M Tris–Cl pH 8, 1 % (v/v) Triton X-100, and 0.0001 M $CaCl_2$, mix 20 μL 1 M Tris–Cl pH 8, 100 μL 10 % (v/v) Triton X-100, 1 μL 0.1 M $CaCl_2$, and 879 μL water in a 1 mL microcentrifuge tube. Keep on ice.
7. Prepare 10 mL of pH 8 Tris–Cl rinse buffer per brain: for working concentrations of 0.02 M Tris–Cl pH 8 and 0.0001 M $CaCl_2$, mix 220 μL Tris–Cl pH 8, 11 μL 0.1 M $CaCl_2$, and 9.769 mL water in a 15 mL conical tube. Keep on ice.

3.1.2 Homogenization and Subcellular Fractionation

1. Place a whole brain on ice until thawed (~25 min).
2. Mix inhibitors into homogenization and centrifugation buffers. Add 3 mL of the homogenization buffer to the tissue grinder and 7 mL of centrifugation buffer to one ultracentrifuge tube.
3. Place the thawed brain into tissue grinder and manually homogenize with 16 strokes (up/down = 1 stroke). Save a small aliquot of the brain homogenate if desired.
4. Transfer the brain homogenate to the ultracentrifuge tube and pipette the solution several times to rinse residual homogenate from the transfer pipette.
5. Use the remaining centrifugation buffer and a 10 mL stereological pipette to rinse the remaining brain homogenate from the tissue grinder.
6. Seal the ultracentrifuge tube with laboratory film and invert it 20 times by hand.
7. Place the ultracentrifuge tube into a swinging bucket upon a manual balance and slowly add overlay buffer (~8 mL) with a transfer pipette until balanced approximately 1 cm below the rim.
8. Ultracentrifuge at 4 °C and 100,000 × *g* for 3 h using the lightest brake setting.
9. Remove the top myelin layer and extract the synaptosomal fraction from the 1/1.25 M sucrose interface (1–4 mL) using a transfer pipette and transfer it to an empty ultracentrifuge tube. Save a small aliquot of the synaptosomes if desired. Store the synaptosomal fraction at −80 °C or proceed with the subcellular fractionation (see Note 8).
10. Mix the inhibitors into pH 6 Tris–Triton buffer and add the buffer to the synaptosomes at a 10:1 volume. Seal the ultracentrifuge tube with laboratory film and invert 20 times by hand.
11. Affix the ultracentrifuge tube to the Labquake® tube shaker/rotator and incubate it for 30 min.

12. Place the ultracentrifuge tube in a swinging bucket and balance it with pH 6 Tris–Cl rinse buffer.
13. Ultracentrifuge at 4 °C and 40,000 × *g* for 30 min using the lightest brake setting.
14. Decant the supernatant (or save if desired) and place the ultracentrifuge tube upside down so residual supernatant drains out. Dry the inside of the tube with a task wipe being careful not to touch the intact-synapse pellet.
15. Use a transfer pipette to gently rinse the intact-synapse pellet with pH 6 Tris–Cl rinse buffer (3 × 1 mL). Add buffer slowly so the pellet does not detach from the ultracentrifuge tube. Remove the buffer with a 1,000 μL pipette tip.
16. Add 1 mL of pH 8 Tris–Triton buffer to the ultracentrifuge tube, break up the pellet with the 1,000 μL pipette tip, and transfer the intact-synapse pieces to a 1.5 mL micro-ultracentrifuge tube.
17. Use an insulin syringe and 3 plunges (up/down = 1) to solubilize the intact-synapse pieces.
18. Affix the micro-ultracentrifuge tube to the Labquake® tube shaker/rotator and incubate for 30 min. Save an aliquot of the intact-synapse fraction if desired (see Note 9).
19. Micro-ultracentrifuge at 4 °C and 40,000 × *g* for 30 min.
20. Remove the presynaptic supernatant (or save if desired) and rinse the PSD pellet with pH 8 Tris–Cl rinse buffer (3 × 1 mL). Microcentrifuge at 4 °C and 9,000 × *g* for 5 min between rinses (see Note 9).
21. Directly proceed to Subheading 3.2.

3.2 Protein Purification

After the addition of methanol, this protocol may be performed at room temperature.

1. Add 500 μL of water to the PSD pellet.
2. Solubilize on ice with the ultrasonic liquid homogenizer (10 × 0.5 s pulses at 10 % power).
3. Add 640 μL of methanol and vortex (3 × 1 s).
4. Add 160 μL of chloroform and vortex (3 × 1 s) (see Note 10).
5. Microcentrifuge at 4 °C and 9,000 × *g* for 10 min (see Note 11).
6. Attach a gel-loading pipette tip to the vacuum trap and suction off the aqueous top layer while being careful not to disturb the protein interface.
7. Add 640 μL of methanol and gently invert the tube 5 times by hand.
8. Microcentrifuge at 4 °C and 9,000 × *g* for 10 min.

9. Suction out all of the liquid.
10. Add 640 μL of methanol and gently invert tube 5 times by hand.
11. Microcentrifuge for 10 min at 9,000 × *g* and 4 °C.
12. Suction out all of the liquid.
13. Repeat steps 1–12 (see Note 12).
14. Add 400 μL of methanol and pulverize the PSD pellet via ultrasonic homogenization (10 × 0.5 s pulses at 10 % power).
15. Vortex, immediately split in fourths, and store the aliquots at −80 °C.

3.3 Protein Solubilization and Quantitation

1. Prepare 0.2 % (w/v) *Rapi*Gest™: add 500 μL of 0.05 M AmBic pH 7.8 to 1 mg *Rapi*Gest™. Keep on ice.
2. Microcentrifuge one PSD aliquot for 10 min at 9,000 × *g* and 4 °C.
3. Suction out methanol and dry for 5 min at room temperature.
4. Add 50 μL of 0.2 % (w/v) *Rapi*Gest™ and save the *Rapi*Gest™ for later use (see Note 13).
5. Solubilize the pellet on ice with the ultrasonic liquid homogenizer (10 × 0.5 s pulses at 10 % power). Keep on ice.
6. Determine protein concentration with the microplate spectrophotometer and the DC™ Protein Assay according to manufacturer's instructions (see Note 14).
7. Directly proceed to Subheading 3.4.

3.4 Protein Digestion

1. Prepare 0.1 μg/μL modified trypsin endoproteinase: add 200 μL of 0.05 M AmBic pH 7.8 to 20 μg of modified trypsin. Keep on ice (see Note 15).
2. Add 100 μg of solubilized PSD protein in a total volume of 83 μL 0.2 % (w/v) *Rapi*Gest™ (see Note 16).
3. Heat at 100 °C for 5 min in the heat block.
4. Cool to room temperature and microcentrifuge for 1 min at 1,000 × *g*.
5. Add 1.0 μL of 0.5 M DTT (for a 0.005 M final concentration) to reduce disulfide bonds.
6. Incubate at 60 °C for 30 min in the Thermomixer®.
7. Cool to room temperature and microcentrifuge for 1 min at 1,000 × *g*.
8. Add 3.0 μL of 0.5 M IAA (for a 0.015 M final concentration) to alkylate reduced thiols.
9. Incubate at room temperature in the dark for 30 min.
10. Add 1.0 μL of 0.1 M $CaCl_2$ (for a 0.001 M final concentration).

11. Add 10.0 μL of modified trypsin endoproteinase for a 1:100 trypsin to protein ratio.
12. Incubate for 18 h at 37 °C in the Thermomixer®.
13. Cool to room temperature and microcentrifuge for 1 min at 1,000 × *g*.
14. Acidify samples with 2 μL 6 N HCl (for a 0.12 M final concentration).
15. Incubate for 45 min at 37 °C while shaking at 400 rpm in the Thermomixer® (see Note 17).
16. Microcentrifuge 3 × 45 min at 9,000 × *g* and 4 °C to pellet *Rapi*Gest™ by-products. Transfer to new microcentrifuge tubes between spins while being careful not to transfer precipitant (see Note 18).
17. Directly proceed to Subheading 3.5 (see Note 19).

3.5 Liquid Chromatography and Mass Spectrometry

3.5.1 Reverse-Phase Column

1. Cut a 90 cm piece of 75 μm ID fused silica capillary tubing.
2. Use the alcohol lamp to remove the polyimide coating from a 4 cm section at its center.
3. Pull a 5 μm tip at the exposed site using the laser micropipette puller.
4. Mix Aqua C18 reverse-phase with HPLC grade methanol in a microcentrifuge tube, vortex, and load into the high-pressure packing bomb.
5. Pack the fused silica column with 30 cm of reverse-phase at 1,000 psi. If methanol is not leaving the column's tip, gently score the tip with the ceramic scribe (see Note 20).
6. Trim the flat end of the column to 40 cm.

3.5.2 Liquid Chromatography

1. Prepare 100 mL of mobile phase A: add 0.1 mL of 96 % formic acid and 2 mL of HPLC grade acetonitrile to 97.9 mL of HPLC grade water.
2. Prepare 100 mL of mobile phase B: add 0.1 mL of 96 % formic acid to 99.9 mL of HPLC grade acetonitrile.
3. Prepare the mobile phase to run at 200 nL/min and step from 0 to 10 % B over 60 min followed by another 60 min step to 30 % B. Include a column equilibration event (10× bed volume) at the beginning of the gradient and a high organic wash at the end.
4. Attach the column to the EASY-nLC 1000 chromatography system via the heated nanoelectrospray ionization source and set at 40 °C.

3.5.3 Mass Spectrometry

1. Place the column in line with the heated capillary source of the Orbitrap Elite Mass Spectrometer.
2. Configure the Orbitrap Elite Mass Spectrometer to collect high-resolution (R = 60,000 at m/z 400) broadband mass spectra

(m/z 400–1,400) for the top-10 most abundant precursor ions. Set collision energy for CID to 35 (arbitrary units) and the dynamic exclusion window to 15 s with a list of 100 ions.

3.5.4 Sample Dilution and Runs

1. Prepare 10 mL of 0.1 % (v/v) formic acid: add 10 μL of 96 % formic acid to 9.99 mL of HPLC grade water.
2. Dilute the sample from 1 to 0.2 μg/μL in 0.1 % (v/v) formic acid.
3. Dilute the bovine standard to 20 fmol/μL in 0.1 % (v/v) formic acid.
4. Load the diluted sample and bovine standard into glass autosampler vials and place in the refrigerated autosampler tray.
5. Loading the reverse-phase column with a volume equivalent to 1 μg of sample and load 100 fmol of the bovine standard every fifth injection. Perform data-dependent acquisition using the Orbitrap Elite Mass Spectrometer (see Note 21).

3.6 Label-Free Differential Analysis

1. Use the MakeMS2 program (18) to convert the raw mass spectrometry data (RAW files) into compact MS1 (cMS1) and compact MS2 (cMS2) text files.
2. Upload cMS1 and cMS2 files to the Beowulf LINUX cluster.
3. Run the Hardklör (19) program to determine monoisotopic mass and charge state of overlapping precursor peptides. Next, use Bullseye (20) to reassign Hardklor deconvoluted precursor mass spectra to their MS/MS identifications.
4. Analyze acquired MS/MS spectra using SEQUEST (21, 22) against real and randomized mouse-derived proteome databases. Search data using tryptic enzyme specificity allowing for one missed cleavage site. Set a fixed modification of 57.021464 on cysteine to account for carboxyamidomethylation during alkylation and set the peptide mass tolerance value according to instrument mass accuracy (see Note 22).
5. Use Percolator (23) software to determine peptide–spectrum match (PSM) false discovery rate (FDR).
6. Create a workspace file in Topograph. Upload Percolator results (.perc.xml file), SEQUEST searches from the real database (SQT files) the mouse derived protein database (in FASTA format), and un-extracted mass spectrometry data (RAW files). Remove isotope tracers and keep static modification for cysteine carboxyamidomethylation. Analyze chromatograms while having Topograph look for peptides in all samples. Keep default mass accuracy and chromatogram extension values and recalculate results one addition time after all chromatograms have been generated.
7. View results grouped by protein and/or cohort.
8. Export results in a Comma Separated Value (CSV) format and analyze according to specific experimental needs.

4 Notes

1. If there are more samples than tissue grinders, clean with a mild detergent between uses. To decrease drying time, rinse the tissue grinder with a volatile liquid like methanol and air dry. Do not dry glass or pestle with a task wipe since this may lead to keratin contamination of sample.
2. Do not use a protease inhibitor cocktail that contains AEBSF because it may cause differential mass shifts in response to covalent protein modifications (24). Moreover, the use of protease inhibitors can attenuate endoproteinase digestion, so we recommend against their use when interested in soluble subcellular fractions (e.g., presynaptic fraction).
3. Make 500 mL 2 M sucrose, 100 mL 0.1 M $CaCl_2$ and 100 mL 0.1 M $MgCl_2$, 100 mL Tris–Cl pH 6 and 50 mL 10 % (v/v) Triton X-100. With sterile filtration and 4 °C storage, it is not necessary to make fresh 2 M sucrose before homogenizations and fractionations that take place on different days. To efficiently solubilize Triton X-100 in water, tape a 50 mL conical containing 10 % (v/v) Triton X-100 to a Labquake® tube shaker/rotator and incubate for approximately 30 min.
4. Make 100 mL 0.05 M AmBic pH 7.8 and 1 mL of both 0.5 M DTT and 0.5 M IAA. Store DTT and IAA in 10 μL aliquots and do not freeze thaw.
5. While Topograph (25) was designed to analyze protein turnover data, it can also be used for protein quantification in label-free experiments (26). Topograph performs retention time alignments based on MS/MS identification events enabling the quantification of precursor ion spectra in samples that do not share MS/MS identifications. This software requires high-resolution precursor ion (i.e MS1) spectra.
6. Sample preparation is limited by ultracentrifuge capacity. With a 6-swinging bucket rotor, it is possible to fractionate upto 6 brains in 1 day (See Note 8).
7. pH 6 Tris–Triton buffer should be added to synaptosomes at a 10:1 volume, respectively. Accordingly, working concentrations for this buffer were calculated for 44 mL total volume. If less than 4 mL of synaptosomes are extracted, there will be leftover buffer. Try to collect synaptosomes in the smallest volume possible to reduce contamination by non-synaptosomal proteins.
8. Add pH 6 Tris–Triton buffer to synaptosomes at a 10:1 volume. Since the ultracentrifuge tubes hold approximately 30 mL, if more than 2.5 mL of synaptosomes are collected, split synaptosomes between two ultracentrifuge tubes and recombine after protein purification. Doubling the number of ultracentrifuge tubes may overwhelm rotor capacity.

9. If sample is saved, purify it according to Subheading 3.2. Count the quantity of the soluble protein as the water component of the purification protocol and keep water, methanol, and chloroform at a 3:4:1 ratio, respectively.
10. After adding chloroform and vortexing, the solution will turn cloudy white, as methanol and chloroform are not miscible at these ratios.
11. After centrifugation, a sizable protein "wafer" will form at the interface between chloroform (bottom) and water/methanol (top).
12. We find that sequential protein purification improves liquid chromatography and peptide identification by mass spectrometry.
13. Be careful not to over dilute the sample. If protein concentration is too low, digestion reagents will require adjustment (see Note 16).
14. Undiluted, protein concentration will be outside the dynamic range of a 0.2–1.6 mg/mL BSA standard curve. Determine protein concentration after diluting the sample by 1/5 in 0.2 % (w/v) *Rapi*Gest™.
15. Other endoproteinase such as proteinase K or chymotrypsin may be used for neuroproteomic characterization of integral membrane proteins.
16. Reagent and endoproteinase dilutions are calculated based on a final protein concentration of 1 μg/μL (100 μg of protein in 100 μL of solution). This is achieved by mathematically dividing 83 μL (100 μL total volume substracted by reagent volumes) by the protein concentration determined in Subheading 3.3. If the concentrated protein is diluted to less than 1.2 μg/μL, adjust reagent and/or endoproteinase volumes to final concentrations stated in the protocol.
17. Solution will become cloudy white as acidification causes water immiscible *Rapi*Gest™ by-products to precipitate out of solution.
18. This microcentrifugation regiment is rather extreme, but if all *Rapi*Gest™ by-products are not removed, the reverse-phase column may clog.
19. We recommend immediate analysis by mass spectrometry; however, peptides may be stored at 4 °C for several days or at −20 °C indefinitely. After −20 °C storage, the digested mixture may turn a light yellow color.
20. While holding onto the column, gently loosen the ferrule. The reverse-phase will settle, so tap the column on the bottom of the microcentrifuge tube a few times until Aqua C18 is viewed traveling up the column. Continue tapping until the column is packed.

21. The bovine standard is used as an analytical metric of retention time reproducibility and ionization efficiency. We use Skyline software (27) to monitor peptide peak areas and retention time from reconstructed mass chromatograms between sample sets. Our mass spectrometer and Skyline methods are available upon request.
22. SEQUEST will export search results as SQT text files. Two sets of SQT files will be generated corresponding to forward and randomized database searches.

Acknowledgments

This work was supported by the following grants: HD41697-10 (SPG), GM103533/AA016171(MJM), and AA016653/AA016171 (CCW).

References

1. Sheng M (2001) Molecular organization of the postsynaptic specialization. Proc Natl Acad Sci USA 98:7058–7061
2. Bayés A, van de Lagemaat LN, Collins MO et al (2011) Characterization of the proteome, diseases and evolution of the human postsynaptic density. Nat Neurosci 14:19–21
3. Szumlinski KK, Kalivas PW, Worley PF (2006) Homer proteins: implications for neuropsychiatric disorders. Curr Opin Neurobiol 16:251–257
4. MacCoss MJ, Matthews DE (2005) Quantitative MS for proteomics: teaching a new dog old tricks. Anal Chem 77:294A–302A
5. Cotman CW (1974) Isolation of postsynaptic densities from rat brain. J Cell Biol 63:441–455
6. Keen JH, Willingham MC, Pastan IH (1979) Clathrin-coated vesicles: isolation, dissociation and factor-dependent reassociation of clathrin baskets. Cell 16:303–312
7. Phillips GR, Huang JK, Wang Y et al (2001) The presynaptic particle web. Neuron 32:63–77
8. Phillips GR, Anderson TR, Florens L et al (2004) Actin-binding proteins in a postsynaptic preparation: Lasp-1 is a component of central nervous system synapses and dendritic spines. J Neurosci Res 78:38–48
9. Phillips GR, Florens L, Tanaka H, Khaing ZZ, Fidler L, Yates JR, Colman DR (2005) Proteomic comparison of two fractions derived from the transsynaptic scaffold. J Neurosci Res 81:762–775
10. Hahn C-G, Banerjee A, Macdonald ML et al (2009) The post-synaptic density of human postmortem brain tissues: an experimental study paradigm for neuropsychiatric illnesses. PLoS One 4:e5251
11. Ramos-Ortolaza DL, Bushlin I, Abul-Husn N, Annangudi SP, Sweedler J, Devi LA (2010) Quantitative neuroproteomics of the synapse. Methods Mol Biol 615:227–246
12. Wessel D, Flügge UI (1984) A method for the quantitative recovery of protein in dilute solution in the presence of detergents and lipids. Anal Biochem 138:141–143
13. Yu Y-Q, Gilar M, Lee PJ, Bouvier ESP, Gebler JC (2003) Enzyme-friendly, mass spectrometry-compatible surfactant for in-solution enzymatic digestion of proteins. Anal Chem 75:6023–6028
14. Speers AE, Blackler AR, Wu CC (2007) Shotgun analysis of integral membrane proteins facilitated by elevated temperature. Anal Chem 79:4613–4620
15. Farias SE, Kline KG, Wu CC (2010) Quantitative improvements in peptide recovery at elevated chromatographic temperatures from microcapillary liquid chromatography-mass spectrometry analyses of brain using selected reaction monitoring. Anal Chem 82:3435–3440
16. Finney GL, Blackler AR, Hoopmann MR, Canterbury JD, Wu CC, MacCoss MJ (2008) Label-free comparative analysis of proteomics mixtures using chromatographic alignment of high-resolution μLC–MS data. Anal Chem 80:961–971
17. Neilson KA, Ali NA, Muralidharan S et al (2011) Less label, more free: approaches in label-free quantitative mass spectrometry. Proteomics 11:535–553
18. McDonald WH, Tabb DL, Sadygov RG et al (2004) MS1, MS2, and SQT—three unified,

compact, and easily parsed file formats for the storage of shotgun proteomic spectra and identifications. Rapid Commun Mass Spectrom 18:2162–2168

19. Hoopmann MR, Finney GL, MacCoss MJ (2007) High-speed data reduction, feature detection, and MS/MS spectrum quality assessment of shotgun proteomics data sets using high-resolution mass spectrometry. Anal Chem 79:5620–5632
20. Hsieh EJ, Hoopmann MR, MacLean B, MacCoss MJ (2010) Comparison of database search strategies for high precursor mass accuracy MS/MS data. J Proteome Res 9:1138–1143
21. Eng JK, McCormack AL, Yates JR (1994) An approach to correlate tandem mass spectral data of peptides with amino acid sequences in a protein database. J Am Soc Mass Spectrom 5:976–989
22. Moore RE, Young MK, Lee TD (2002) Qscore: an algorithm for evaluating SEQUEST database search results. J Am Soc Mass Spectrom 13:378–386
23. Käll L, Canterbury JD, Weston J, Noble WS, MacCoss MJ (2007) Semi-supervised learning for peptide identification from shotgun proteomics datasets. Nat Methods 4:923–925
24. Schuchard MD, Mehigh RJ, Cockrill SL et al (2005) Artifactual isoform profile modification following treatment of human plasma or serum with protease inhibitor, monitored by 2-dimensional electrophoresis and mass spectrometry. Biotechniques 39:239–247
25. Hsieh EJ, Shulman NJ, Dai DF, Vincow ES, Karunadharma PP, Pallanck L, Rabinovitch PS, MacCoss MJ (2012) Topograph, a software platform for precursor enrichment corrected global protein turnover measurements. Mol Cell Proteomics 11:1468–1474
26. Bateman NW, Goulding SP, Shulman N, Gadok AK, Szumlinski KK, MacCoss MJ, Wu CC. Maximizing peptide identification events in proteomic workflows utilizing data-dependent acquisition. Under review, Molecular & Cellular Protoemics
27. MacLean B, Tomazela DM, Shulman N et al (2010) Skyline: an open source document editor for creating and analyzing targeted proteomics experiments. Bioinformatics 26: 966–968

Chapter 23

Fractionation of Peptides by Strong Cation-Exchange Liquid Chromatography

King C. Chan and Haleem J. Issaq

Abstract

Chromatography in all its formats plays an important role in proteomics research. The search for protein biomarkers for different diseases has been an active area of research. The dynamic concentration range and the large number of proteins in a proteome require the development of multidimensional separation strategies to allow for the identification of the largest number of proteins. Strong cation-exchange chromatography (SCX) has been used extensively for the fractionation of proteins and peptides. This chapter provides a detailed description for the SCX fractionation of complex proteome samples.

Key words Protein, Peptides, Fractionation, Mass spectrometry, Strong cation-exchange chromatography

1 Introduction

There have been increased efforts in the last 15 years in the search for protein biomarkers for different diseases. The samples that are used the most are blood (serum and plasma), urine, and tissues. The search procedure is straightforward (Fig. 1); proteins are extracted from the biological specimen, dissolved in a buffer, analyzed by gel electrophoresis or enzymatically digested, fractionated by chromatography, separated by reversed-phase liquid chromatography (LC), and detected and identified by mass spectrometry (MS).

The first step in the analysis of a proteome is to decide if the analysis should be carried out at the protein or the peptide level. Proteins and peptides can be fractionated by ion-exchange LC. The primary advantage of initially digesting the proteins into peptides is solubility. Peptides generally have a wider range of solubility than proteins, and mass spectrometers, in spite of all the technological advances, are still best at identifying peptides rather than intact proteins. Therefore, proteome samples need to be broken down into peptides prior to MS analysis. A quick scan of the literature

Ming Zhou and Timothy Veenstra (eds.), *Proteomics for Biomarker Discovery: Methods and Protocols*, Methods in Molecular Biology, vol. 1002, DOI 10.1007/978-1-62703-360-2_23,

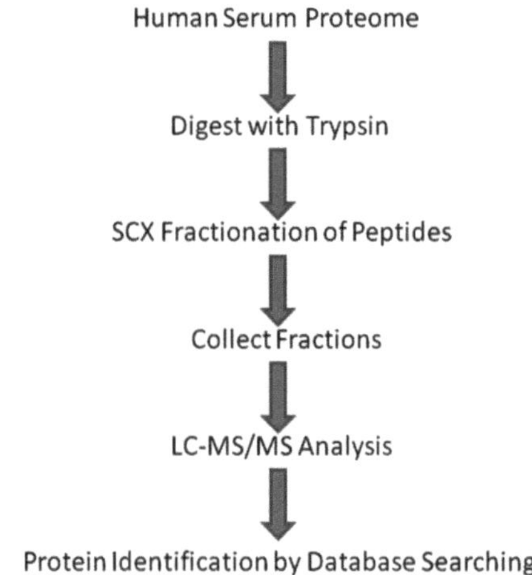

Fig. 1 Diagram of the proteome fractionation procedure

(PubMed) reveals that most solution-based proteome studies, except protein profiling, are conducted at the peptide level. The digestion of serum, cell, tissue, urine, or plasma proteome would results in thousands up to one million peptides. Analysis of such a large number of peptides by LC/MS is well beyond the capability of any mass spectrometer and would exceed the capacity of the system. Therefore to be able to analyze such a complex mixture and achieve meaningful results, the mixture has to be simplified. This can be achieved by fractionating the mixture using a separation technique.

Fractionation is the first step in the multidimensional separation of proteins/peptides of a proteome, and it will allow the analyst to (a) detect and identify the largest number of proteins/peptides by MS and (b) utilize the dynamic range of the system. Fractionation of the complex tryptic protein digestate has been done using different chromatographic and electrophoretic procedures such as affinity, reversed-phase, ion-exchange, size-exclusion, and hydrophilic interaction chromatography, isoelectric focusing, and gel electrophoresis. The fractionation by liquid chromatography can be done on-line or off-line. The off-line approach allows the analyst to (a) use any column type and size and (b) collect as many fractions as needed and (c) is not limited by the injected volume, and (d) it is not destructive, which means that the collected fractions are preserved for further analysis if needed. The discussion in this chapter is limited to off-line fractionation of protein digests using strong cation-exchange chromatography (SCX).

Dr. John Yates at Scripps Research Institute introduced the multidimensional protein identification technology (MudPIT). The procedure used SCX for the fractionation of a protein digest on-line with mass spectrometry using an HPLC biphasic column packed with reversed-phase material followed by strong cation-exchange particles (1). At the same year Issaq et al. (2) and Janini et al. (3) used off-line reversed-phase chromatography for the fractionation of protein digests. Since then SCX has been used as the first separation dimension in the fractionation of peptides in multidimensional separations, while the second separation dimension in most published studies is reversed-phase chromatography in which the mobile phase solvents are more suitable for the MS than SCX mobile phases (4).

2 Materials

2.1 Chemicals

1. Sequencing-grade modified trypsin (Promega).

2.2 Buffers

1. SCX–LC buffer A: 25 % (v/v) acetonitrile (CH_3CN) in H_2O.
2. SCX–LC buffer B: 25 % (v/v) CH_3CN containing 0.5 M NH_4HCO_2, pH 3.

2.3 Apparatus

1. HPLC system (Agilent, model 1100).
2. SCX column: 1 × 150 mm, 5 μm, 300 Å, Polysulfoethyl A (PolyLC, Inc.).
3. UV laser-induced fluorescence detector.
4. Fraction collector 96-well format (ISCO: model Foxy Jr.).

3 Methods

3.1 Digestion

1. Add trypsin to the protein sample to create an enzyme-to-protein ratio of 1:50 (see Note 1).
2. Incubate sample for 16 h at 37 °C.
3. Lyophilize samples and resuspend in 0.1 % formic acid, 20 % acetonitrile.
4. Pass sample solution through a C-18 solid-phase extraction mini-column (ZipTip) to remove any salt from the sample prior to SCX fractionation.
5. Elute peptides off the column using methanol.
6. Dry the peptides, and re-dissolve the peptides in 100 μL of 0.1 % formic acid/25 % acetonitrile.

3.2 SCX Fractionation

1. Mobile phase A is 25 % CH_3CN.
2. Mobile phase B is 25 % CH_3CN, 0.5 M NH_4HCO_2, pH 3.0.
3. Inject 90 μL of sample onto the column.

4. Set up method to deliver mobile phases at a flow rate of 0.05 mL/min.
5. Use a multistep gradient as follows (see Note 2): 2 % mobile phase B for 2 min, followed by 32 % B in 60 min, then 100 % B in 13 min and maintained at 100 % B for 11 min (see Note 3).
6. Detect eluting peptides off column using UV laser-induced fluorescence detector with excitation and emission wavelengths set at 266 and 340 nm, respectively (see Note 4).
7. Collect fractions off the column in 1 min intervals.
8. Lyophilize each collected fraction and reconstituted in 30 μL of 0.1 % TFA prior to LC–MS/MS analysis.

4 Notes

1. Trypsin is commonly used for these types of experiments; however, other proteolytic enzymes, such as chymotrypsin, can be used to prepare the sample.
2. It is important to desalt the sample after digestion and prior to fractionation. Excess salt will interfere with the quality of the SCX fractionation as illustrated in Fig. 2.
3. The gradient provided in the method is recommended; however, depending on the complexity of the sample, a different gradient may be utilized (Fig. 3).
4. While a laser-induced fluorescence detector is recommended due to its sensitivity, an UV detector may also be used.

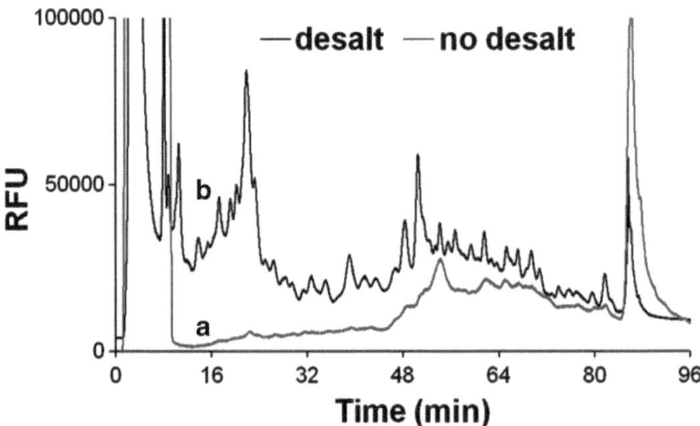

Fig. 2 Chromatograms of a serum sample before (**a**) and after (**b**) desalting

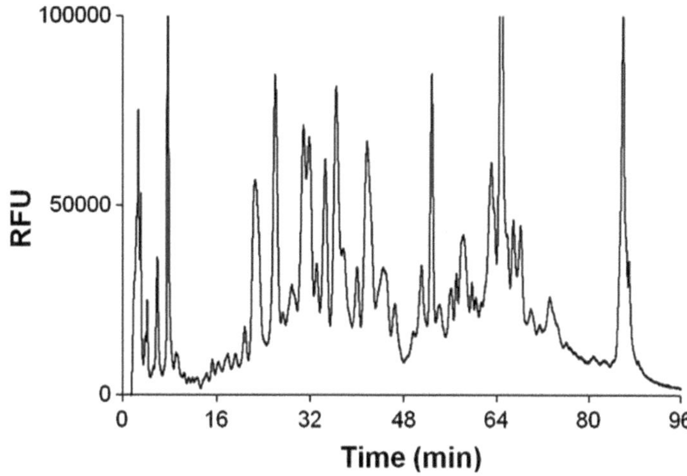

Fig. 3 Strong cation-exchange chromatogram of serum proteome

References

1. Link AJ, Eng J, Schieltz DM et al (1999) Direct analysis of protein complexes using mass spectrometry. Nat Biotechnol 17: 676–682
2. Issaq HJ, Chan KC, Janini GM, Muschik GM (1999) A simple two-dimensional high performance liquid chromatography/high performance capillary electrophoresis set-up for the separation of complex mixtures. Electrophoresis 20:1533–1537
3. Janini GM, Chan KC, Conrads TP, Issaq HJ, Veenstra TD (2004) Two-dimensional liquid chromatography-capillary zone electrophoresis-sheathless electrospray ionization-mass spectrometry: evaluation for peptide analysis and protein identification. Electrophoresis 25:1973–1980
4. Issaq HJ, Conrads TP, Janini GM, Veenstra TD (2002) Methods for fractionation, separation and profiling of proteins and peptides. Electrophoresis 23:3048–3061

ERRATUM

Chapter 22

Label-Free Differential Analysis of Murine Postsynaptic Densities

Scott P. Goulding, Michael J. MacCoss, and Christine C. Wu

Ming Zhou and Timothy Veenstra (eds.) *Proteomics for Biomarker Discovery*, Methods in Molecular Biology, vol. 1002, DOI: 10.1007/978-1-62703-360-2, pp. 295–309, © Springer Science+Business Media, LLC 2013

DOI 10.1007/978-1-62703-360-2_24

The publisher regrets that in the original publication there was an error. Note 16 of Chapter 22 should read as follows:

Reagent and endoproteinase dilutions are calculated based on a final protein concentration of 1 ug/uL (100 ug of protein in 100 uL of solution). This is achieved by mathematically dividing 100 ug (83 uL is the total volume substracted by reagent volumes) by the protein concentration determined in Subheading 3.3. If the concentrated protein is diluted to less than 1.2 ug/uL, adjust reagent and/or endoproteinase volumes to final concentrations stated in the protocol.

The online version of the original chapter can be found at http://dx.doi.org/10.1007/978-1-62703-360-2_22

INDEX

Ming Zhou and Timothy Veenstra (eds.), *Proteomics for Biomarker Discovery: Methods and Protocols*, Methods in Molecular Biology, vol. 1002, DOI 10.1007/978-1-62703-360-2, © Springer Science+Business Media, LLC 2013

MIX
Papier aus verantwortungsvollen Quellen
Paper from responsible sources
FSC® C105338

If you have any concerns about our products,
you can contact us on
ProductSafety@springernature.com

In case Publisher is established outside the EU,
the EU authorized representative is:
Springer Nature Customer Service Center GmbH
Europaplatz 3, 69115 Heidelberg, Germany

Printed by Libri Plureos GmbH
in Hamburg, Germany